STRUCTURE AND BONDING is issued at irregular intervals, according to the material received. With the acceptance for publication of a manuscript, copyright of all countries is vested exclusively in the publisher. Only papers not previously published elsewhere should be submitted. Likewise, the author guarantees against subsequent publication elsewhere. The text should be as clear and concise as possible, the manuscript written on one side of the paper only. Illustrations should be limited to those actually necessary.

Manuscripts will be accepted by the editors:

Professor Dr. *J. D. Dunitz*   Laboratorium für Organische Chemie der Eidgenössischen Hochschule
CH-8006 Zürich, Universitätsstraße 6/8

Professor Dr. *P. Hemmerich*   Universität Konstanz, Fachbereich Biologie
D-7750 Konstanz, Postfach 733

Professor *J. A. Ibers*   Department of Chemistry, Northwestern University
Evanston, Illinois 60201/USA

Professor Dr. *C. K. Jørgensen*   51, Route de Frontenex,
CH-1207 Genève

Professor *J. B. Neilands*   University of California, Biochemistry Department
Berkeley, California 94720/USA

Sir *Ronald S. Nyholm*, FRS †

Professor Dr. *D. Reinen*   Institut für Anorganische Chemie der Universität
Marburg
D-3550 Marburg, Gutenbergstraße 18

Professor *R. J. P. Williams*   Wadham College, Inorganic Chemistry Laboratory
Oxford OX1 3QR/Great Britain

SPRINGER-VERLAG

D-6900 Heidelberg 1
P. O. Box 1780
Telephone (06221) 4 91 01
Telex 04-61 723

D-1000 Berlin 33
Heidelberger Platz 3
Telephone (0311) 82 20 01
Telex 01-83319

SPRINGER-VERLAG
NEW YORK INC.

175, Fifth Avenue
New York, N. Y. 10010
Telephone 673-2660

STRUCTURE
AND BONDING

Volume 13

Editors: J. D. Dunitz, Zürich
P. Hemmerich, Konstanz · J. A. Ibers, Evanston
C. K. Jørgensen, Genève · J. B. Neilands, Berkeley
Sir Ronald S. Nyholm†, London · D. Reinen, Marburg · R. J. P. Williams, Oxford

With 70 Figures

Springer-Verlag
Berlin Heidelberg GmbH 1973

ISBN 978-3-540-06125-0   ISBN 978-3-540-38289-8 (eBook)
DOI 10.1007/978-3-540-38289-8

Originally published by Springer-Verlag Berlin · Heidelberg · New York in 1973

Library of Congress Catalog Card Number 67-11280.

Contents

Structural Systematics in Actinide Fluoride Complexes *

R. A. Penneman and R. R. Ryan

University of California, Los Alamos Scientific Laboratory
Los Alamos, New Mexico 87544/USA

A. Rosenzweig

Department of Geology, University of New Mexico
Albuquerque, New Mexico 87106/USA

Table of Contents

1. Introduction

During the last twenty years of active investigation, a large number of crystal structures have been determined for compounds of the actinides with fluorine. It is the purpose of this paper to survey the structural characteristics displayed by these compounds. Among the d-transition elements, the number of d-electrons has a significant effect on bonding, although this effect is minimum in their fluoro complexes (1). Evidence of f-orbital participation in actinide fluoride bonding has been sought (2); however, this survey will show that the 5f-electrons have little directional bonding effects in determining actinide-fluoride complex structures.

---

* This work was performed under the auspices of the U.S. Atomic Energy Commission.

R. A. Penneman, R. R. Ryan and A. Rosenzweig

Actinide cations appear to behave in a largely ionic manner; radius and charge play a predominant role in the determination of structure and coordination types (3).

The actinides (5f elements, $Z = 90$—103) offer to inorganic and structural chemists a unique opportunity for study, since these elements display a multiplicity of valences (Fig. 1), as well as fixed valences for

	+2	+3	+4	+5	+6	+7
Ac		X				
Th			X			
Pa			X	X		
U		X	X	X	X	
Np		X	X	X	X	O
Pu		X	X	X	X	O
Am	(X)	X	X	⊕	⊕	
Cm		X	⊕			
Bk		X	⊕			
Cf	(X)	X	⊕			
Es	(X)	X				
Fm		X				
Md	X	X				
No	X	X				
Lr		X				

O Circles indicate valence achieved with oxygen
⊕ Oxygen and fluorine
(X) Valence rare

Fig. 1. Actinide valences

upwards of a dozen consecutive elements. This latter aspect they share with the lanthanide (4f) series, which, however, has a more monotonous chemistry, being predominantly trivalent with but few divalent and tetravalent species. The higher valence states of the actinides seem to be achieved only in compounds with the highly electronegative ligands oxygen and fluorine. The highest valence (+7) occurs with oxygen (4, 5, 6); for americium the +5 and +6 valences occur only in compounds containing oxygen or both oxygen and fluorine (7, 8, 9) (Fig. 1). It is

2

further noted that, excepting obvious chemical restrictions, the highest oxidation states occur early in the series, while the stability of lower valence states predominates in the last half of the series with the divalent state occuring as early as californium (9a). However, there is evidence for Am(II) at low temperature in a CaF_2 matrix (10) and recent work at Oak Ridge and Los Alamos has established anhydrous divalent americium halides (10a).

1.1. Trends in Actinide Size and Valence

Cation sizes have reached a maximum with the actinide ions and high coordination numbers are demanded. The first actinide of a particular charge is the largest elemental cation of that charge (thorium, for example, is the largest M^{4+} ion). Cell dimensions for oxides containing the M^{4+} actinides are known for nine successive elements (Th^{4+} through Cf^{4+}) (11, 12); for M^{3+} the data now extend through Es_2O_3 (13). Within the 5f series, ionic sizes become demonstrably smaller with increasing atomic number for the tri- and tetravalent ions. A similar pattern is shown by the M^{5+} species, with Pa^{5+} being the largest (14), followed by U^{5+}, etc. (15). Pentavalent bismuth is the largest of the Group V, non-transition elements in this oxidation state, but it is still smaller than U^{5+} ($NaBiF_6$ is isostructural with rhombohedral (16) $NaUF_6$, but of smaller cell dimensions (17)). When we compare the lanthanide (4f) and actinide (5f) series with respect to relative trivalent ion sizes, we find that those of the latter series are larger and that, for example, trivalent californium is the size of the trivalent lanthanide, europium (11), three elements before the 4f counterpart of californium. Although there are fewer data on tetravalent lanthanides to compare, we find the same general size displacement of the tetravalent 5f ions with respect to their 4f analogs, Ce^{4+} being smaller than U^{4+} and Tb^{4+} than Cf^{4+}.

In Fig. 2, the actinide element (in its various valences) is written beneath each last row element if it displays that group valence. Uranium, for example, is shown in the last row in its customary position between protactinium and neptunium; additionally, it is written beneath actinium to indicate that uranium has a tri-positive state and under thorium, since uranium has a tetra-positive state, etc. This has the effect of pivoting the bottom row of the periodic table at each actinide, bringing those elements to its right beneath it. The pivot element displays the maximum ion size for that particular valence. General group chemical similarity is displayed within a column. However, the reduction in ion size as the atomic number increases may cause differences such as change in coordination, which results in the appearance of analogous compounds which are isostoichiometric but not isostructural.

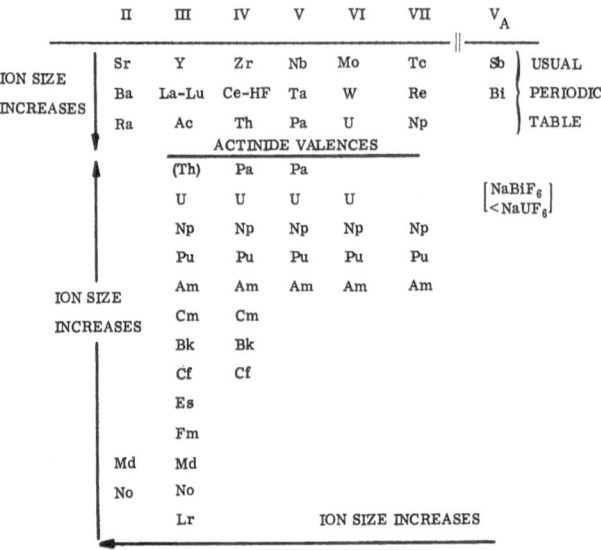

	II	III	IV	V	VI	VII	V$_A$	
ION SIZE	Sr	Y	Zr	Nb	Mo	Tc	Sb	USUAL
INCREASES	Ba	La–Lu	Ce–HF	Ta	W	Re	Bi	PERIODIC
	Ra	Ac	Th	Pa	U	Np		TABLE

ACTINIDE VALENCES

(Th)	Pa	Pa	
U	U	U	U
Np	Np	Np	Np	Np
Pu	Pu	Pu	Pu	Pu
Am	Am	Am	Am	Am
Cm	Cm		
Bk	Bk		
Cf	Cf		
Es			
Fm			
Md	Md		
No	No		
	Lr		

$\begin{bmatrix} NaBiF_6 \\ < NaUF_6 \end{bmatrix}$

ION SIZE INCREASES (vertical, left)

ION SIZE INCREASES (horizontal, bottom)

Fig. 2. Modified periodic table, showing valence location of actinides and size trends

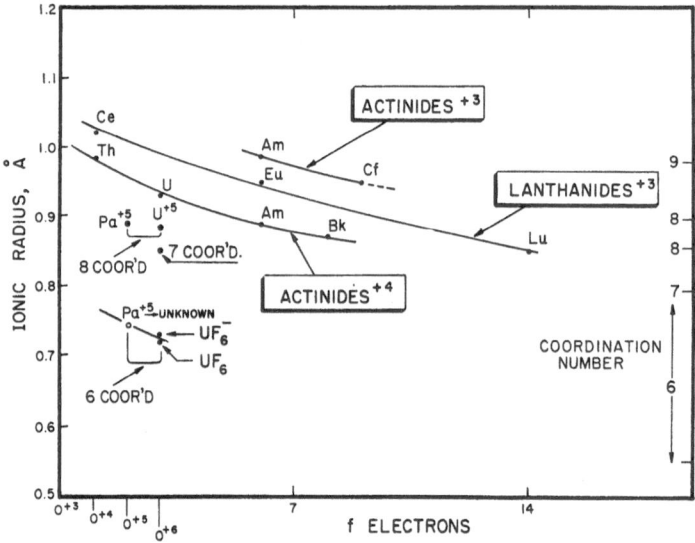

Fig. 3. Ionic radii of some actinides and lanthanides in various valences

1.2. Coordination Number and Polyhedra

In Fig. 3 are shown some ionic radii for M^{3+}, M^{4+}, and M^{5+} oxidation states. Ionic radii vary, depending on choice of counter ion radius, model compounds, coordination number, etc. For a recent review see *Shannon* and *Prewitt* (*18*). Most of the radii used in Fig. 3 are those of *Zacharia-sen* (*19*) and of *Peterson* and *Cunningham* (*11*), who compared similar lanthanide and actinide compounds. Ions of both series are obviously large and of a size sufficient to preclude lower than six coordination with F^- and to make six coordination a rarity. In the review by *Muetterties* and *Wright* it was noted that the actinide and lanthanide fluorides favor high coordination numbers (*20*). Excellent illustrations of high coordination number polyhedra are found in their paper. In his book, *D. Brown* devotes a chapter to fluorides and oxyfluorides of the lanthanides and actinides (*20a*).

Table 1

Octahedron	6	0.55 Å
Capped octahedron	7	0.79
Antiprism	8	0.86
Dodecahedron	8	0.89
Cube	8	0.97
Tricapped trigonal prism	9	0.97

The coordination polyhedra, coordination number, and maximum size of coordinated metal ion which will just allow F—F hard sphere contacts are given in Table 1. Coordination numbers of 6—9 are thus expected for actinide fluorides, (see Fig. 3) with 6 being attained only for valences greater than 4+. Eight coordination spans the largest range of radius ratios, and extends to the nine-coordination radius ratio. Coordination of the majority of actinide ions to fluoride ion (1.33 Å) yields radius ratios within this range. We expect highest coordination numbers to be observed at the actinide element where a given charge type first occurs, since it is the largest ion of its type. As the size diminishes with atomic number, we can anticipate changes to occur such as diminution of coordination number, and recurrence of structures previously observed for the smaller d-transition elements. For example, K_2UF_7 and K_2TaF_7 are isostructural with coordination number 7 for the pentavalent element, but K_2PaF_7 has a different structure and coordination number of 9 for protactinium (V) (*14*).

In 1948 *Zachariasen* (*21*), studying a number of alkali actinide fluorides, noted that the cell volumes of these compounds could be attributed to the volume of the fluoride ions and the larger alkali ions alone, with the actinide ions occupying interstices between the anions. In his examples he assigns a volume of 18 Å³ for F^-, 7 Å³ for Na^+, and 21 Å³ for K^+, etc. This relationship is borne out in many structures of related compounds which have since been determined. The tri- and tetravalent actinide ions are quite large (similar to Na^+), and the fluorine volume deduced from measured cell volumes includes a contribution from the actinide. The difference in actinide size and coordination would produce variation in effective F^- volume, but this variation will be small because of the generally large coordination number of the actinide. Thus, known stoichiometry leads to a useful estimation of volume/formula wt, which is essentially independent of structure, as are other functions of volume, such as the Lorentz-Lorenz molar refractivity (22).

These observations lead one to suspect that the fluoride ions and larger alkali ions in these compounds are arranged in an approximation of a close packed array. These suspicions are easily confirmed by an inspection of the available structural data. The details of the close packed array are strongly influenced by the alkali ion size. The smallest can be accommodated in the interstices of a fluoride anion array, but as the cation size increases the array will be progressively distorted, and ultimately the alkali may become a part of the close packing scheme. These two extremes are exemplified by the following: In $LiUF_6$ ($LiSbF_6$ type (23) the alkali and actinide are relatively small and occupy octahedral sites in an array of fluorides which approximates a hexagonal close-packed array. In the compounds KUF_6 and $RbUF_6$ the alkali ions are ten coordinate and the actinide eight-coordinate (24), but the model for the packing is difficult to describe. In $CsUF_6$ the U^{5+} again occupies an octahedral site but the large Cs^+ is twelve-coordinated; the cesium and fluorine, in this case, make up a distorted cubic close-packed array (15). In Na_3PaF_8 (25) the fluorines are less densely packed (cube corners) and both Na^+ and Pa^{5+}, though much smaller than fluorine, become part of the distorted cubic close-packed array. Further distortions are caused by such phenomena as hydrogen bonding and cation-cation repulsions. A striking example of the first of these is seen in the pair of compounds Rb_2UF_6 and $(NH_4)_2UF_6$ (see section on eight-coordination). The second type is discernable in structures having fluoride polyhedra which are linked together; in these distortion occurs invariably. This is a result of the actinide-actinide repulsion which tends to pull the shared fluorines together giving them a shorter than normal 'contact' distance and a longer fluoride-actinide bond. A similar effect can often be seen even in the case of "isolated" polyhedra and arises from cation packing effects. For example, an alkali cation opposite an edge will cause a shortening of the F—F distance on that edge, and a displacement of the kernel actinide away from that edge.

1.3. Depolymerization by Fluoride; Variable Valence in Fluoride Compounds

To achieve high coordination number, the binary fluorides MF_x of the actinides are polymeric, with fluorines being shared between two or more actinides (an obvious exception to this is presented by the low-boiling MF_6 fluorides). In the usual case, the MF_x unit in the lattice is part of a chain or of a three dimensional polymer. For example, eight coordinated UF_4 has all fluorides shared. However, it can be *depolymerized* by addition of alkali fluoride, successively passing through a series of complexes terminating at the discrete anionic species UF_8^{4-}. In chain structures, the distance between nearest neighbor actinides is about 4 Å; in structures having isolated anions they are much farther apart. In the case of K_5ThF_9 further "dilution" occurs and the structure contains ThF_8^{4-} ions as well as F^- ions which are "free" in the sense of not being bound to thorium. There are cases in which "dilution" by free fluoride causes no further depolymerization: $(NH_4)_3ThF_7$ and $(NH_4)_4ThF_8$ both have chains of 9-coordinate polyhedra with the latter compound having "free" fluoride. From this viewpoint, the great number of complex compounds formed between the actinide tetrafluorides and the alkali fluorides is more readily understood. The various fluorides form a series in which the ratio of MF_4: alkali fluoride goes from large positive values, *e.g.* 6:1 as in KTh_6F_{25}, to fractional values, *e.g.* 1:5 as in K_5ThF_9, frequently yielding compounds having unusual mole ratios. The situation is frequently complicated by the existence of several polymorphs.

With the metals of the first transition series, the maximum coordination number of higher oxidation states is six, and this is so firmly fixed that in their fluoride complexes the oxidation state of the metal in question can be fixed by controlling the mol fraction of alkali metal present (26). Thus, the fluorination of a vanadium salt in the presence of a one, two, or three mol ratio of potassium ion, yields KV^VF_6, $K_2V^{IV}F_6$, or $K_3V^{III}F_6$. The same tendency is shown, but to a lesser degree, by metals of the second transition series, as exemplified by $KRuF_6$ and K_2RuF_6. For an unusual example in the third series, note that OsF_6 is known, OsF_7 is not stable, but heptavalent osmium is found as six-coordinated $OsOF_5$ (27).

With the actinides this just mentioned tendency is minimized. Actinide fluoride complex structures can display the same coordination number over a range of valence states, i.e. $Na_2U^{VI}F_8$ and $Na_3U^VF_8$ in which the actinide coordination is cubic. However, this behavior is unusual. The weak directional influence of the f-electrons, coupled with the larger size of actinide ions allows them to accommodate to the several different coordination polyhedra possible with higher coordination num-

7

bers. Change in valence of the kernel actinide is frequently accompanied by a change in coordination number. However, this may leave the heavy atom positions essentially unchanged. This is illustrated by the structures of Na_2UF_8 and Na_3UF_7, which are derived from Na_3UF_8 as follows:

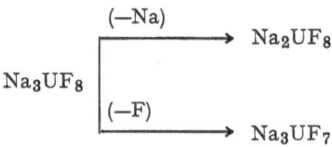

Thus, Na_2UF_8 has sodium ion vacancies and Na_3UF_7 has fluoride ion vacancies. An additional example is provided by a series of uranium compounds in which the parent fcc cell of ~ 9.2 Å can accommodate $3\,KF \cdot UF_3$, $3\,KF \cdot UF_4$, or $3\,KF \cdot UF_5$, and whose powder patterns are essentially indistinguishable (28). Additional substances with similar unit cells are listed by *Wyckoff* (29).

2. Structural Characteristics of Actinide Fluoride Complexes

2.1. Six-coordinated Actinide-fluoride Complexes

The structural chemistry of six-coordinated fluoro complexes of the d-transition elements is the subject of an entire review paper by *Babel* (1). In contrast, six coordination is rare in actinide fluoro-complexes. The relatively large sizes of the actinide ions (Fig. 3) suggest that low coordination numbers should be found only in the fluoride complexes of the higher oxidation states. The radius ratio, r^+/r^-, predicted for the lower limit of stability of octahedral coordination is $\sqrt{2} - 1$, which corresponds to a positive ion radius of 0.55 Å in coordination with fluoride (1.33 Å).

Of the very few known structure types displaying six-coordinated actinides, only two, UF_6 and $CsUF_6$, have been determined with sufficient accuracy to give trustworthy values for the effective ionic radii of the actinide ions. In UF_6 and $CsUF_6$ the uranium ions (U^{6+} and U^{5+}) have radius ratios with fluorine well above the lower limit of stability for octahedral coordination but still lower than the lower limit predicted for the 8-coordinated square antiprism (0.645) or the 7-coordinated capped octahedron (0.592). However, both U^{6+} and U^{5+} are sufficiently larger than 0.55 Å so that F—F distances in the octahedron cannot all be van der Waals' contact distances, and thus, the MF_6 octahedra are susceptible to distortion.

Isolated octahedral groups are found in orthorhombic UF_6 (*30*). Within the accuracy of the single crystal X-ray experiment the coordination was not demonstrated to be significantly distorted from that of a regular octahedron. The average U—F distance of 2.05 Å gives an effective U^{6+} radius of 0.72 Å.

There are several examples of isolated MF_6^- ions which are slightly distorted from ideal octahedral geometry (Table 2). In the rhombohedral $CsUF_6$ (*15*), (see Fig. 4) U^{5+} has six fluorine neighbors at the corners of an octahedron which is slightly elongated (stated as flattened in Ref. (*15*)) along the three-fold axis (symmetry D3*d*). The angle between the trigonal axis of the UF_6^- group and a U—F bond is 53° 23', smaller than the value 54° 44' for a regular octahedron. The U—F distance is 2.057 Å, giving an effective U^{5+} radius of 0.73 Å. The cesium ion in $CsUF_6$ is twelve-coordinated, having six fluorines at 3.10 Å and six more F neighbors at a slightly longer distance (3.17 Å) (*15*). Cesium is located on the trigonal axis and separates two UF_6^- polyhedra. The compounds $CsNpF_6$ (*31*) and $CsPuF_6$ (*32*) are isostructural with $CsUF_6$. These structures are of the $KOsF_6$ (*33*) type, which was determined only from powder data.

6 COORDINATION

TRIGONAL PRISM   OCTAHEDRON

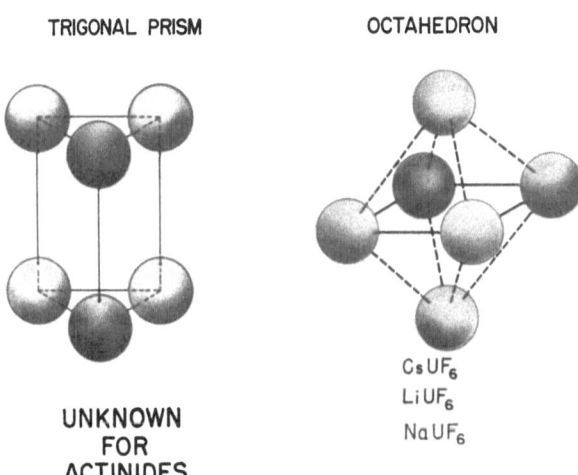

$Cs\,UF_6$
$Li\,UF_6$
$Na\,UF_6$

UNKNOWN
FOR
ACTINIDES

Fig. 4. Polyhedra for six-coordinated actinides

The X-ray powder patterns of $LiUF_6$ (*34*) and of α-$NaUF_6$ (*16*) indicate that they are both isostructural with rhombohedral ($R\bar{3}$)

Table 2. Six-coordinated compounds

Compound[a]	Space Grp	Lattice Constants[b]			Distortion[c]	Ref.
LiSbF$_6$	R$\bar{3}$	5.18	13.60		none	(23)
α-NaUF$_6$	R$\bar{3}$	5.596	15.526		trigonal	(16)
LiUF$_6$	R$\bar{3}$	5.262	14.295		trigonal	(34)
NaSbF$_6$	Fm3m	8.201			none	(38)
β-NaUF$_6$	Fm3m	8.608			trigonal	(34)
NOUF$_6$	Ia3	10.464			trigonal	(36, 42)
CsUF$_6$	R$\bar{3}$	8.021	8.430		trigonal	(15)
CsNpF$_6$	R$\bar{3}$	8.017	8.386		trigonal	(37)
CsPuF$_6$	R$\bar{3}$	8.006	8.370		trigonal	(32)
α-UF$_5$	I4/m	6.525	4.472		tetragonal	(43)
UF$_6$	Pnma	9.916	8.960	5.245	none?	(154)
NpF$_6$	Pnma	9.909	8.997	5.202		(154)
PuF$_6$	Pnma	9.912	8.942	5.206		(154)

[a] Underlined compounds are structure types, determined by single crystal techniques.
[b] In this and subsequent tables, lattice constants are given for the several systems as follows: cubic — a; tetragonal and hexagonal — a, c; rhombohedral — a, c or a$_{rh}$, α; orthorhombic — a, b, c; monoclinic — a, b, c; triclinic — a, b, c, α, β, γ.
[c] Distortion from octahedral, O$_h$, symmetry.

LiSbF$_6$ (23). The single crystal study of LiSbF$_6$ shows the Li$^+$ coordination number to be six (in contrast to the coordination number of twelve for Cs$^+$ in rhombohedral CsUF$_6$) and the SbF$_6^-$ ion to be a regular octahedron. However, ESR measurements made by *Drifford, Dianoux, Rigny,* and *Plurien* (35, 36) on the series of rhombohedral MUF$_6$ compounds (M = Cs, Na, and Li) indicate that they all show a significant distortion (elongation) along the trigonal axis. The β-modification of NaUF$_6$ is face-centered cubic and reported to be isostructural with NaSbF$_6$ (16, 34). NaSbF$_6$ has recently been shown, by NMR (37) as well as single crystal techniques (38), to contain at room temperature an SbF$_6^-$ ion on a site with O$_h$ symmetry in the space group F$_{m3m}$. Again, work performed by *Drifford et al.,* on the cubic phase, β-NaUF$_6$, using ESR techniques indicates a significant trigonal distortion of the UF$_6^-$ ion (36) (along the cubic diagonal).

It was demonstrated recently that NaUF$_6$ (rh) converts cleanly to NaUF$_6$ (fcc) upon heating at 55 °C for 24 hours (39). The conversion of NaUF$_6$ (fcc) to NaUF$_6$ (rh) is sluggish; the lines of the fcc phase persist after the sample is held at -5 °C for 24 hours, but the three strongest lines of the rhombohedral phase are clearly visible. Since cubic NaUF$_6$ undergoes a phase change (at least partially) to the rhombohedral form at temperatures only slightly below room temperature and since the French ESR spectra of fcc NaUF$_6$ were taken at liquid nitrogen temperatures, it is not inconceivable that their sample was not entirely in the cubic form. It is also possible that, at room temperature, the distorted UF$_6^-$ disorders (with but slight displacements of the atoms) in such a way as to average to O$_h$ site symmetry. Such a model allows both the X-ray pattern to have perfect F$_{m3m}$ symmetry *and* the UF$_6^-$ to be distorted on the ESR time scale. Further, it would carry the interesting implication that the distortions of the UF$_6^-$ ion in the rhombohedral structures are not entirely due to crystal packing effects and that the UF$_6^-$ ion may have an inherent tendency toward trigonal distortion.

The original report on NOUF$_6$ suggested a pseudo-cubic cell of a = 5.18 Å (40). Preliminary single crystal precession photos of NOUF$_6$ indicated a 10.5 Å cell (41). In a neutron diffraction study of NOUF$_6$ by Madame *Charpin* (42), cubic cells of 10.464 Å at room temperature and 10.336 Å at 4.2 °K were found. In this case also the ESR data are consistent with a trigonal distortion of the UF$_6^-$ octahedron (35, 36).

On the basis of powder X-ray diffraction data, *Zachariasen* (43) proposed a structure for α-UF$_5$. In this structure the octahedrally coordinated UF$_6$ groups share opposite corners to form chains, thus distorting the octahedron along a four-fold axis (D4h symmetry). The U—F distance for the unshared fluorines is 2.18 Å and the average distance is 2.20 Å. This yields a minimum effective U^{5+} radius of 0.85 Å. This value is considerably larger than in the well-determined $CsUF_6$ structure, and the radius ratio is just above the lower limit for 8-coordination.

2.2. Seven-coordinated Actinide-fluoride Complexes

Of the several seven-coordination polyhedra possible, there are two markedly different types which have been established (Fig. 5). The first includes both the monocapped trigonal prism (symmetry C_{2v}) (Fig. 5a) and the monocapped octahedron (symmetry C_{3v}) which are closely related. They require a central atom with a minimum radius ratio of about 0.59 or a radius of about 0.79 Å with fluorine. Though of different symmetry, only a slight shift of atoms is required to change one polyhedron into the other. The second type is the pentagonal bipyramid (Fig. 5b) which requires a minimum radius ratio of about 0.70 or a central ion radius of 0.93 Å with fluorine. Note, in this case the central ion is large enough to lead to the expectation of eight coordination.

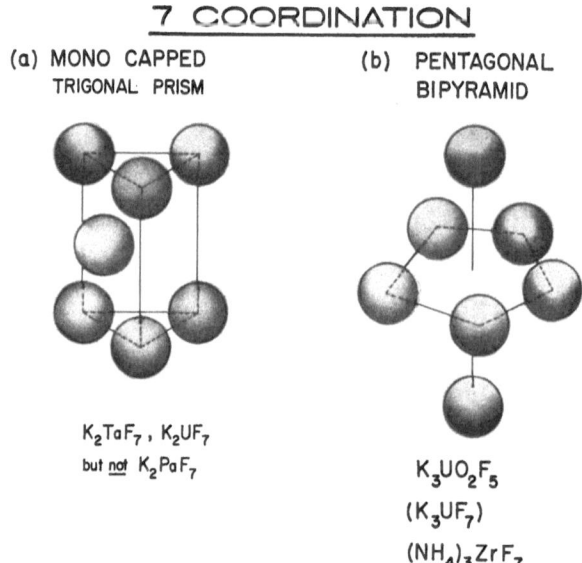

7 COORDINATION

(a) MONO CAPPED TRIGONAL PRISM

(b) PENTAGONAL BIPYRAMID

K_2TaF_7, K_2UF_7
but not K_2PaF_7

$K_3UO_2F_5$
(K_3UF_7)
$(NH_4)_3ZrF_7$

Fig. 5. Polyhedra for seven-coordinated actinides (capped octahedron not shown)

A number of actinide-fluoride complexes of the type R_2MF_7 (not including K_2PaF_7) (14) are isostructural with K_2NbF_7 and K_2TaF_7 whose structures were first deduced by *Hoard* (44). The structure of K_2NbF_7 was later refined by *Brown* and *Walker* by neutron diffraction

(45). The coordination polyhedron of the transition metal ion is described as a monocapped trigonal prism with an average M—F distance of 1.945 Å and rather close approaches (2.36–2.41 Å) for the F—F contact distances *(45)*. Although the polyhedron is described as a capped trigonal prism, it is distorted toward a capped octahedron. The Nb^{5+} site has no crystallographically required symmetry in this structure, but the polyhedron nearly has a mirror plane. The following actinide compounds have powder patterns corresponding to that of K_2NbF_7: K_2UF_7, Rb_2UF_7, $(NH_4)_2UF_7$, Cs_2UF_7 *(34)*, Rb_2NpF_7 *(37)*, Rb_2PuF_7 *(32)*. No parameters are available which allow the calculation of the ionic radii of actinide ions in this structure type. (Table 3).

Zachariasen (46) determined, by single crystal techniques, the structure of $K_3UO_2F_5$ to be tetragonal with pronounced pseudo-cubic character. The uranyl ion was found to be linear with five equatorial fluorines forming a pentagonal bipyramidal $UO_2F_5^{3-}$ ion. Further, he prepared K_3UF_7 in two phases, one cubic and the other tetragonal *(47)*. He deduced that the ordered tetragonal phase was essentially isostructural with $K_3UO_2F_5$ (with F replacing O and the bonds lengthening). The estimated U—F distances have an average value of 2.26 Å, giving an estimated radius for U^{4+} of 0.93 Å. Later work suggested that preparations of K_3UF_7 melted in the presence of air could contain sufficient uranyl ion to yield the tetragonal structure. When made on a larger scale in the absence of air, and with carefully purified materials, this ordered, tetragonal phase of K_3UF_7 ($K_3UO_2F_5$-type) was not found *(48)*. However, a blue-green orthorhombic phase *(49)* was identified in addition to the disordered cubic phase found by *Zachariasen (47)*. In the orthorhombic-cubic conversion, the orthorhombic *a* axis becomes the *(110)* direction in the cube *(49)*. The cubic phase of K_3UF_7 gave diffuse diffraction spots, and could be formed only by quenching melts of K_3UF_7; slow cooling yielded the orthorhombic phase.

Zachariasen suggested that cubic K_3UF_7 contained seven-coordinated uranium, having UF_7^{3-} pentagonal bipyramids similar to $K_3UO_2F_5$ (Fig. 5b). His argument was based on his tetragonal structure of $K_3UO_2F_5$ which can be readily altered to give a fcc structure by very slight movement of the K and U atom positions and disordering the fluorine atoms *(47)*.

More definitive work has been done on the isostructural fcc $(NH_4)_3ZrF_7$. Its structure, based on very limited data, was described by *Hampson* and *Pauling (50)* as having isolated $(ZrF_7)^{3-}$ groups in the form of a capped octahedron with some rotational freedom. *Zachariasen (47)* questioned this ZrF_7^{3-} geometry and proposed for it as well as for cubic K_3UF_7 the above-mentioned pentagonal-bipyramidal conformation. Quite recently, a detailed single-crystal structure determination

Table 3. *Seven-coordinated compounds*

Compound [a]	Space Grp	Lattice Constants				Conformation	Ref.
K$_3$UO$_2$F$_5$	I4$_1$/a	9.160	18.167			pentagonal bipyramid	(46)
K$_3$NpO$_2$F$_5$	I4$_1$/a	9.12	18.12				(159)
K$_3$UF$_7$	I4$_1$/a	9.22	18.34			(ordered, may contain uranyl)	(47)
Rb$_3$UF$_7$	tetragonal						(48)
(NH$_4$)$_3$UO$_2$F$_5$	Cm	29.22	9.48	13.51	136°07'	distorted pentagonal bipyramid	(134)
(NH$_4$)$_3$ZrF$_7$	Fm3m	9.419				pentagonal bipyramid (disordered)	(51)
K$_3$UF$_7$	Fm3m	9.22					(47)
Rb$_3$UF$_7$	Fm3m	9.567					(48)
Cs$_3$UF$_7$	Fm3m	9.90					(54)
K$_3$ThF$_7$	Fm3m	~9.2					(49, 52)
Rb$_3$ThF$_7$	Fm3m	9.62					(53)
Cs$_3$ThF$_7$	Fm3m	10.04					(55)
β-UF$_5$	I4̄2d	11.47	5.208			much distorted pentagonal bipyramid with 4 corners shared	(43)
NpF$_5$	I4̄2d						(160)
K$_2$NbF$_7$	P2$_1$/c	5.846	12.693	8.515	90.0°	capped trigonal prism ~C2v	(45)
K$_2$TaF$_7$	P2$_1$/c	5.85	12.67	8.50	~90°		(44)
K$_2$UF$_7$	P2$_1$/c						(32, 34)
Rb$_2$UF$_7$	P2$_1$/c						(32, 34)
(NH$_4$)$_2$UF$_7$	P2$_1$/c						(32, 34)
Cs$_2$UF$_7$	P2$_1$/c						(32, 34)
Rb$_2$NpF$_7$	P2$_1$/c	6.26	13.42	8.90	~90°		(31)
Rb$_2$PuF$_7$	P2$_1$/c	6.27	13.416	8.84	~90°		(32)

[a] Underlined compounds are the structure types, determined by single crystal techniques.

of $(NH_4)_3ZrF_7$ by *Hurst* and *Taylor* (*51*) indeed confirmed the presence
of disordered $(ZrF_7)^{3-}$ groups. A refinement, including population factors
for statistically occupied sites, allows a slightly distorted pentagonal-
bipyramid to be placed in the unit cell. However, their structure contains
F—F bond contacts as short as 2.15 Å. Since this is much shorter than
2.4—2.5 Å observed between F—F pairs *shared* between polyhedra containing highly charged ions, it seems unlikely that F—F contacts could
be as short as 2.15 Å in isolated ZrF_7^{3-} polyhedra coordinated by singly
charged ammonium cations.

A number of analogous actinide compounds have corresponding face-
centered cubic powder patterns and are clearly related structurally. In
order of increasing cell edge (9.2—10.0 Å) these are: K_3UF_7 (*47*), K_3ThF_7
(*52*), Rb_3UF_7 (*48*), Rb_3ThF_7 (*53*), Cs_3UF_7 (*54*), and Cs_3ThF_7 (*55*).
Compounds of different stoichiometry but having apparently similar fcc
structures include: K_3UF_6 (*28*), K_3UF_8 (*28, 34*), K_3PaF_8 (*57*), Rb_3PaF_8
(*56*) and Cs_3PaF_8 (*57*). However, no structural details are known for
these compounds.

2.3. Eight Coordination

Eight coordination frequently occurs in lanthanide and actinide fluoride
complexes. The preponderance of these octacoordinate compounds can
be assigned to one of two idealized coordination polyhedra, namely the
square antiprism, symmetry (D_{4d}), or the dodecahedron (D_{2d}) with
triangular faces. Of the remaining idealized eight-coordinate polyhedra,
some are not represented and others have but few representatives in this
class of compounds.

An attempt has been made by *Hoard* and *Silverton* (*58*) to quantify
the energy relationship between the square antiprism and the dodecahedron considering coulombic repulsions and non-bonding repulsions
to be the dominating factors in determining the polyhedron type. The
main conclusion of their study was that no significant difference exists
between the ground state energies of the two types of polyhedra. Although
the dangers inherent in such a simple model are self evident, the experimental evidence supports the conclusion that these two polyhedra are
roughly equivalent in energy; the choice between them, in the actinide
fluoride structures, is apparently decided by the packing forces. Another
important, but not surprising, conclusion of the *Hoard* and *Silverton*
work is that the cube is considerably higher in energy than either of the
two more usual octa-coordination types. The relative interligand repulsions in various polyhedra have been tabulated by *King* (*59*).

The antiprism can be characterized by two parameters, i.e. ℓ/s, the
ratio of the length of the side of a triangular face to that of the edge of
the square face, and θ, the angle between the $\bar{8}$ axis and the vector from

15

the central ion to one of the ligands (see Fig. 6a). *Hoard* and *Silverton* established the "most favorable" values for these parameters, based on their model, to be $\ell/s = 1.057$ and $\theta = 57.3°$. The hard sphere model for the antiprism demands $\ell = s$ and $\theta = 59.25°$.

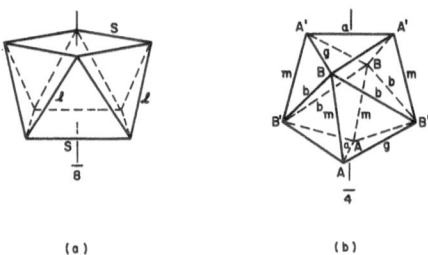

(a)   (b)

Fig. 6. Polyhedra for eight coordination showing distinguishing parameters. 6a. square antiprism, 6b. dodecahedrno

Three parameters are necessary to characterize the dodecahedron with triangular faces (see Fig. 6b). *Hoard* and *Silverton* chose for these parameters: θ_A, the angle between the M—A bond vector and the $\bar{4}$ axis of the polyhedron; θ_B, the corresponding angle made with the M—B bond vector; and the ratio of the bond lengths, M—A vector/M—B vector. The "most favorable" values for these parameters are $\theta_A = 35.2°$, $\theta_B = 73.5°$ and M—A/M—B = 1.03. The corresponding values for the hard sphere model are M—A = M—B, $\theta_A = 36.9°$ and $\theta_B = 69.5°$.

It is of interest to note that it takes only a small distortion of the antiprism to bring it to the dodecahedral conformation; in crystal structures for which the site symmetry of the central ion is low, it is often a moot question as to which is the better idealized geometry for the description of the polyhedron. It has been the practice to use the parameters chosen by *Hoard* and *Silverton* to distinguish between the two polyhedra. This procedure has recently been criticized by *Lippard* and *Russ* (60) who propose that different parameters be used in order to make this decision. The parameters they suggest are θ_{LR}, the angle between "best" planes through the atoms labeled AABB and through B'B'A'A' (interpenetrating trapezoidal planes, Fig. 6b, and Fig 7a) and $\langle d \rangle$ (see Table 4) the average deviation of the ligand atoms from these planes. It should be noted that these parameters, while helpful, amount essentially to a rejection test and do not provide sufficient criteria for assigning a distorded polyhedron to one of the idealized polyhedra. We have however tabulated them along with the more conventional parameters for several structures for which accurate structural data are available. A different and perhaps more objective criterion for testing the degree of deviation

8   COORDINATION

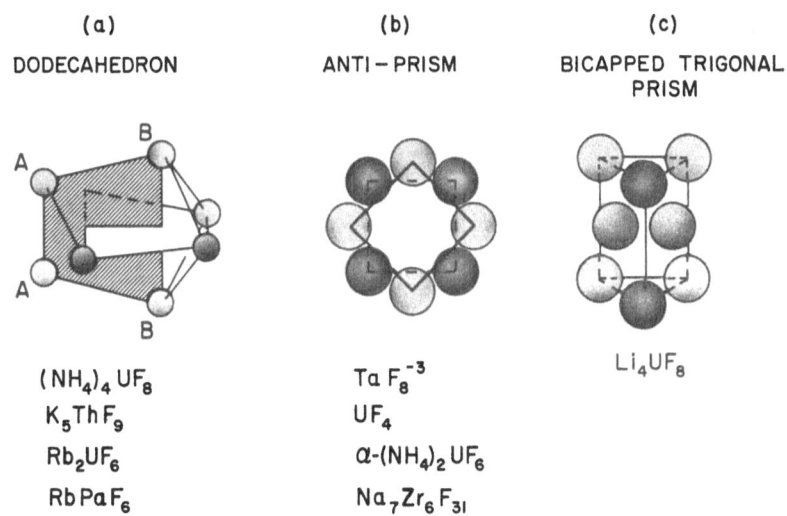

(a)	(b)	(c)
DODECAHEDRON	ANTI – PRISM	BICAPPED TRIGONAL PRISM

$(NH_4)_4 UF_8$ $Ta F_8^{-3}$ $Li_4 UF_8$

$K_5 ThF_9$ UF_4

$Rb_2 UF_6$ $\alpha\text{-}(NH_4)_2 UF_6$

$Rb Pa F_6$ $Na_7 Zr_6 F_{31}$

Fig. 7. Some examples of eight-coordination polyhedra

of a polyhedron from an idealized geometry is to find the best least squares fit of idealized polyhedra to the actual atom positions. A comparison of the sum of the squares of the residuals for the geometries to be tested then provides a meaningful parameter for making the choice of a suitable idealized geometry. In Table 4 UF_8^{4-} is assigned to the antiprism type by the authors in the original paper (61) although they stated it to be intermediate between the antiprism and the dodecahedron. The rms deviations (Å) for the distance of the atoms from their idealized positions in the case of $(NH_4)_4UF_8$ are 0.094 for the antiprism and 0.054 for the dodecahedron and would favor the latter configuration.

Isolated Polyhedra. The influence of the alkali cation on the coordination of the actinide ion makes it difficult to speak of "isolated polyhedra". Three dimensional structures have been determined for four actinide compounds which can be characterized as having isolated coordination polyhedra in the sense that there is no sharing of fluorine between the actinides. Of these four compounds, two [i.e., K_5ThF_9 (62) and $(NH_4)_4UF_8$ (67)] can properly be described as containing dodecahedra with triangular faces. That packing effects are important in these structures is apparent from the nature of the distortions. For example, in K_5ThF_9 the potassium atom closest to an edge of the polyhedron causes a shortening of that F—F distance; the concomittant shift of the thorium away from the edge produces longer than average Th—F distances. In fact, close examination of fluoride structures reveals that it is generally

Table 4. *Classification of Eight-coordinated Polyhedra*

Compound	X-ray symm	ℓ/s	M-A/M-B	θ	θ_A	θ_B	θ LR	$\langle d_0 \rangle$[a]	$\langle d_A \rangle$[b]	original authors' designation
Isolated Polyhedra										
$(NH_4)_4UF_8$ (61)	2	1.05	1.02	57.1	36.8	72.1	88°11′	0.12	0.18	A, (D_0)[c]
$K_4ThF_8 \cdot KF$ (62)	m		1.03		36.3	72.6	90	0		D_0
Na_3PaF_8 (25)	4/mmm	—	—	—	—	—	—	—	—	C
Li_4UF_8 (63)	m									bicapped trig prism
Chain Structures[d]										
α-$(NH_4)_2CeF_6$* (75)	2	1.04	1.00	56.2	40.2	71.8	84.5°	0.30	0.02	A
Rb_2UF_6 (80)	mm		1.04		39.5	69.9	90	0		D_0
$RbPaF_6$ (24)	222		1.12		30.9	77.6	90	0.02		D_0
γ-Na_2UF_6** (65)	mmmm	—	—	—	—	—	—	—	—	C
Three Dimensional Polymers										
UF_4 (77)									$\begin{cases} 0.01 \\ 0.11 \end{cases}$	A, A
$Na_7Zr_6F_{31}$*** (79) dodecahedron, D_0	D_{2d}	0.98	1.03	58.6	35.2	73.5	90	0	0.01	A
antiprism, A	D_{4d}	1.06		57.3					0	

* Isostructural with α-$(NH_4)_2UF_6$.

** Based on powder data.

*** The many actinide compounds isostructural with this compound are listed in Table 5.

[a] $\langle d_0 \rangle$ is the mean distance of the atoms from the two perpendicular planes through the dodecahedron, intersecting in the $\bar{4}$ axis.

[b] $\langle d_A \rangle$ is the mean distance of the atoms from the two parallel planes containing the square edges of the antiprism.

[c] As shown in the text, UF_8^{4-} fits parameters for the dodecahedron better than an antiprism.

[d] The simplest chain, a dimer, occurs in $(NH_4)_3CeF_7 \cdot H_2O$, which has no known actinide analog (81).

true that distortions in the polyhedra can be understood in terms of the cation packing, with radius ratio effects dominating.

The dodecahedron requires a larger radius ratio than the antiprism (i. e., 0.665 versus 0.645), and it is tempting to say that it is favored over the antiprism as a coordination type for the larger (with respect to the d-transition series) actinide elements. The lack of a sufficiently large number of well-determined structures makes such a suggestion tenuous at this time. However, there is some support for this thesis among the one dimensional polymers, as explained in the following section.

The coordination of Li_4UF_8 has been described as a triangular prism (Fig. 7c) with pyramids on two of the prism faces (63). This polyhedron (idealized symmetry C_{2v}) has been suggested to be of slightly higher energy than either the dodecahedron or antiprism. If this is true, then the reason for its stabilization must be looked for in the packing forces in the crystal. The dominating force is probably the approach of a ninth fluorine to the third rectangular face of the polyhedron, making this structure transitionary to the nine coordinate, tricapped trigonal prism to be discussed in the next section. In Li_4UF_8 this ninth fluorine lies at a distance of $3.30 \pm .03$ Å from the U^{4+} versus the average of $2.29 \pm .02$ Å for the other eight distances. To date, the bicapped trigonal prism has only been shown to exist in circumstances where a ninth atom approaches the third face (noted as "frustrated" nonacoordination in ref. 20)[1].

The eight-fold coordination type with the ligands situated at the corners of a cube maximizes non-bonded repulsions (20, 58, 59, 64), and it, therefore, is not an appealing coordination type. The cube is, indeed, unknown as a structure type among the molecular coordination compounds of the d-transition elements although it is prevalent among inonic compounds with the CaF_2 and $CsCl$ lattice types.

The single well-established example of cubic coordination in a complex fluoride is provided by the recent single crystal X-ray study of the coordination of PaF_8^{3-} in Na_3PaF_8 (25). The site symmetry for the PaF_8^{3-} ion is 4/mmm with F—F distances of 2.47 Å parallel to the four fold axis and 2.60 Å perpendicular to this axis, giving the ion a significant tetragonal distortion (with equal Pa—F distances). The Na_3PaF_8 structure type (which apparently includes Na_3UF_8 and Na_3NpF_8) can be visualized as the result of stacking two fluorite cells to make the tetragonal cell, placing the Pa atom at the origin, one Na atom half way up the tetragonal axis and the other two sodium atoms in the faces of the "cubic" cell (Fig. 8). Zachariasen first proposed this model for the structure of Na_3UF_7 in which seven fluorines are statistically distributed among the eight possible sites (65).

---

[1]) For a contrary view, see Ref. (173).

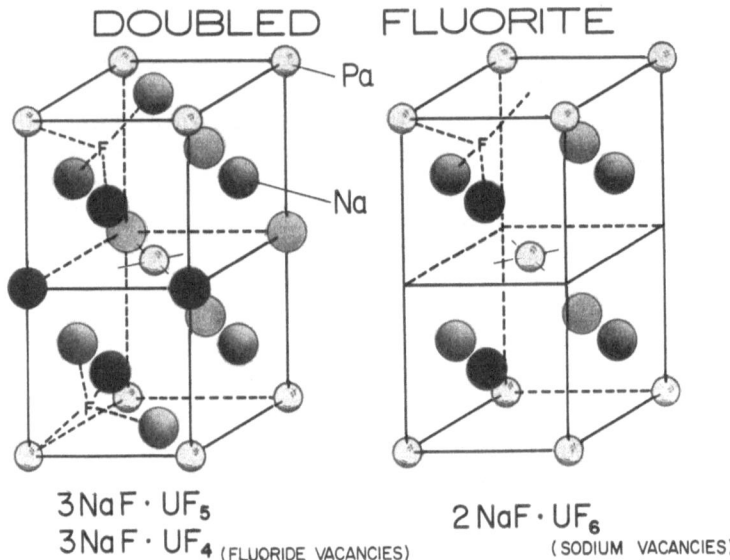

Fig. 8. The doubled fluorite cell as a model for Na_3PaF_8, Na_3UF_7, Na_2UF_8

The powder diffraction data of *Ruedorff* and *Leutner* (66) for Na_3UF_8 indicate it to be isostructural with Na_3PaF_8. If this is true, the fluorine atoms are required to be equivalent by symmetry. However, *Fukushima* and *Hecht* (67) find from NMR studies that the fluorines in Na_3UF_8 are non-equivalent and therefore the UF_8^{3-} unit has symmetry lower than D_{4h}. Yet, the X-ray evidence on Na_3PaF_8 shows the thermal ellipsoids for the fluorines to be normal, with no suggestion of statistical disorder at room temperature. Interestingly, *Fukushima* and *Hecht* found no evidence of a first order phase transition in Na_3UF_8 on lowering the temperature. Distortion in Na_3UF_8 was also observed by *Rigny*, *Dianoux* und *Plurien* in their magnetic susceptibility study. Their calculations agreed better with trigonal than with tetragonal distortion of UF_8^{3-} (36).

The proposed structure (68) of Na_2UF_8 is similar to that of Na_3UF_8 except that the Na has been removed from the tetragonal axis (see Fig. 8). Again, *Dianoux* and *Rigny* (69) find from NMR data that the UF_8^{2-} group only approximates D_{4h} symmetry, but is actually lower.

Instead of tetragonal symmetry as found for Na_3PaF_8 (and possibly for Na_3UF_8 and Na_3NpF_8) the compounds M_3UF_8 containing the larger alkali cations ($M = K$, Rb, NH_4, and Cs) and M_3PaF_8 ($M = K$, Rb, Cs) have cubic ($Fm3m$) symmetry with unit cells containing four formula weights. The cell edge for K_3PaF_8 is 9.235 Å (57), a value similar to the $Fm3m$ cell of K_3UF_7 descussed in the previous section (47). Placement of

20

fluorines in 32-fold positions with $x \approx 1/8$ results in a cubic PaF_8^{3-} arrangement with O_h symmetry but the Pa—F and K—F distances would be only about 2.0 Å, indicating that this arrangement cannot be correct. A preferable alternative is that these structures are related to the cubic $(NH_4)_3ZrF_7$—K_3UF_7 type with eight instead of seven statistically distributed fluorines. For example, a dodecahedron can be approximated starting with a seven coordinated model of the above type (see Fig. 5b), having one fluorine above and below an equatorial band of five fluorines. Removing one of the equatorial fluorines leaves four in the form of a trapezoid; two fluorines are added in line with the non-equatorial fluorines, with their contact approximately in the location of the one originally removed from the equatorial position. This yields a second trapezoid normal to the first. Minor adjustments will then permit the eight fluorines to be distributed in sites of the F_{m3m} space group. Slight displacement of alkali ions from their ideal F_{m3m} sites is also required as done by *Zachariasen* (47).

Volume considerations rule out the inefficient cubic arrangement of eight fluorines. For example, Rb_3PaF_8 has a cell edge (56) of 9.6 Å which gives a volume per formula weight of 221 Å³. Subtracting 3×28 Å³ for the three rubidium ions (21) leaves 17.2 Å for each of the eight fluorines. In the case of K_3PaF_8 the volume (57) per fluorine is less than 17 Å³ if we take 21 Å³ for K^+. The structure is clearly tightly packed. Nonetheless there are many examples of compounds belonging to this fcc class, not only with the tetravalent and pentavalent actinides, but also including the tri- and tetravalent lanthanides as well. For example, $M_3CeF_{6,7}$ (71), $M_3PrF_{6,7}$ (72), (M = Rb, Cs) and $Cs_3NdF_{6,7}$ (73) $Cs_3DyF_{6,7}$ (73) and Cs_3TbF_7 (74).

Structures containing linear chains. Of the several octacoordinate structures which contain linear chains of actinide fluoride polyhedra, only three have been accurately determined using three dimensional X-ray data. All of these structures have been found to contain polyhedra of either the dodecahedron or antiprism type.

One can connect dodecahedra across a, b or m type edges (see Fig. 6). Edges of the b type are probably not favored for connecting polyhedra because they are not "contact" distances. (It is generally observed that two atoms in a bridge are at a distance shorter than their normal Vd W contact distance.) There are two structures indicated in Table 4 to be linear chains of dodecahedra. The first of these $RbPaF_6$, utilizes edges of type a for linking. One can see from examination of Table 4 that the angles in this PaF_8 dodecahedron are greatly distorted from the most favorable dodecahedron calculated by Hoard and Silverton. The reason for this is, of course, the strong Pa-Pa repulsion across the a edge, which tends to reduce θ_A and at the same time allows the θ_B angle to increase.

Table 5. *Eight-coordinated compounds*

Compound[a]	Space Grp	Lattice Constants		Conformation	Ref.
Na_3PaF_8	I4/mmm	5.493	10.970	cube D4h	(25)
Na_3UF_8	I4/mmm	5.470	10.940		(25, 66)
Na_3NpF_8	I4/mmm	5.443	10.837		(25)
$LaOF$	P$\bar{4}$2$_1$m	4.091	5.837	cube D4h	(142)
$PuOF$		4.05	5.72		(142)
Na_3UF_7	I4/mmm	5.46	10.92	cube D4h statistically occupied	(65)
Na_3PuF_7	I4/mmm	5.46(?)	10.92(?)	by 7F's	(156)
$CsUF_7$	Fm3m	5.51		cube D4h statistically occupied by 7 F's	(158)
α-K_2UF_6	Fm3m	5.934		cubic, with disordered cations	(21)
α-K_2ThF_6	Fm3m	5.994			(21)
α-K_2NpF_6	Fm3m	5.905			(159)
α-Na_2UF_6	Fm3m	5.565			(21)
α-Na_2ThF_6	Fm3m	5.687			(97)
$CfOF$	Fm3m	5.561		cube Oh with disordered anions	(141)
$AcOF$		5.943			(142)
Na_2UF_8	I4/mmm	5.27	11.20	cubic (not D4h)	(68) (69)

		a	b	c			Ref.
K_5ThF_9 Rb_5ThF_9	$Cmc2_1$	7.848	12.840	10.785		dodecahedron	(62) (54)
$(NH_4)_4UF_8$	$C2/c$	13.126	6.692	13.717	121.3°	dodecahedron (distorted)	(61)
$(NH_4)_4PaF_8$	$C2/c$	13.18	6.71	13.22	117.17°		(82)
$(NH_4)_4NpF_8$	$C2/c$						(83)
$(NH_4)_4PuF_8$	$C2/c$						(84)
$(NH_4)_4AmF_8$	$C2/c$						(85)
$(NH_4)_4CeF_8$	$C2/c$	13.02	6.66	13.61	121.3°		(86)
UO_2F_2	$R\bar{3}m$	4.21	15.69			bicapped (O), octahedron (F), D3d	(87)
NpO_2F_2		4.178	15.80				(8)
PuO_2F_2		4.154	15.84				(88, 8)
AmO_2F_2		4.136	15.85				(8)
$(NH_4)_3CeF_7 \cdot H_2O$	$P2_1/n$	11.064	12.104	7.134	95.70°	dodecahedron shared edge dimer	(81, 86)
$\gamma\text{-}Na_2UF_6$	$Immm$	5.56	4.01	11.64		distorted cube sharing edges to form chain	(65)
$\gamma\text{-}(NH_4)_2UF_6$	$Pbca$	16.17	14.07	11.69		alternating antiprisms and dodecahedra forming chain	(38)
Rb_2UF_6	$Cmcm$	6.998	12.098	7.669		dodecahedron sharing edges to form chain	(80, 89)
Rb_2NpF_6	$Cmcm$	6.986	12.068	7.628			(89)
Rb_2PuF_6	$Cmcm$	6.971	12.033	7.602			(89)
Rb_2AmF_6	$Cmcm$	6.962	12.001	7.579			(89, 90)
Rb_2CmF_6	$Cmcm$	6.931	11.996	7.567			(89)

Table 5 (continued)

Compound[a]	Space Grp.	Lattice Constants			Conformation	Ref.
RbPaF$_6$	Cmma	8.048	12.025	5.861	dodecahedron sharing edges to form chain	(24)
RbUF$_6$	Cmma	8.06	11.89	5.82		(56)
KPaF$_6$	Cmma	7.98	11.54	5.64		(56)
KUF$_6$	Cmma	7.96	11.46	5.61		(56)
NH$_4$PaF$_6$	Cmma	8.03	11.90	5.84		(56)
NH$_4$UF$_6$	Cmma	8.03	11.89	5.83		(56)
CsPaF$_6$	Cmma	8.06	12.56	6.14		(56)
α-(NH$_4$)$_2$CeF$_6$	Pbcn	7.026	12.098	7.439	antiprism sharing edges to form chain	(75)
α-(NH$_4$)$_2$UF$_6$						(76, 84)
KAmO$_2$F$_2$	R$\bar{3}$m	6.78	α = 36.25°		bicapped octahedron linked to form sheets (O's form cap)	(91)
KNpO$_2$F$_2$	R$\bar{3}$m	6.80	36.32			(159)
RbNpO$_2$F$_2$	R$\bar{3}$m	6.814	36.3			(8)
RbPuO$_2$F$_2$	R$\bar{3}$m	6.796	36.3			(8)
RbAmO$_2$F$_2$	R$\bar{3}$m	6.789	36.25			(8)
NH$_4$PuO$_2$F$_2$	R$\bar{3}$m	6.817	36.25			(8)
Li$_4$UF$_8$[b]	Pnma	9.960	9.883	5.986	bicapped trigonal prism	(63)
YF$_3$[c]	Pnma	6.353	6.850	4.393	bicapped trigonal prism (nearly tricapped)	(92)
BkF$_3$	Pnma	6.70	7.09	4.41		(11)
CfF$_3$	Pnma	6.653	7.041	4.395		(93)

Compound						Description	Ref.
UF_4	C2/c	12.73	10.75	8.48	126.33°	antiprism linked in 3-dimensions by sharing all corners	(77)
ThF_4	C2/c	12.90	10.93	8.58	126.4°		(94, 95)
PaF_4	C2/c	12.83	10.82	8.45	126.4°		(82, 95, 96)
NpF_4	C2/c	12.64	10.70	8.36	126.4°		(95, 97)
PuF_4	C2/c	12.59	10.69	8.29	126.0°		(95, 97)
AmF_4	C2/c	12.56	10.58	8.25	125.9°		(95, 98)
CmF_4	C2/c	12.51	10.61	8.20	125.8°		(95)
BkF_4	C2/c	12.47	10.58	8.17	125.9°		(95)
CfF_4	C2/c						(99)
$Na_7Zr_6F_{31}$	R$\bar{3}$	13.807	9.429			antiprism sharing 4 corners and one edge to form 3-dimentional network	(79)
$Na_7Th_6F_{31}$	R$\bar{3}$	14.96	9.912				(100, 101)
$K_7Th_6F_{31}$	R$\bar{3}$	{15.293	10.449				(70)
		15.32	10.49				(52)
$Rb_7Th_6F_{31}$	R$\bar{3}$	15.39	10.73				(54, 101)
$Na_7Pa_6F_{31}$	R$\bar{3}$	14.81	9.85				(82)
$K_7Pa_6F_{31}$	R$\bar{3}$	15.18	10.47				(82)
$Rb_7Pa_6F_{31}$	R$\bar{3}$	15.49	10.75				(82)
$Na_7U_6F_{31}$	R$\bar{3}$	14.72	9.84				(21, 100, 101)
$K_7U_6F_{31}$	R$\bar{3}$	15.09	10.38				(21, 48, 102, 101)
$Rb_7U_6F_{31}$	R$\bar{3}$	15.49	10.42				(48, 102)
$(NH_4)_7U_6F_{31}$	R$\bar{3}$	15.40	10.49				(76, 84)
$Na_7Np_6F_{31}$	R$\bar{3}$	{14.60	9.728				(103)
		14.64	9.785				(104)
$K_7Np_6F_{31}$	R$\bar{3}$	14.99	10.31				(103)
$Rb_7Np_6F_{31}$	R$\bar{3}$	15.24	10.62				(103)
$Na_7Pu_6F_{31}$	R$\bar{3}$	{14.55	9.741				(104, 21)
		14.52	9.704				(103)

Table 5 (continued)

Compound[a]	Space Grp.	Lattice Constants		Conformation	Ref.
$K_7Pu_6F_{31}$	$R\bar{3}$	14.93	10.28		(103, 21)
$Rb_7Pu_6F_{31}$	$R\bar{3}$	15.21	10.61		(103)
$(NH_4)_7Pu_6F_{31}$	$R\bar{3}$	15.18	10.36		(84)
$Na_7Am_6F_{31}$	$R\bar{3}$	14.48	9.665		(104)
$K_7Am_6F_{31}$	$R\bar{3}$	14.938	10.293		(105)
$Na_7Cm_6F_{31}$	$R\bar{3}$	14.41	9.661		(104)
$K_7Cm_6F_{31}$	$R\bar{3}$	14.89	10.254		(105)
$Na_7Bk_6F_{31}$	$R\bar{3}$				(99)
$Na_7Cf_6F_{31}$	$R\bar{3}$				(99)

[a] Underlined compounds are structure types, determined by single crystal techniques.
[b] In Li_4UF_8 a ninth fluorine is at a distance of 3.3 Å versus the average of 2.29 Å for the other eight.
[c] In YF_3 the ninth fluorine is at 2.60 Å versus 2.3 Å; Y is thus intermediate between 8 and 9 coordinated.
[d] For additional examples of bicapped trigonal prismatic coordination see Ref. (173).

The many compounds which are isostructural with $RbPaF_6$ are listed in Table 5.

The compounds Rb_2UF_6 (80) and α-$(NH_4)_2UF_6$ (75) exhibit closely related chain structures. The Rb-compound consists of infinite chains of $[UF_8]$ dodecahedra linked by edges of type m; the NH_4-containing compound consists of chains of antiprisms linked by s edges on opposite square faces. The uranium positions as well as the nitrogen and rubidium positions are equivalent in the two structures. The fluorines in $(NH_4)_2UF_6$ have been distorted from the dodecahedral arrangement in Rb_2UF_6 so as to make maximum use of the H---F hydrogen bonding. This behavior has been taken to indicate that the dodecahedron provides the energetically lower ground state configuration for this structure type (75).

Besides the α-phase, $(NH_4)_2UF_6$ exhibits at least three higher temperature phases (76), β, γ, δ. All of these phases have U—U distances of ca. 4 Å and therefore contain UF_8 chains. Preliminary results indicate, for example, that γ-$(NH_4)_2UF_6$, the room temperature stable phase, contains chains of alternate antiprisms and dodecahedra (38). The common edges are apparently the ℓ edges for the antiprisms and the a edges for the dodecahedra.

A linear chain of polyhedra with a distorted cubic arrangement has been proposed for γ-Na_2UF_6, on the basis of powder data (65). This interesting proposal is plausible, especially in light of the established structure of Na_3PaF_8, but verification by single crystal techniques would certainly be worthwhile.

Polyhedra linked in 3-dimensions. Complete three dimensional structural studies have been done on eight-coordinated UF_4 (77) which is of the ZrF_4 structure type (78) and on $Na_7Zr_6F_{31}$ (79) which is isostructural with all of its tetravalent actinide analogs (Table 5). Both structures can be described as three dimensional polymers. In such compounds the M—M repulsion becomes so important that the actinide coordination polyhedra are severely distorted from any idealized coordination type. Both structures can probably be best described, however, as containing the antiprismatic coordination type.

The structure of $Na_7Zr_6F_{31}$ typifies the structure of all of the known 7:6 alkali fluoride/MF_4 compounds of which there are many examples known for alkali metals heavier than lithium (see Table 5) (79). These structures contain one formula per unit cell in the space group R $\bar{3}$. Each Zr (or actinide) is surrounded by eight F atoms arranged as a square antiprism. These antiprisms share corners to enclose a large cubo-octahedral cavity which is occupied by one disordered fluorine atom. *Thoma* has made predictions of the existence of many fluoro complexes having this 7:6 ratio based on the M^+/M^{4+} radius ratio (153). α-NH_4UF_5 has a powder pattern indistinguishable from that of $(NH_4)_7U_6F_{31}$ (84), and is therefore presumed to be an NH_4F-deficient form of this structure type. An excellent review of Eight-coordination Chemistry is that of *Lippard* (*172*).

2.4. Nine Coordination

As has been pointed out in the introduction, the structures of many of the actinide fluoride complexes can be described as some distortion of a model based on closest packing of spheres. These relationships are most obvious for the nine-coordinate structures since, for example, hexagonal closest packing provides a natural way for the packing of the nine coordinate "tricapped trigonal prism" polyhedron with only a small distortion of the closest packed arrangement. It is therefore useful to consider the packing model as well as the coordination polyhedron type when describing these structures. The structures to be discussed in this section are closely related, as can be seen by comparison of their cell constants (see Table 6). The relationship between the cell constants is a reflection of the packing, and this is pointed out in the following discussion wherever it is felt to be illustrative.

Although several different coordination types have been established among the eight-coordinate actinide fluorides, only one polyhedron type has been discovered among the nine-coordinate compounds. This polyhedron is the tricapped trigonal prism (symmetry D_{3h}) which can be thought of as being the result of placing atoms in the three rectangular faces of a triangular prism (Fig. 9). A second, less symmetrical polyhedron,

9 COORDINATION

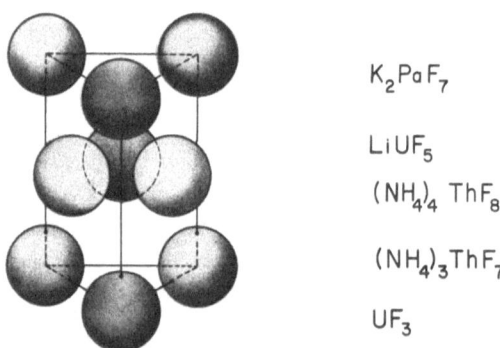

K_2PaF_7

$LiUF_5$

$(NH_4)_4ThF_8$

$(NH_4)_3ThF_7$

UF_3

Fig. 9. The tricapped trigonal prism and some compounds showing this coordination

the monocapped square antiprism may be derived from the tricapped prism with only very small displacements of the atoms and would therefore be expected to occur occasionally among the nine-coordinated actinides. One might expect, since the actinide ions are the largest representatives of their charge types, and since the tricapped trigonal prism, requiring a radius ratio of 0.732, can comfortably accommodate positive ions with a radius of ca. 97 Å, that nine coordinate structures would be very common among the actinide fluorides. This is indeed the case, especially among the lower oxidation states where, for example, an isostructural series exists from AcF_3 to BkF_3 in which the actinide ion is nine-coordinate, and the structure is that of LaF_3. As the series is ascended, the actinide (III) radius decreases so that at Bk a dimorphic structure appears for BkF_3 which corresponds to the YF_3 structure in which the coordination is between eight and nine, with one of the fluorines significantly more distant from the central ion than the other eight (92). As might be expected, CfF_3 and the heavier lanthanides have this same structure.

Nonacoordination is rare among elements of the lower transition series, especially for isolated ions containing monatomic ligands. One might have expected the larger actinides to provide some examples of such ions. The fact is that no example of such an ion has been found among the actinide fluorides, and the only known compounds with a 5:1 alkali fluoride to actinide fluoride ratio occur with thorium e.g. K_5ThF_9 which forms a "double salt" structure with eight-coordinated thorium and an isolated fluoride ion (62). Such a result is obviously caused by the inability of the structure to stabilize an ion with a charge of minus five. An attempt to prepare an isolated nine-coordinated ion with a smaller negative charge by choosing as the central ion an actinide in a higher oxidation state (such as Pa^{5+}) would probably not succeed because of its relatively smaller ionic radius. However, investigations of tetravalent actinide fluoride systems with cations larger than Cs^+ might prove fruitful. The simplest nonacoordinate structures thus far found are those in which the actinide fluoride part of the structure is a linear chain, although a dimeric unit (39) having this coordination may exist in a compound of approximate composition $7 NH_4F \cdot 2 ThF_4 \cdot H_2O$.

Structures have been established for compounds having chains formed by connecting polyhedra by edges and by triangular faces. An example of this latter type of chain is afforded by β_2-Na_2ThF_6 and β_1-K_2UF_6 and their isostructural compounds (investigated only by powder diffraction techniques). In this type of chain the tricapped trigonal prisms share the triangular faces perpendicular to the three fold axis of the ideal polyhedron to form infinite chains. These structures can be constructed by considering the fluorines to be hexagonal closest packed with the stacking

Table 6. *Nine-coordinated compounds*

Compound[a]	Space Grp.	Lattice Constants		Conformation	Ref.
LaF$_3$	P$\bar{3}$c1	7.185	7.351	irregular tricapped trigonal prism	(106, 106a)
AcF$_3$	P$\bar{3}$c1	7.41	7.55		(107)
UF$_3$	P$\bar{3}$c1	7.181	7.348		(19, 133)
NpF$_3$	P$\bar{3}$c1	7.129	7.288		(19)
PuF$_3$	P$\bar{3}$c1	7.093	7.254		(19)
AmF$_3$	P$\bar{3}$c1	7.044	7.225		(108)
CmF$_3$	P$\bar{3}$c1	6.999	7.179		(108)
BkF$_3$[b]	P$\bar{3}$c1	6.97	7.14		(11)
CaThF$_6$	P$\bar{3}$c1	6.994	7.171		(109)
SrThF$_6$	P$\bar{3}$c1	7.162	7.324		(109)
BaThF$_6$	P$\bar{3}$c1	7.419	7.516		(109)
PbThF$_6$	P$\bar{3}$c1	7.280	7.404		(109)
EuThF$_6$	P$\bar{3}$c1	7.124	7.360		(109)
CdThF$_6$	P$\bar{3}$c1	6.963	7.109		(109)
CaUF$_6$[c]	P$\bar{3}$c1	{ 6.928	7.127		(109)
	P$\bar{3}$c1	{ 3.997	7.103	(Na$_3$As structure type?)	(133)
SrUF$_6$	P$\bar{3}$c1	7.122	7.293		(109)
BaUF$_6$	P$\bar{3}$c1	7.403	7.482		(109)
PbUF$_6$	P$\bar{3}$c1	7.245	7.355		(109)
CaNpF$_6$	P$\bar{3}$c1	6.918	7.100		(109)
SrNpF$_6$	P$\bar{3}$c1	7.109	7.260		(109)
BaNpF$_6$	P$\bar{3}$c1	7.374	7.450		(109)
PbNpF$_6$	P$\bar{3}$c1	7.212	7.360		(109)
CaPuF$_6$	P$\bar{3}$c1	6.918	7.097		(109)
SrPuF$_6$	P$\bar{3}$c1	7.091	7.255		(109)

Compound	Space group	a	b	c	angle	Structure	Ref
$LiUF_5$	$I4_1/a$	14.884	6.547			tricapped trigonal prism sharing edges and corners to form network	(110)
$\underline{LiThF_5}$	$I4_1/a$	15.10	6.60				(111, 101)
$LiPaF_5$	$I4_1/a$	14.97	6.576				(82)
$LiNpF_5$	$I4_1/a$	14.80 / 14.75	6.516 / 6.478				(112) / (103)
$LiPuF_5$	$I4_1/a$	14.67 / 14.65	6.479 / 6.486				(112) / (103)
$LiAmF_5$	$I4_1/a$	14.63	6.449				(112)
$LiCmF_5$	$I4_1/a$	14.57	6.437				(112)
U_2F_9	$I\bar{4}3m$	8.471				capped trig prism?, forming framework	(113)
$\underline{NaTh_2F_9}$	$I\bar{4}3m$	8.722					(113)
KU_2F_9	Pnam	8.70	7.035	11.48		capped trig. prism, sharing ends and edges	(114, 21)
KTh_2F_9	Pnam	8.85	7.16	11.62			(21, 52)
KNp_2F_9	Pnam	8.63	7.01	11.43			(21)
KPu_2F_9	Pnam	8.56	6.95	11.33			(21)
$\underline{CsU_2F_9}$	C2/c	15.649	7.087	8.689	118.1°	capped trigonal prism	(115)
$\underline{(Na, Li)_7Th_6F_{31}}$	P$\bar{3}$c1	9.9056		13.282		in part tricapped trigonal prism linked to ThF_{10} polyhedron in 3-dimensions	(116, 168)
$\underline{Na_7Th_6F_{31}}$	P$\bar{3}$c1						(168)
$\underline{(NH_4)_3ThF_7}$	Pnma	13.944	7.928	7.041		tricapped trigonal prism, 4 shared fluorines	(117)

Table 6 (continued)

Compound[a]	Space Grp.	Lattice Constants				Conformation	Ref.
K_2PaF_7	C2/c	13.760	6.742	8.145	125.17°	tricapped trigonal prisms, sharing edges to form chain	(14)
Rb_2PaF_7	C2/c						(56)
$(NH_4)_2PaF_7$	C2/c						(56)
Cs_2PaF_7	C2/c	14.937	7.270	8.266	125.32°		(118)
$(NH_4)_4ThF_8$	P1̄	8.477 88.38°	8.364 96.08°	7.308 106.33°		tricapped trigonal prism sharing edges to form chain. One F not bonded to Th	(119)
β_1-K_2UF_6	P6̄2m	6.553	3.749			tricapped trigonal prism sharing ends to form chain	(65, 120)
β_1-K_2ThF_6	P6̄2m	6.579	3.824				(65)
Rb_2ThF_6	P6̄2m	6.85	3.83				(121)
β_1-K_2NpF_6	P6̄2m	6.56	3.73				(159)
β_1-Cs_2PuF_6	P6̄2m	7.185	4.11				(155)
β_2-Na_2ThF_6	P32	5.99	3.81			tricapped trigonal prism sharing ends to form chain	(65, 21)
β_2-Na_2UF_6	P32	5.95	3.73				(65, 21)
β_2-K_2UF_6	P32	6.53	4.04				(65, 21)
$NaNdF_4$	P6̄	6.100	3.711			tricapped trigonal prism sharing ends to form chain	(138)
$NaPuF_4$[d]	P6̄	6.119	3.752				(138, 65, 21, 167)
$NaUF_4$[d]	P6̄	6.167	3.77				(138, 101)
$NaAmF_4$[d]	P6̄	6.109	3.731				(150)
$KNaThF_6$	P3̄	6.307	7.891			tricapped trigonal prisms sharing ends	(101, 131)
$RbNaThF_6$	P3̄						(101)

$RbTh_3F_{13}$	$P2_1ma$	8.649	8.176	7.4453	tricapped trig. prism, forming columns by sharing alternate edges, the columns joined by shared corners	(130)
KTh_3F_{13}	?					(52)
$CsTh_3F_{13}$	?					(55)
RbU_3F_{13}	?					(48)
$NH_4U_3F_{13}$	?					(84)
$NH_4Pu_3F_{13}$	?					(84)
CsU_6F_{25}	$P6_3/mmc$	8.242	16.412		tricapped trigonal prism, sharing edges and corners to form double rings of six polyhedra each	(128)
KTh_6F_{25}						(129)
$CsTh_6F_{25}$						(129)
KU_6F_{25}						(129)
RbU_6F_{25}						(129)
CsU_6F_{25}						(129)
$\alpha\text{-}KTh_6F_{25}$	$R\bar{3}m$	8.313	25.262		tricapped trigonal prism, sharing edges and corners to form double rings of six polyhedra each	(129)
$RbTh_6F_{25}$						(129)

[a] Underlined compounds are structure types, determined by single crystal techniques.
[b] BkF_3 and CfF_3 occur in the YF_3 structure type and are described in Table 5.
[c] $CaUF_6$ may be of the Na_3As structure type, $P6_3/mmc$, rather than the LaF_3 type, see Ref. (133).
[d] $NaF \cdot MF_3$ compounds are likely isostructural with $NaNdF_4$, Ref. (138).
[e] In $(NH_4)_7Th_6F_{31} \cdot H_2O$ dimeric units are found, in which two thoriums share three fluorines, and each thorium is further coordinated by a puckered equatorial ring of six fluorines (39).

33

sequence ABAB----. In both of these structures the unit cell may be chosen so that each layer has three fluorines and six holes. The actinide ion is placed in one of these holes in, let us say, the A layer and chosen as the origin. In β_1-K_2UF_6, K^+ is placed in two holes in the B layer at coordinates 1/3, 2/3, 1/2 and 2/3, 1/3, 1/2 (65, 120). This introduction of a large cation results in a considerable expansion of the layer and a corresponding compaction of the interlayer spacing. Thus the axial ratio, c/a', decreases relative to the ideal hcp axial ratio from 1.62 to ca. 1.00. In β_2-Na_2ThF_6, which has a different space group, the Na^+, though smaller than K^+, occupies an octahedral site between the A and B layers (65). Thus the ion is smaller, the hole longer, and the resultant distortion of close packing is less. The axial ratio, c/a', is now ~1.12.

Compounds containing linear chains of polyhedra connected by edges are: K_2PaF_7 (14), $(NH_4)_4ThF_8$ (119) and $(NH_4)_3ThF_7$ (117). Of these, the first two compounds form a chain by sharing an edge formed by connecting a vertex above the square face to a corner of the trigonal face in the prism. The second shared edge is related to the first by what would be a two-fold rotation axis in the idealized polyhedron, in such a way that the edge is formed by a corner of the opposite trigonal face and the vertex of an adjacent prism face. In the resultant chain, the trigonal axis of the idealized polyhedra is inclined to the chain axis. It is interesting to note that in $(NH_4)_4ThF_8$ one of the fluorines is not included in the first coordination sphere of thorium and that the compound forms a "double salt" structure which could probably be more properly written as $(NH_4)_3ThF_7 \cdot NH_4F$. The second type of chain is formed by $(NH_4)_3ThF_7$ which utilizes the same type of edge of the polyhedron, but the two chaining edges are related by what would be a mirror parallel to the three fold axis in the idealized polyhedron (117). In this arrangement the trigonal axis of the idealized polyhedron is normal to the chain axis.

The structure of the uranium compound in the series of isostructural compounds $LiMF_5$ where M = Th, U, --- Cm has been accurately determined (110). The coordination polyhedron of the U atom can be described in terms of the usual tricapped trigonal prism. In this case the polyhedron shares eight corners with adjacent polyhedra with one fluorine atom free, completing the nine coordination. The lithium atom fills a slightly distorted octahedral hole.

The compounds KM_2F_9 with M = Th, U, Np and Pu are all isostructural, with accurate X-ray data being available for the uranium compound (114). The tricapped trigonal prisms are packed into a complicated three dimensional network sharing edges and corners. The K^+ has six fluorines at a distance of 2.7 Å and two at 3.09 Å to form a distorted cube.

The unit cells of U_2F_9 and $NaTh_2F_9$ are closely related and structures have been proposed by Zachariasen on the basis of powder pattern data (113). The structure of U_2F_9 was solved by consideration of fluorine packing and coordination requirements for uranium. While the structure has not been subjected to the test of accurate intensity data, the coordination number of uranium is nine (in the usual tricapped trigonal prism) and the fluorine packing is only slightly distorted from hcp. The uranium atoms are equivalent, being in the special position with coordinates x, x, x in the space group $I\bar{4}3m$. The crystallographic equivalence of the uranium atoms and the black color of the crystals suggest that the uranium atoms resonate between the valence states of four and five. The proposed structure (113) of $NaTh_2F_9$ is the same as that for U_2F_9 with the exceptions that the Na atoms fill a distorted octahedral hole at 1/2, 0, 0 and that the actinide ion has the fixed valence of 4+.

In the compound CsU_2F_9, whose structure has recently been carefully determined, the U^{4+} coordination involves a very distorted tricapped trigonal prism (115). In this case, however, eight of the neighboring fluorines are shared with adjacent polyhedra to form a layered polyion. The ninth corner of the coordination polyhedron is effectively occupied by only one half a fluorine in a statistically occupied set. Thus the coordination number might be said to be 8.5.

The long isostructural series of compounds MF_3 (where M = La, Ac, U, ---Bk) and the series CMF_6 (where M = Th, U or Np and C ranges over a large selection of divalent cations) were originally thought to have the space group $P6_3/mcm$. The structure for LaF_3 as proposed in this space group gives an hcp packing for fluorine with the heavy metal ion packed in a site with five nearest neighbors and site symmetry $\bar{6}m2$ (a trigonal bipyramid). *Mansmann* (106) and *Zalkin, Templeton,* and *Hopkins* (106a) later recognized that the cell symmetry is trigonal (instead of hexagonal) and refined the structure in the space group $P\bar{3}c1$ and found it to be significantly distorted from the previous structure. The refined structure can be envisioned as a distortion of the earlier structure in which the trigonal bipyramid has been distorted in such a way as to allow closer approach of four additional fluorine atoms to give the heavy element ion a coordination of nine. In the compounds CMF_6 the actinide and divalent cation are presumed to be disordered on the La sites of the LaF_3 structure (or the Na_3As structure (133)). In the YF_3 structure type (92), which applies to BkF_3 and CfF_3, the actinide coordination is between eight and nine, with one of the fluorines being significantly more distant from the actinide than the other eight.

$Na_7Th_6F_{31}$ apparently displays two structure types, the rhombohedral $Na_7Zr_6F_{31}$ which is metastable, and a trigonal ($P3c1$) structure (168). The latter structure forms at least a partial solid solution in which

Table 7. *Summary*

Transition Metal Ion	Compound	Structure Type	Coordination Number	Reference
Y^{+3}	YF_3	YF_3	8	(92)
La^{+3}	LaF_3	LaF_3	9	(106, 106a)
	$LaOF$	$LaOF$ (tetragonal)	8	(142)
Nd^{+3}	$NaNdF_4$	$NaNdF_4$	9	(138)
Ac^{+3}	AcF_3	LaF_3	9	(107, 19)
	$AcOF$	$CfOF$	8	(142)
U^{+3}	Zr_3UF_{15}			(101)
	K_3UF_6	K_3UF_7 (?) (cubic)	6 (?)	(28)
	Rb_3UF_6	K_3UF_7 (?) (cubic)	6 (?)	(101)
	Cs_3UF_6	K_3UF_7 (?) (cubic)	6 (?)	(101)
	$NaUF_4$	$NaNdF_4$	9	(101, 138)
	UF_3	LaF_3	9	(19, 133)
Np^{+3}	NpF_3	LaF_3	9	(19)
Pu^{+3}	$NaPuF_4$	$NaNdF_4$	9	(65, 21, 138, 167)
	$KPuF_4$	$KLaF_4$ (cubic)	8	(167)
	KPu_2F_7	Cubic		(167)
	PuF_3	LaF_3	9	(19)
	$PuOF$	$LaOF$ (tetragonal)	8	(142)
Am^{+3}	$NaAmF_4$	$NaNdF_4$	9	(150)
	$KAmF_4$	$KLaF_4$ (cubic)	8	(167)
	KAm_2F_7	Cubic		(167)
	AmF_3	LaF_3	9	(108)
Cm^{+3}	CmF_3	LaF_3	9	(108)
Bk^{+3}	BkF_3	YF_3	8	(11)
	BkF_3	LaF_3	9	(11)
Cf^{+3}	CfF_3	YF_3	8	(93)
	$CfOF$	$CfOF$	8	(141)
Zr^{+4}	$(NH_4)_3ZrF_7$	K_3UF_7 (cubic)	7	(47, 51)
	$Na_7Zr_6F_{31}$	$Na_7Zr_6F_{31}$	8	(79)
	$\gamma\text{-}Na_2ZrF_6$	$\gamma\text{-}Na_2ZrF_6$	7	(135)
Ce^{+4}	$(NH_4)_4CeF_8$	$(NH_4)_4UF_8$	8	(86)
	$(NH_4)_3CeF_7 \cdot H_2O$	$(NH_4)_3CeF_7 \cdot H_2O$	8	(86, 81)
	$\alpha\text{-}(NH_4)_2CeF_6$	$\alpha\text{-}(NH_4)_2CeF_6$	8	(75)

Table 7 (continued)

Transition Metal Ion	Compound	Structure Type	Coordination Number	Reference
Th^{+4}	K$_5$ThF$_9$	K$_5$ThF$_9$	8	(62, 52)
	Rb$_5$ThF$_9$	K$_5$ThF$_9$?	(8)	(54)
	Na$_4$ThF$_8$	cubic		(100, 124, 21)
	(NH$_4$)$_4$ThF$_8$	(NH$_4$)$_4$ThF$_8$	9	(119)
	Na$_7$Th$_2$F$_{15}$	Na$_7$U$_2$F$_{15}$		(100)
	K$_7$Th$_2$F$_{15}$	β-Rb$_7$Th$_2$F$_{15}$		(101)
	α-Rb$_7$Th$_2$F$_{15}$			(54)
	β-Rb$_7$Th$_2$F$_{15}$	β-Rb$_7$Th$_2$F$_{15}$		(54)
	Li$_3$ThF$_7$	Li$_3$UF$_7$		(111)
	K$_3$ThF$_7$	K$_3$UF$_7$ (cubic)	7	(52, 49)
	Rb$_3$ThF$_7$	K$_3$UF$_7$ (cubic)	7	(53)
	(NH$_4$)$_3$ThF$_7$	(NH$_4$)$_3$ThF$_7$	9	(117)
	Cs$_3$ThF$_7$	K$_3$UF$_7$ (cubic)	7	(55)
	α-Na$_2$ThF$_6$	α-K$_2$ThF$_6$	8	(97)
	β_2-Na$_2$ThF$_6$	β_2-Na$_2$ThF$_6$	9	(65, 21, 100)
	δ-Na$_2$ThF$_6$	δ-Na$_2$ThF$_6$ (hex)	?	(100, 124, 21)
	α-K$_2$ThF$_6$	α-K$_2$ThF$_6$	8	(21, 97)
	β_1-K$_2$ThF$_6$	β_1-K$_2$UF$_6$	9	(52, 65)
	Rb$_2$ThF$_6$	β_1-K$_2$UF$_6$	9	(121)
	Cs$_2$ThF$_6$	Cs$_2$UF$_6$		(55)
	KNaThF$_6$	KNaThF$_6$	9	(101, 131)
	RbNaThF$_6$	KNaThF$_6$	9	(101)
	CaThF$_6$	LaF$_3$	9	(109)
	SrThF$_6$	LaF$_3$	9	(109)
	BaThF$_6$	LaF$_3$	9	(109)
	PbThF$_6$	LaF$_3$	9	(109)
	EuThF$_6$	LaF$_3$	9	(109)
	CdThF$_6$	LaF$_3$	9	(109)
	Na$_3$Th$_2$F$_{11}$			(100, 124)
	Na$_7$Th$_6$F$_{31}$	Na$_7$Zr$_6$F$_{31}$	8	(100, 101, 105)
	Na$_7$Th$_6$F$_{31}$	(Na, Li)$_7$Th$_6$F$_{31}$	9, 10	(168)
	(Na, Li)$_7$Th$_6$F$_{31}$	(Na, Li)$_7$Th$_6$F$_{31}$	9, 10	(116)
	K$_7$Th$_6$F$_{31}$	Na$_7$Zr$_6$F$_{31}$	8	(52, 70)
	Rb$_7$Th$_6$F$_{31}$	Na$_7$Zr$_6$F$_{31}$	8	(54, 101)
	LiThF$_5$	LiUF$_5$	9	(111, 101)
	α-NaThF$_5$			(100, 124)
	β-NaThF$_5$			(100, 124)
	CsThF$_5$	CsUF$_5$?		(55)
	NaBeTh$_3$F$_{15}$			(125)
	Cs$_2$Th$_3$F$_{14}$	Cs$_2$U$_3$F$_{14}$		(125)
	LiTh$_2$F$_9$			(111)
	NaTh$_2$F$_9$	NaTh$_2$F$_9$	9	(113, 21, 100)
	KTh$_2$F$_9$	KU$_2$F$_9$	9	(21, 52)
	CsTh$_2$F$_9$			(55)
	α-KTh$_3$F$_{13}$	?		(52)
	β-KTh$_3$F$_{13}$	?		(52)

Table 7 (continued)

Transition Metal Ion	Compound	Structure Type	Coordination Number	Reference
	γ-KTh_3F_{13}	?		(52)
	$RbTh_3F_{13}$	$RbTh_3F_{13}$	9	(53, 130)
	$CsTh_3F_{13}$	$CsTh_3F_{13}$	9	(55)
	$LiTh_4F_{17}$	LiU_4F_{17} ?		(111)
	α-KTh_6F_{25}	α-KTh_6F_{25}	9	(129)
	KTh_6F_{25}	CsU_6F_{25}	9	(126, 129)
	$RbTh_6F_{25}$	α-KTh_6F_{25}	9	(54, 129)
	$CsTh_6F_{25}$	CsU_6F_{25}	9	(55, 129)
	ThF_4	UF_4	8	(94, 95, 113)
	$ThF_4 \cdot xH_2O$			(169)
Pa^{+4}	$(NH_4)_4PaF_8$	$(NH_4)_4UF_8$	8	(82)
	$Na_7Pa_6F_{31}$	$Na_7Zr_6F_{31}$	8	(82)
	$K_7Pa_6F_{31}$	$Na_7Zr_6F_{31}$	8	(82)
	$Rb_7Pa_6F_{31}$	$Na_7Zr_6F_{31}$	8	(82)
	$LiPaF_5$	$LiUF_5$	9	(82)
	PaF_4	UF_4	8	(82, 95, 96)
U^{+4}	Li_4UF_8	Li_4UF_8	8 (+)	(122, 63)
	$(NH_4)_4UF_8$	$(NH_4)_4UF_8$	8	(61)
	$Na_7U_2F_{15}$			(122, 28, 100)
	Li_3UF_7			(122)
	Na_3UF_7	Na_3PaF_8	7 atoms in 8-site	(65, 21, 100, 122)
	K_3UF_7	(orthorhombic)		(49, 48)
	K_3UF_7	K_3UF_7 (tetragonal)	7	(47)
	K_3UF_7	K_3UF_7 (cubic)	7	(28, 47)
	Rb_3UF_7	K_3UF_7 (cubic)	7	(48)
	Rb_3UF_7	K_3UF_7 (tetrag.) ?	7	(48)
	Cs_3UF_7	K_3UF_7 (cubic)	7	(54)
	α-Na_2UF_6	α-K_2ThF_6	8	(21, 97)
	β_2-Na_2UF_6	β_2-Na_2ThF_6	9	(65, 21, 100, 121)
	γ-Na_2UF_6	γ-Na_2UF_6	8	(65, 21, 100, 122)
	δ-Na_2UF_6	δ-Na_2ThF_6 (hex)	?	(122, 100, 124, 21)
	α-K_2UF_6	α-K_2ThF_6	8	(21, 97)
	β_1-K_2UF_6	β_1-K_2UF_6	9	(120, 65)
	β_2-K_2UF_6	β_2-Na_2ThF_6	9	(65, 21)
	Rb_2UF_6	Rb_2UF_6	8	(48, 89, 80, 90)
	α-$(NH_4)_2UF_6$	α-$(NH_4)_2CeF_6$	8	(76, 84)
	β-$(NH_4)_2UF_6$			(76, 84)
	γ-$(NH_4)_2UF_6$	γ-$(NH_4)_2UF_6$	8	(84, 76, 38)
	δ-$(NH_4)_2UF_6$			(76, 84)
	Cs_2UF_6	Cs_2UF_6		(54)
	$KNaUF_6$	$KNaThF_6$		(101)
	$RbNaUF_6$	$KNaThF_6$		(101)
	$CaUF_6$	LaF_3 (Na_3As)	9	(109, 133)

Table 7 (continued)

Transition Metal Ion	Compound	Structure Type	Coordination Number	Reference
	$SrUF_6$	LaF_3 ?	9	(109)
	$BaUF_6$	LaF_3 ?	9	(109)
	$PbUF_6$	LaF_3 ?	9	(109)
	$MnUF_6 \cdot 3\,H_2O$	monoclinic		(146)
	$ZnUF_6 \cdot 5\,H_2O$	orthorhombic		(146)
	$Na_5U_3F_{17}$			(122, 100)
	$Na_7U_6F_{31}$	$Na_7Zr_6F_{31}$	8	(21, 100, 102, 122)
	$K_7U_6F_{31}$	$Na_7Zr_6F_{31}$	8	(21, 48, 102)
	$Rb_7U_6F_{31}$	$Na_7Zr_6F_{31}$	8	(48, 102, 105)
	$(NH_4)_7U_6F_{31}$	$Na_7Zr_6F_{31}$	8	(76, 84)
	$LiUF_5$	$LiUF_5$	9	(110)
	$RbUF_5$	$CsUF_5$?		(48)
	$\alpha\text{-}NH_4UF_5$	$Na_7Zr_6F_{31}$?		(84)
	$\beta\text{-}NH_4UF_5$			(84)
	$\gamma\text{-}NH_4UF_5$			(84)
	$CsUF_5$	$CsUF_5$		(54)
	$NaBeU_3F_{15}$	$NaBeTh_3F_{15}$		(125)
	$Rb_2U_3F_{14}$			(48)
	$Cs_2U_3F_{14}$	monoclinic		(125)
	NaU_2F_9			(122, 100)
	KU_2F_9	KU_2F_9	9	(114, 21, 48)
	CsU_2F_9	CsU_2F_9	9	(115)
	RbU_3F_{13}	$RbTh_3F_{13}$?	9	(48)
	$NH_4U_3F_{13}$			(84)
	LiU_4F_{17}			(122)
	KU_6F_{25}	CsU_6F_{25}	9	(21, 129)
	RbU_6F_{25}	CsU_6F_{25}	9	(126, 48, 129)
	CsU_6F_{25}	CsU_6F_{25}	9	(128)
	UF_4	UF_4	8	(95, 77)
	$UF_4 \cdot 1\frac{1}{2}\,H_2O$			(170)
	$U(C_5H_5)_3F$			(171)
Np^{+4}	$(NH_4)_4NpF_8$	$(NH_4)_4UF_8$	8	(83)
	Na_2NpF_6	$\delta\text{-}Na_2UF_6$		(103)
	$\alpha\text{-}K_2NpF_6$	$\alpha\text{-}K_2ThF_6$	8	(159)
	$\beta_1\text{-}K_2NpF_6$	$\beta_1\text{-}K_2UF_6$	9	(159)
	Rb_2NpF_6	Rb_2UF_6	8	(89)
	$CaNpF_6$	LaF_3	9	(109)
	$SrNpF_6$	LaF_3	9	(109)
	$BaNpF_6$	LaF_3	9	(109)
	$PbNpF_6$	LaF_3	9	(109)
	$Na_7Np_6F_{31}$	$Na_7Zr_6F_{31}$	8	(103, 104)
	$K_7Np_6F_{31}$	$Na_7Zr_6F_{31}$	8	(103, 105)
	$Rb_7Np_6F_{31}$	$Na_7Zr_6F_{31}$	8	(103, 105)
	$LiNpF_5$	$LiUF_5$	9	(112, 103)

Table 7 (continued)

Transition Metal Ion	Compound	Structure Type	Coordination Number	Reference
	KNp_2F_9	KU_2F_9	9	(21)
	NpF_4	UF_4	8	(95, 97)
Pu^{+4}	Li_4PuF_8			(161, 162)
	$(NH_4)_4PuF_8$	$(NH_4)_4UF_8$	8	(84)
	Na_3PuF_7	Na_3UF_7	7 atoms in 8-site	(156)
	Na_2PuF_6	$\begin{cases}\delta\text{-}Na_2UF_6 \\ \beta_2\text{-}Na_2ThF_6\end{cases}$	9	(103, 156) (21, 137)
	Rb_2PuF_6	Rb_2UF_6	8	(89)
	$\beta_1\text{-}Cs_2PuF_6$	$\beta_1\text{-}K_2UF_6$	9	(155)
	Cs_2PuF_6	$\gamma\text{-}(NH_4)_2UF_6(?)$	8 (?)	(155)
	$(NH_4)_2PuF_6$			(84)
	$CaPuF_6$	LaF_3	9	(109)
	$SrPuF_6$	LaF_3	9	(109)
	$Na_7Pu_6F_{31}$	$Na_7Zr_6F_{31}$	8	(21, 103, 104)
	$K_7Pu_6F_{31}$	$Na_7Zr_6F_{31}$	8	(21, 103, 105)
	$Rb_7Pu_6F_{31}$	$Na_7Zr_6F_{31}$	8	(103, 105)
	$(NH_4)_7Pu_6F_{31}$	$Na_7Zr_6F_{31}$	8	(84)
	$LiPuF_5$	$LiUF_5$	9	(112, 103)
	$\alpha\text{-}NH_4PuF_5$	$\alpha\text{-}NH_4UF_5$		(84)
	$\beta\text{-}NH_4PuF_5$	$\beta\text{-}NH_4UF_5$		(84)
	$\gamma\text{-}NH_4PuF_5$	$\gamma\text{-}NH_4UF_5$		(84)
	KPu_2F_9	KU_2F_9	9	(21)
	$CsPu_2F_9 \cdot 3\,H_2O$			(137)
	$NH_4Pu_3F_{13}$			(84)
	PuF_4	UF_4	8	(95, 97)
Am^{+4}	$(NH_4)_4AmF_8$	$(NH_4)_4UF_8$	8	(85)
	Rb_2AmF_6	Rb_2UF_6	8	(89, 90)
	$Na_7Am_6F_{31}$	$Na_7Zr_6F_{31}$	8	(104)
	$K_7Am_6F_{31}$	$Na_7Zr_6F_{31}$	8	(105)
	$LiAmF_5$	$LiUF_5$	9	(112)
	AmF_4	UF_4	8	(95, 98)
Cm^{+4}	Rb_2CmF_6	Rb_2UF_6	8	(89)
	$Na_7Cm_6F_{31}$	$Na_7Zr_6F_{31}$	8	(104)
	$K_7Cm_6F_{31}$	$Na_7Zr_6F_{31}$	8	(105)
	$LiCmF_5$	$LiUF_5$	9	(112)
	CmF_4	UF_4	8	(95)
Bk^{+4}	$Na_7Bk_6F_{31}$	$Na_7Zr_6F_{31}$	8	(99)
	BkF_4	UF_4	8	(123, 95)
Cf^{+4}	$Na_7Cf_6F_{31}$	$Na_7Zr_6F_{31}$	8	(99)
	CfF_4	UF_4	8	(99)

Table 7 (continued)

Transition Metal Ion	Compound	Structure Type	Coordination Number	Reference
$U^{+4.5}$	U_2F_9	U_2F_9	9	(113)
Nb^{+5}	K_2NbF_7	K_2NbF_7 (K_2TaF_7)	7	(45)
Sb^{+5}	$LiSbF_6$	$LiSbF_6$	6	(23)
	$NaSbF_6$	$NaSbF_6$	6	(38)
Ta^{+5}	K_2TaF_7	K_2TaF_7	7	(44)
Pa^{+5}	Li_3PaF_8	tetragonal	8 (?)	(57)
	Na_3PaF_8	Na_3PaF_8	8	(25)
	K_3PaF_8	K_3UF_7 (cubic)?	8 (?)	(10, 50, 28, 57)
	Rb_3PaF_8	K_3UF_7 (cubic)?	8 (?)	(56)
	Cs_3PaF_8	K_3UF_7 (cubic)?	8 (?)	(57)
	K_2PaF_7	K_2PaF_7	9	(14)
	Rb_2PaF_7	K_2PaF_7	9	(56)
	$(NH_4)_2PaF_7$	K_2PaF_7	9	(56)
	Cs_2PaF_7	K_2PaF_7	9	(118, 56, 57)
	$LiPaF_6$			(56)
	$NaPaF_6$			(56)
	$KPaF_6$	$RbPaF_6$	8	(56)
	$RbPaF_6$	$RbPaF_6$	8	(56, 24)
	NH_4PaF_6	$RbPaF_6$	8	(56)
	$CsPaF_6$	$RbPaF_6$	8	(56)
	PaF_5	β-UF_5	7	(96)
	Pa_2OF_8	U_2F_9		(96)
U^{+5}	Na_3UF_8	Na_3PaF_8	8	(25, 66)
	K_3UF_8	K_3UF_7 (cubic)?	8 (?)	(28, 34)
	Rb_3UF_8	K_3UF_7 (cubic)?	8 (?)	(34, 69)
	$(NH_4)_3UF_8$	K_3UF_7 (cubic)?	8 (?)	(34, 69)
	Cs_3UF_8	K_3UF_7 (cubic)?	8 (?)	(34, 69)
	Ag_3UF_8	cubic		(143)
	Tl_3UF_8	cubic		(143)
	K_2UF_7	K_2TaF_7	7	(32, 34)
	Rb_2UF_7	K_2TaF_7	7	(32, 34)
	$(NH_4)_2UF_7$	K_2TaF_7	7	(32, 34)
	Cs_2UF_7	K_2TaF_7	7	(32, 34)
	$N_2H_6UF_7$			(165, 166)
	$LiUF_6$	$LiSbF_6$	6	(34)
	β-$NaUF_6$	$NaSbF_6$ (cubic)	6	(34, 16)
	α-$NaUF_6$	$LiSbF_6$ (rhombohedral)	6	(16)
	KUF_6	$RbPaF_6$	8	(16, 56)
	$RbUF_6$	$RbPaF_6$	8	(16, 56)
	NH_4UF_6	$RbPaF_6$	8	(16, 56)

Table 7 (continued)

Transition Metal Ion	Compound	Structure Type	Coordination Number	Reference
	$CsUF_6$	$CsUF_6$	6	(15, 16, 33)
	$NOUF_6$	$KSbF_6$?	6	(40, 36, 42)
	$N_2H_6(UF_6)_2$			(166)
	$AgUF_6$	tetragonal		(143)
	$CoU_2F_{12} \cdot 4\,H_2O$	triclinic		(145)
	$NiU_2F_{12} \cdot 4\,H_2O$	triclinic		(145)
	$CuU_2F_{12} \cdot 4\,H_2O$	triclinic		(145)
	$\alpha\text{-}UF_5$	$\alpha\text{-}UF_5$	6	(43)
	$\beta\text{-}UF_5$	$\beta\text{-}UF_5$	7	(43)
	$UF_5 \cdot XeF_6$	monoclinic	8 (?)	(136, 164)
Np^{+5}	Na_3NpF_8	Na_3PaF_8	8	(25)
	Rb_2NpF_7	K_2TaF_7	7	(31)
	$CsNpF_6$	$CsUF_6$	6	(31)
	NpF_5 (?)	$\beta\text{-}UF_5$	7	(160)
	$KNpO_2F_2$	$KAmO_2F_2$	8	(159)
	$RbNpO_2F_2$	$KAmO_2F_2$	8	(8)
Pu^{+5}	Rb_2PuF_7	K_2TaF_7	7	(32)
	$CsPuF_6$	$CsUF_6$	6	(32)
	$RbPuO_2F_2$	$KAmO_2F_2$	8	(8)
	$NH_4PuO_2F_2$	$KAmO_2F_2$	8	(8)
Am^{+5}	$KAmO_2F_2$	$KAmO_2F_2$	8	(91)
	$RbAmO_2F_2$	$KAmO_2F_2$	8	(8)
U^{+6}	Na_3UF_9			(148)
	Ag_3UF_9			(148)
	Na_2UF_8	Na_2UF_8	8	(68, 69, 144)
	K_2UF_8			(68)
	Rb_2UF_8 (?)			(148)
	$Na_3U_2F_{15}$			(149)
	$K_3U_2F_{15}$			(148)
	$NaUF_7$			(68)
	KUF_7			(68)
	NH_4UF_7			(127, 147)
	$CsUF_7$	cubic	7 atoms in 8-site?	(158)
	$CsUF_7$	tetragonal	?	(127, 158)
	UF_6	UF_6	6	(30, 154)
	$K_3UO_2F_5$	$K_3UO_2F_5$	7	(46)
	$(NH_4)_3UO_2F_5$	$(NH_4)_3UO_2F_5$	7	(134)
	$Cs_3UO_2F_5$	cubic		(163)
	$K_5(UO_2)_2F_9{}^a)$	monoclinic		(151, 152)
	UO_2F_2	UO_2F_2	8	(8, 87, 87a)

Table 7 (continued)

Transition Metal Ion	Compound	Structure Type	Coordination Number	Reference
Np^{+6}	NpF_6	UF_6	6	(154)
	$K_3NpO_2F_5$	$K_3UO_2F_5$	7	(159)
	NpO_2F_2	UO_2F_2	8	(8)
Pu^{+6}	PuF_6	UF_6	6	(154)
	PuO_2F_2	UO_2F_2	8	(88, 8)
Am^{+6}	AmO_2F_2	UO_2F_2	8	(8)

a) A series of compounds $M(UO_2)_2F_5$ (M = K, Rb, Cs) were prepared but no single crystal structures were reported (152).

Li^+ replaces Na^+. The P3c1 structure, determined on a preparation in which $Na:Li \approx 1:1$ has two types of actinide coordination polyhedra; the tricapped trigonal prism and a ten-coordinated polyhedron. These are linked to one another to form a three-dimensional polyion.

2.5. List of Observed Stoichiometries by Element and Structure Type

Table 7 is a summary table and includes some compounds on which definitive structural information is not available.

3. Optical Characteristics

3.1. Molar Refractivity

Structural details concerning coordination polyhedra are still best elucidated by single crystal X-ray and neutron diffraction studies, although it is clear that the newer techniques of NMR and ESR have application. The structural variety displayed by the complexes between the actinide fluorides and the alkali fluorides has been amply demonstrated in the foregoing sections. Even with their different structures, such complexes display common characteristics which are independent of such packing details. Prominent among such general properties are unit cell volume and refractivity volume, both of which are additive functions of composition.

In this section attention is focused on the refractivity volume. In order to calculate this quantity, it is necessary to have refractive index and density data. The density can of course be calculated from X-ray data and the unit cell contents. However, X-ray data are not a requirement. The measured density together with refractive index data are sufficient, and both types of measurements (139) can frequently be made using the microscope[2]). Examination of crystals using a petrographic microscope is an old technique, but it appears to have fallen out of favor with inorganic and structural chemists. Thus, it is not widely known that the molar refractivity of fluoride complexes is a sensitive measure of their composition.

The quantitative use of refractive index data involves the refractivity volume. The quantity to be calculated is the Lorentz-Lorenz molar refractivity[3]).

For typical actinide fluoride complexes the refractive indices range from ~1.5 to ~1.7. The resulting Lorentz-Lorenz function, $\frac{n^2-1}{n^2+2}$, then has values of several tenths (0.3—0.4). This factor, since it multiplies the molar volume directly, can be regarded as that portion of the molar volume which has the property of additivity, as measured by its effect on visible light. It is this refractivity volume which is an additive function of unit cell composition; it is essentially independent of symmetry, coordination number and whether fluoride is shared between actinides, bound to an individual actinide, or "free" in the lattice.

It has been found (22) that the molar refractivities of a series of complex fluorides containing a common alkali fluoride are a linear function of composition (Fig. 10). That is, for such a series as $NaF \cdot UF_4$, $2 NaF \cdot UF_4$, $3 NaF \cdot UF_4$, etc., the refractivities can be expressed as a constant, R_{UF_4}, plus an incremental increase in refractivity, R_{NaF}, so that the total

---

[2]) The density can be determined by pycnometry, or often estimated with sufficient accuracy be flotation techniques. The refractive index of small crystals may be measured conveniently by the immersion method. In the discussion to follow, the geometric mean index of refraction is used.

[3]) The Lorentz-Lorenz molar refractivity, R, is obtained from Eq. (1), relating the geometric mean refractive index, n, the formula weight, M, and the density, d. Where the unit cell dimensions and number of formula units per cell areknown, the second equation is useful, avoiding the unnecessary addition of the formula weight.

$$R = \frac{n^2-1}{n^2+2} \left(\frac{M}{d} \right) \tag{1}$$

$$R = \frac{n^2-1}{n^2+2} \left(\frac{\text{cell volume } (\text{Å}^3) \times 0.6025}{\text{formula units per cell}} \right) \tag{2}$$

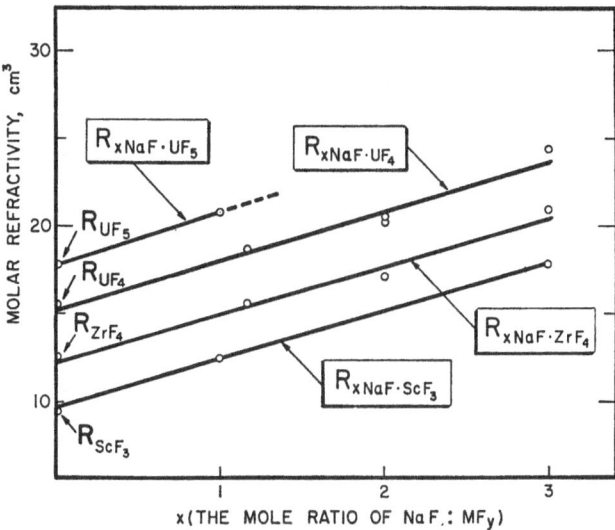

Fig. 10. The molar refractivity of sodium fluoride complexes of tri-, tetra-, and pentavalent elements versus composition

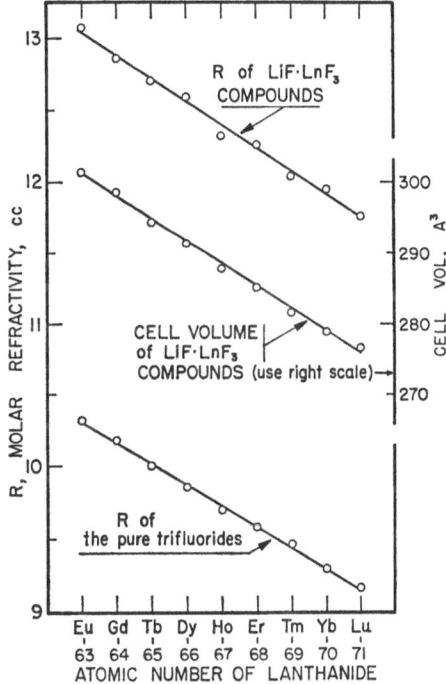

Fig. 11. The molar refractivity and cell volume of lithium fluoride complexes with the smaller lanthanide trifluorides. Courtesy of the Am. Chem. Soc.

refractivity for a given compound is $R_{UF_4} + xR_{NaF}$, where x is the number of moles of NaF per mole of UF_4. If the actinide is changed from U to Zr, for example, a second line is obtained, parallel to the first, and displaced by the difference in the intrinsic molar refractivity values of the tetrafluorides of U and Zr. A change in valence of the kernel element yields a parallel line similarly displaced in absolute value, i. e. the $ScF_3 \cdot xNaF$ and $UF_3 \cdot xNaF$ curves. Similar linear behavior is displayed when the molar refractivities of complex fluorides between LiF and lanthanide trifluorides (132) are shown, see Fig. 11.

Formation of the $LiF \cdot MF_3$ compounds with the lanthanides is size discriminatory (132), and stops abruptly with trivalent lanthanide ions larger than Eu^{3+}. Because of the shift in 5f ion size equivalence versus the 4f series, as previously mentioned, the beginning of $LiF \cdot$ actinide trifluoride compound formation is expected at californium(III) and likely not at the Bk^{3+} radius (or larger). Both the larger M^{3+} ions of americium and curium failed to yield $LiF \cdot MF_3$ compounds as expected (132).

By statistical tests of over 60 compounds it was demonstrated (22) that the empirical slopes, which represent the incremental increase in refractivity per mole of alkali fluoride in such complexes, are in fact adequately represented by the sum of Fajans' aqueous refractivity values for a given alkali metal ion and fluoride ion (140). Since the slope is known, a single point would determine the line for a series of unknown complexes. The predictive value, and use for determining composition by simple methods are obvious. For example, the empirical refractivity per mole of KF is 5 cm^3 which is such a large fraction of typical refractivity values (about 30 cm^3) that even a crude refractive index determination can usually fix the formula within 0.1 to 0.2 mol.

Many of the fluoride complexes are anhydrous, being prepared from melts, and even hydrates are not numerous among those prepared from aqueous solution. In the case of hydrates, the refractivity of water must be allowed for. An example of a hydrate, $(NH_4)_3CeF_7 \cdot H_2O$ obtained in the $NH_4F-CeF_4-H_2O$ system (81, 86), is shown in Fig. 12. When the refractivity of H_2O is subtracted, the point for the yet unknown anhydrous $(NH_4)_3CeF_7$ compound falls on the line established by the known anhydrous ammonium fluoride-cerium tetrafluoride complexes.

This linearity of refractivity is little, if any, perturbed by the amount of alkali fluoride in the lattice, coordination number of actinide, whether the fluorides are all shared between highly charged ions (as in eight coordinated UF_4), unshared as in UF_8^{4-}, or in part unbonded to the highly charged ion as in $5 KF \cdot ThF_4$. This must reflect the essential ionic nature of the lattice, and the reluctance of fluoride ion to deform. Actually, in such a series of compounds, the coordination is high and remains nearly constant, although it may go from eight (in ThF_4) to nine (as in

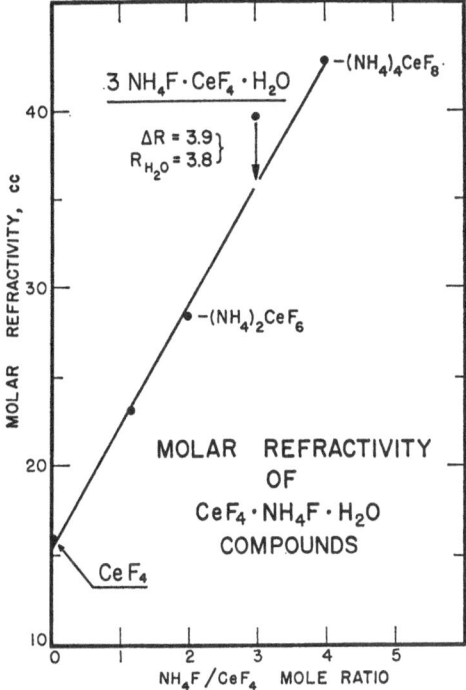

Fig. 12. The molar refractivity of hydrated and anhydrous ammonium fluoride-cerium tetrafluoride complexes

$(NH_4)_3ThF_7)$. The initial value of the refractivity of ThF_4, for example, is higher per fluoride than that of subsequently added fluoride and reflects the maximum polarization of fluoride in its complexes with thorium. Beginning with that base value for the refractivity of ThF_4, the effect on the refractivity of its complexes as the number of moles of alkali fluoride is increased is purely additive. This linear result may be thought of as being parallel, in effect at least, to the idea expressed in the beginning of this paper, namely, that the highly polymeric structures of the binary fluorides are progressively depolymerized by the addition of alkali fluoride until individual fluoro-anions are obtained. Further addition of alkali fluoride is then usually rejected by the lattice. Free fluoride in these complexes is known in only two compounds: K_5ThF_9, where it exists together with isolated ThF_8^{4-} polyhedra; and $(NH_4)_4ThF_8$, where it exists with chains of ThF_9 polyhedra.

The ionic nature of fluoride in three isolated fluoro anions was reported by Prof. Buslaev (174) from ESCA measurements on $NaTaF_6$, Na_2TaF_7, and Na_3TaF_8. No detectable difference was observed in the inner and valence electron levels of Na, Ta, or F in these htree complexes.

4. References

1. *Babel, D.:* Struct. Bonding, *3*, 1 (1967).
2. *Kettle, S. F. A., Smith, A. J.:* J. Chem. Soc. (A) 688 (1967).
3. *Pauling, L.:* J. Am. Chem. Soc. *49*, 765 (1927); Z. Krist *67*, 377 (1928); J. Am. Chem. Soc. *51*, 1010 (1929).
4. *Krot, N. N., Gel'man, A. D.:* Dokl. Akad. Nauk SSSR *177*, 1, 124 (1967).
5. *Keller, C., Seiffert, H.:* Inorg. Nucl. Chem. Letters *5*, 1, 51 (1969); Angew. Chem. Intern. Ed. Engl. *8* (4), 279 (1969).
6. *Zaitseva, V. P.:* Dokl. Akad. Nauk SSSR *188*, 826 (1969).
7. *Keller, C.:* Adv. Chem. Series 71, Lanthanide/Actinide Chemistry, Am. Chem. Soc. p. 228, 1967.
8. *Keenan, T. K.:* Inorg. Nucl. Chem. Letters *4*, 381 (1968); Inorg. Chem. *4*, 1500 (1965).
9. *Penneman, R. A., Keenan, T. K.:* The Radiochemistry of Americium and Curium, National Academy of Sciences, NS 3006, 1960.
9a. *Baybarz, R. D., Peterson, J. R.:* private communication, 1971, Oak Ridge Nat'l. Lab.
10. *Edelstein, N., Easley, W., McLaughlin, R.:* J. Chem. Phys. *44*, 3130 (1966).
10a. *Baybarz, R. D., Asprey, L. B., Strouse, C. E., Fukushima, E.:* J. Inorg. Nucl. Chem. *34*, 3427 (1972).
11. *Peterson, J. R., Cunningham, B. B.:* J. Inorg. Nucl. Chem. *30*, 1775 (1968); Inorg. Nucl. Chem. Letters *3*, 327 (1967).
12. *Baybarz, R. D., Haire, R. G., Fahey, J. A.:* J. Inorg. Nucl. Chem. *34*, 557 (1972). On the Californium Oxide System.
13. *Haire, R. G., Baybarz, R. D.:* private communication.
14. *Brown, D., Kettle, S. F. A., Smith, A. J.:* J. Chem. Soc. (A) 1429 (1967).
15. *Rosenzweig, A., Cromer, D. T.:* Acta Cryst. *23*, 865 (1967).
16. *Sturgeon, G. D., Penneman, R. A., Kruse, F. H., Asprey, L. B.:* Inorg. Chem. *4*, 748 (1965).
17. *Mastin, S. H., Penneman, R. A.:* unpublished work, LASL (1970).
18. *Shannon, R. D., Prewitt, C. T.:* Acta Cryst. B *25*, 925 (1969).
19. *Zachariasen, W. H.:* In: The Actinide Elements NNES Ch. 18, p. 769. New York: McGraw Hill 1954.
20. *Muetterties, E. L., Wright, C. M.:* Quart. Rev. Chem. Soc. *21*, 109 (1967).
20a. *Brown, D.:* In: Halides of the Lanthanides and Actinides. London: Wiley and Sons 1968.
21. *Zachariasen, W. H.:* J. Am. Chem. Soc. *70*, 2147 (1948). — *Ellinger, F. H., Zachariasen, W. H.:* J. Phys. Chem. *58*, 405 (1954).
22. *Penneman, R. A.:* Inorg. Chem. *8*, 1379 (1969).
23. *Burns, J. H.:* Acta Cryst. *15*, 1098 (1962).
24. — *Levy, H. A., Keller, O. L. Jr.:* Acta Cryst. B *24*, 1675 (1968).
25. *Brown, D., Easey, J. F., Rickard, C. E. F.:* J. Chem. Soc. (A) 1161 (1969).
26. *Klemm, W.:* J. Inorg. Nucl. Chem. *8*, 532 (1958).
27. *Bartlett, N., Trotter, J.:* J. Chem. Soc. A 543 (1968).
28. *Thoma, R. E., Friedman, H. A., Penneman, R. A.:* J. Am. Chem. Soc. *88*, 2046 (1966).
29. *Wyckoff, R. W. G.:* Crystal Structures, Ed. 2, Vol. 3, pp. 325, 426. New York: Interscience Publishers 1965.
30. *Hoard, J. L., Stroupe, J. D.:* paper 45, USAEC TID 5290 Book I, 325—49 (1958).

31. *Asprey, L. B., Keenan, T. K., Penneman, R. A., Sturgeon, G. D.:* Inorg. Nucl. Chem. Letters 2, 19 (1966).
32. *Penneman, R. A., Sturgeon, G. D., Asprey, L. B., Kruse, F. H.:* J. Am. Chem. Soc. 87, 5803 (1965).
33. *Hepworth, M. A., Jack, K. H., Westland, G. J.:* J. Inorg. Nucl. Chem. 2, 79 (1956).
34. *Penneman, R. A., Sturgeon, G. D., Asprey, L. B.:* Inorg. Chem. 3, 126 (1964).
35. *Drifford, M., Rigny, P., Plurien, P.:* Phys. Letters 27 A, 620 (1968). — *Rigny, P., Plurien, P.:* J. Phys. Chem. Solids 28, 2589 (1967).
36. *Rigny, P., Dianoux, A. J., Plurien, P.:* private communication 1971.
37. *Fukushima, E.:* Acta Cryst. A 27, 65 (1971).
38. *Ryan, R. R.:* unpublished work, LASL (1970), submitted to chem. komm.
39. — *Penneman, R. A.:* unpublished work, LASL (1971).
40. *Geichman, J. R., Smith, E. A., Trond, S. S., Ogle, P. R.:* Inorg. Chem. 1, 661 (1962).
41. *Rosenzweig, A.:* unpublished work, LASL (1967).
42. *Charpin, P.:* SACLAY, Neutron Diffraction of NOUF$_6$, private communication (1970).
43. *Zachariasen, W. H.:* Acta Cryst. 2, 296 (1949).
44. *Hoard, J. L.:* J. Am. Chem. Soc. 61, 1252 (1939).
45. *Brown, G. M., Walker, L. A.:* Acta Cryst. 20, 220 (1966).
46. *Zachariasen, W. H.:* Acta Cryst. 7, 783 (1954).
47. — Acta Cryst. 7, 792 (1954).
48. *Thoma, R. E., Insley, H., Landau, B. S., Friedman, H. A., Grimes, W. R.:* J. Am. Ceram. Soc. 41, 538 (1958).
49. *Burns, J. H.:* ORNL-3262, Reactor Chem. Div., 1/31/62, pp. 15—16.
50. *Hampson, G. G., Pauling, L.:* J. Am. Chem. Soc. 60, 2702 (1938).
51. *Hurst, H. J., Taylor, J. C.:* Acta Cryst. B 26, 417 (1970).
52. *Asker, W. J., Segnit, E. R., Wylie, A. W.:* J. Chem. Soc. 4470 (1952).
53. *Dergunov, E. P., Bergman, A. G.:* Dokl. Akad. Nauk. SSSR 60, 391 (1948).
54. ORNL 2548 Phase Diagrams of Nuclear Reactor Materials (1959), pp. 76, 92.
55. *Thoma, R. E., Carlton, T. S.:* J. Inorg. Nucl. Chem. 17, 88 (1961).
56. *Asprey, L. B., Kruse, F. H., Rosenzweig, A., Penneman, R. A.:* Inorg. Chem. 5, 659 (1966).
57. *Brown, D., Easey, J. F.:* J. Chem. Soc. A 254 (1966).
58. *Hoard, J. L., Silverton, J. V.:* Inorg. Chem. 2, 235 (1963).
59. *King, R. B.:* J. Am. Chem. Soc. 92, 6455 (1970).
60. *Lippard, S. J., Russ, B. J.:* Inorg. Chem. 7, 1686 (1968).
61. *Rosenzweig, A., Cromer, D. T.:* Acta Cryst. B 26, 38 (1970).
62. *Ryan, R. R., Penneman, R. A.:* Structure of K$_5$ThF$_9$, Acta Cryst. B 27, 829 (1971).
63. *Brunton, G.:* J. Inorg. Nucl. Chem. 29, 1631 (1967).
64. *Kepert, D. L.:* J. Chem. Soc. 4736 (1965).
65. *Zachariasen, W. H.:* Acta Cryst. 1, 265 (1948).
66. *Ruedorff, W., Leutner, H.:* Liebigs Ann. Chem. 632, 1 (1960).
67. *Fukushima, E., Hecht, H. G.:* J. Chem. Phys. 54, 4341 (1971).
68. *Malm, J. G., Selig, H., Siegel, S.:* Inorg. Chem. 5, 130 (1966).
69. *Dianoux, A. J., Rigny, P.:* J. Phys. Radium 29, 791 (1968).
70. *Brunton, G.:* Crystal Structure of K$_7$Th$_6$F$_{31}$, Acta Cryst. B 27, 2290 (1971).
71. *Hoppe, R., Rödder, K. M.:* Z. Anorg. Allgem. Chem. 313, 154 (1961).
72. — *Liebe, W.:* Z. Anorg. Allgem. Chem. 313, 221 (1961).

73. *Asprey, L. B., Cunningham, B. B.:* Unusual Oxidation States of Some Actinide and Lanthanide Elements, pp. 267—302. Book Chapter in Progr. Inorg. Chem. 2, edited by *F. Albert Cotton.* New York: Interscience Publishers, Inc. 1960.
74. *Hoppe, R., Rödder, K. M.:* Z. Anorg. Allgem. Chem. *312*, 277 (1961).
75. *Ryan, R. R., Larson, A. C., Kruse, F. H.:* Inorg. Chem. *8*, 33 (1969).
76. *Penneman, R. A., Kruse, F. H., George, R. S., Coleman, J. S.:* Inorg. Chem. *3*, 309 (1964).
77. *Larson, A. C., Roof, R. B., Cromer, D. T.:* Acta Cryst. *17*, 555 (1964).
78. *Burbank, R. D., Bensey, F. N., Jr.:* USAEC document, K-1280 (1956) see also: Ref. (*29*), Vol. 2, p. 127.
79. *Burns, J. H., Ellison, R. D., Levy, H. A.:* Acta Cryst. B *24*, 230 (1968).
80. *Kruse, F. H.:* J. Inorg. Nucl. Chem. *33*, 1625 (1971).
81. *Ryan, R. R., Penneman, R. A.:* Ammonium Cerium Heptafluoride Hydrate. Acta Cryst. B 27, 1939 (1971).
82. *Asprey, L. B., Kruse, F. H., Penneman, R. A.:* Inorg. Chem. *6*, 544 (1967).
83. *Keenan, T. K.:* unpublished work LASL, (1963).
84. *Benz, R., Douglass, R. M., Kruse, F. H., Penneman, R. A.:* Inorg. Chem. *2*, 799 (1963).
85. *Asprey, L. B., Penneman, R. A.:* Inorg. Chem. *1*, 134 (1962).
86. *Penneman, R. A., Rosenzweig, A.:* Inorg. Chem. *8*, 627 (1969).
87. *Zachariasen, W. H.:* Acta Cryst. *1*, 277 (1948).
87a. *Atoji, M., McDermott, M. J.:* Acta Cryst. B *26*, 1540 (1970).
88. *Alenchikova, I. F., Zaitseva, L. L., Lipis, L. V., Nikolayev, N. S., Fomin, V. V., Chebotarev, N. T.:* Zh. Neorgan. Khim. *3*, 951 (1958).
89. *Keenan, T K.:* Inorg. Nucl. Chem. Letters *3*, 463 (1967).
90. *Kruse, F. H., Asprey, L. B.:* Inorg. Chem. *1*, 137 (1962).
91. *Asprey, L. B., Ellinger, F. H., Zachariasen, W. H.:* J. Am. Chem. Soc. *76*, 5235 (1954).
92. *Zalkin, A., Templeton, D. H.:* J. Am. Chem. Soc. *75*, 2453 (1953).
93. *Peterson, J. R., Copeland, J. C., Temple, D.:* unpublished results quoted in (*11*).
94. *Darnell, A. J., Keneshea, F. J., Jr.:* J. Phys. Chem. *62*, 1143 (1958).
95. *Keenan, T. K., Asprey, L. B.:* Inorg. Chem. *8*, 235 (1969).
96. *Stein, L.:* Inorg. Chem. *3*, 995 (1964).
97. *Zachariasen, W. H.:* Acta Cryst. 2, 388 (1949).
98. *Asprey, L. B.:* J. Am. Chem. Soc. *76*, 2019 (1954).
99. — unpublished work, 1970.
100. *Thoma, R. E., Insley, H., Hebert, G. M., Friedman, H. A., Weaver, C. F.:* J. Am. Ceram. Soc. *46*, 37 (1963).
101. *Brunton, G. D., Insley, H., McVay, T. N., Thoma, R. E.:* ORNL 3761 (1965), Crystallographic Data for Some Metal Fluorides, Chlorides and Oxides.
102. *Harris, L. A.:* ORNL, unpublished work (1958).
103. *Keller, C., Schmutz, H.:* Inorg. Nucl. Chem. Letters 2, 355 (1966).
104. *Keenan, T. K.:* Inorg. Nucl. Chem. Letters 2, 211 (1966).
105. — Inorg. Nucl. Chem. Letters *3*, 391 (1967).
106. *Mansmann, M.:* Z. Krist. *122*, 375 (1965).
106a. *Zalkin, A., Templeton, D. H., Hopkins, T. E.:* Inorg. Chem. *5*, 1466 (1966).
107. *Fried, S., Hagemann, F., Zachariasen, W. H.:* J. Am. Chem. Soc. *72*, 771 (1950).
108. *Asprey, L. B., Keenan, T. K., Kruse, F. H.:* Inorg. Chem. *4*, 985 (1965).
109. *Keller, C., Salzer, M.:* J. Inorg. Nucl. Chem. *29*, 2925 (1967).
110. *Brunton, G.:* Acta Cryst. *21*, 814 (1966).
111. *Harris, L. A., White, G. D., Thoma, R. E.:* J. Phys. Chem. *63*, 1974 (1959).
112. *Keenan, T. K.:* Inorg. Nucl. Chem. Letters 2, 153 (1966).

113. *Zachariasen, W. H.:* Acta Cryst. 2, 390 (1949).
114. *Brunton, G.:* Acta Cryst. B 25, 1919 (1969).
115. *Ryan, R. R., Rosenzweig, A.:* unpublished, Structure of CsU_2F_9 (1971).
116. *Brunton, G., Sears, D. R.:* Acta Cryst. B 25, 2519 (1969).
117. *Penneman, R. A., Ryan, R. R., Kressin, I. K.:* The Crystal structure of Ammonium Heptafluorothorate, Acta Cryst. B 27, 2279 (1971).
118. *Brown, D., Easey, J. F., Holah, D. G.:* J. Chem. Soc. (A) 1979 (1967).
119. *Ryan, R. R., Penneman, R. A., Rosenzweig, A.:* Acta Cryst. B 25, 1958 (1969).
120. *Brunton, G.:* Acta Cryst. B 25, 2163 (1969).
121. *Harris, L. A.:* Acta Cryst. 13, 502 (1960).
122. *Barton, C. J., Friedman, H. A., Grimes, W. R., Insley, H., Moore, R. E., Thoma, R. E.:* J. Am. Ceram. Soc. 41, 63 (1958).
123. *Asprey, L. B., Keenan, T. K.:* Inorg. Nucl. Chem. Letters 4, 537 (1968).
124. *Thoma, R. E., Insley, H., Landau, B. S., Friedman, H. A., Grimes, W. R.:* J. Phys. Chem. 63, 1266 (1959).
125. — J. Am. Ceram. Soc. 43, 608 (1960).
126. *Harris, L. A.:* unpublished work (1958).
127. *Nikolaev, N. S., Sukhoverkhov, V. F.:* Dokl. Akad. Nauk. SSSR 136, 621 (1961).
128. *Brunton, G. D.:* Acta Cryst. B 27, 245 (1971).
129. — The Crystal Structure of α-KTh_6F_{25}, a Polymorph of CsU_6F_{25}, Acta Cryst. B 28, 144 (1972).
130. — The Crystal Structure of $RbTh_3F_{13}$, Acta Cryst. B 27, 1823 (1971).
131. — Acta Cryst. B 26, 1185 (1970).
132. *Thoma, R. E., Brunton, G. D., Penneman, R. A., Keenan, T. K.:* Inorg. Chem. 9, 1096 (1970).
133. *Chebotarev, N. T., Beznosikova, A. V.:* At. Energ. (USSR) 25, 321 (1968).
134. *Brusset, H., Gillier-Pandraud, H., Nguyen-Quy-Dao:* Acta Cryst. B 25, 67 (1969).
135. *Brunton, G. D.:* Acta Cryst. B 25, 2164 (1969).
136. *Slivnik, J., Frlec, B., Zemva, B., Bohinc, M.:* J. Inorg. Nucl. Chem. 32, 1397 (1970).
137. *Alenchikova, I. F., Zaitseva, L. L., Lipis, L. V., Fomin, V. V., Chebotarev, N. T.:* Proc. 2nd Intern. Conf. Peaceful Uses At. Energy 28, 309 (1958).
138. *Burns, J. H.:* Inorg. Chem. 4, 881 (1965).
139. *Chamot, E. M., Mason, C. W.:* Vol. 1, Handbook of Chemical Microscopy, Ed. 3. New York: John Wiley & Sons 1958.
140. *Fajans, K., Lühdemann, R.:* Z. physik. Chem. 29 B, 150 (1935).
141. *Peterson, J. R., Burns, J. H.:* J. Inorg. Nucl. Chem. 30, 2955 (1968).
142. *Zachariasen, W. H.:* Acta Cryst. 4, 231 (1951).
143. *Bougon, R., Plurien, P.:* Compt. Rend. 260, 4217 (1965).
144. *Peka, I., Sedlakova, L., Sykora, F.:* Collection Czech. Chem. Commun. 31, 4449 (1966).
145. *Montoloy, F., Plurien, P.:* Compt. Rend. C 267, 1036 (1968).
146. *Charpin, P., Montoloy, F., Nierlich, M.:* Compt. Rend. C 268, 156 (1969).
147. *Volavsek, B.:* Croat. Chem. Acta 33, 181 (1961).
148. *Martin, H., Albers, A., Dust, H.:* Z. Anorg. Allgem. Chem. 265, 128 (1951).
149. *Nikolaev, N. S., Sadikova, A. T.:* At. Energ. (USSR) 25, 422 (1968).
150. *Keller, C., Schmutz, H.:* Z. Naturforsch. 19b, 1080 (1964).
151. *Staritzky, E., Cromer, D. T., Walker, D. I.:* Anal. Chem. 28, 1355 (1956).
152. *Davidovich, R. L., Buslaev, Yu. A.:* Dokl. Akad. Nauk. (USSR) 191, 355 (1970).
153. *Thoma, R. E.:* Inorg. Chem. 1, 220 (1962).
154. *Schablaske, R.:* ANL-6900, 162 (1964).

155. *Riha, J.:* ANL-7375, 54 (1967).
156. — *Trevorrow, L. E.:* ANL-7425, 52 (1967).
157. *Gerding, T. J., Trevorrow, L. E.:* ANL-7575, 63 (1968).
158. *Sadikova, A. T., Sadikov, G. G., Nikolaev, N. S.:* Sov. At. Energy *26*, 313 (1969).
159. *Thalmayer, C. E., Cohen, D.:* Advan. Chem. Ser. 71, Lanthanide/Actinide Chemistry 256 (1967).
160. *Fried, S., Holloway, J.:* personal communication (1970). In: D. Brown, Ref. *(20a)*.
161. *Steindler, M. J.:* ANL-7438, 108 (1968); J. Inorg. Nucl. Chem. *33*, 2875 (1971).
162. *Riha, J. G., Trevorrow, L. E.:* ANL-7575, 61 (1968); J. Inorg. Nucl. Chem. *33*, 2875 (1971).
163. *Rebenko, A. N., Brusentsev, F. A., Opalovskii, A. A.:* Izv. Sib. Otd. Akad. Nauk. SSSR, Ser. Khim. Nauk. *1*, 136 (1968).
164. *Frlec, B., Charpin, P., Drifford, M., Bohinc, M.;* J. Inorg. Nucl. Chem. (1971) in press.
165. — *Brčić, B. S., Slivnik, J.:* Inorg. Chem. *5*, 542 (1966).
166. — *Hyman, H. H.:* Inorg. Chem. *6*, 2233 (1967).
167. *Schmutz, H.:* Karlsruhe Report KFK-431 (1966).
168. *Thoma, R. E., Brunton, G.:* private communication (1971).
169. *D'Eye, R. W. M., Booth, G. W.:* J. Inorg. Nucl. Chem. *4*, 13 (1957).
170. *Tananaev, I. V., Savchenko, G. S.:* At. Energ. (USSR) *11*, 397 (1962).
171. *Fischer, R. D., Von Ammon, R., Kanellakopulos, B.:* J. Organometal. Chem. *25*, 123 (1970).
172. *Lippard, S. J.:* Eight Coordination Chemistry, pp. 109—193. Book chapter in Prog. Inorg. chem. 8, edited by *F. Albert Cotton.* New York: Interscience Publishers, Inc. 1967.
173. *Porai-Koshits, M A* and *Aslanov, L A.* Some aspects of the stereochemistry of eight-coordinate complexes, Zh. Strukt. Khim. *13*, 266 (1972).
174. Presented at the 4th European Symposium on Fluorine Chemistry, Ljubljana, Yugoslavia, 1972.
    ESCA-Invesigations of Inner and Valence Electron Levels in the Fluorides. *Yu. A. Buslaev, V. I. Nefedov* and *Yu. V. Kokunov*, Academy of Sciences of U.S.S.R., 31 Leninskii Pr., Moscow, USSR.

Received February 22, 1972

Spectra and Energy Transfer of Rare Earths in Inorganic Glasses

Renata Reisfeld

Department of Inorganic and Analytical Chemistry
The Hebrew University of Jerusalem, Israel*

Table of Contents

I. Introduction

Surprisingly little use has been made of the formalism of ligand field theory for the interpretation of phenomena such as energy transfer, absorption and emission spectra of rare earth ions in inorganic glasses. This may be due to the fact that glass is an amorphous substance and, therefore, the possibility of applying symmetry concepts to it may seem impractical. However, as we shall show in this paper, there is a remarkable similarity between the above-mentioned phenomena in glasses and in crystals. It would seem highly improbable that this similarity is merely

* The experimental work performed in this laboratory described here was supported partially by NBS contract (G)–103.

fortuitous. We therefore propose that despite the absence of both long range and cellular disorder as described by *Ziman* (1) the immediate vicinity of all rare earth ions does not significantly vary from one site to another and all sites have a definite microsymmetry. Hence, the optical properties of rare earths in glasses may be analysed in a manner similar to that of impurities in crystals, the principle difference being that the disorder will introduce perturbative effects of these properties, *e.g.* inhomogeneous broadening. In the case of rare earth ions in crystals, it has been well established that these ions may be used as probes of the crystalline symmetry.

The site symmetry can be deduced by the methods of group theory from the number of lines in the emission or absorption spectrum into which the free ion levels are split.

The optical transitions typical of the rare earth ions in crystals and solutions correspond mainly to intra f^N transitions of predominantly electric dipole character. For a free ion, electric dipole transitions between states of the same configuration are strictly parity forbidden and the observed spectra of crystals or solutions result from noncentro-symmetric interactions that lead to a mixing of states of opposite parity.

The various mechanisms of mixing are thoroughly discussed by *Wybourne* (2). One of the most important mechanisms responsible for the mixing is the coupling of states of opposite parity by way of the odd terms in the crystal field expansion of the perturbation potential V, provided by the crystal environment about the ion of interest. The expansion is done in terms of spherical harmonics or tensor operators that transform like spherical harmonics. This can be formulated in a general Eq. (1)

$$V = \sum_{k,\,q,\,i} B_q^k (C_q^k)i \qquad (1)$$

where the summation involving i is over all the elctrons in question, k is the rank of the tensoral operator C, q is the relevant component of that operator $(-k \leq q \leq k)$ and B_q^k are the expansion coefficients. The first term in the expansion has $k = q = 0$ and is spherically symmetric. This term gives the shift of spectral bands to the longer wavelength, as its influence on the upper levels is stronger than it is on the lower, more shielded levels. When only f electrons are involved, the terms in the expansion with $k \leq 6$ are nonzero. In addition, all terms with k odd vanish, hence it is the even terms in the expansion that reflect the splitting. The B_q^k can be regarded as coefficients of expansion to be determined empirically from ΔE, the magnitude of the splitting. A comparison of the spectra of the rare earths and the transition metal ions reveals that the crystal field splitting is of about 100—300 cm^{-1} in the former compared to several thousands cm^{-1} in the latter. The number of levels into which a single level can be split is determined by the site symmetry. Values for different symmetries can be found in the literature (see for instance Ref. (3), p. 60).

The crystal field model may also provide a calculation scheme for the transition probabilities between levels perturbed by the crystal field. Such calculations were made by *Judd* (4) and *Ofelt* (5), who showed that the odd terms in the crystal field expansion can connect the $4f$ configuration with the $5d$ and $5g$ configurations. The result of the calculation for the oscillator strength, due to a forced electric dipole transition between the two states $(l^N \alpha S L J J_z)$ and $(l^N \alpha^1 S L^1 J^1 J_z^1)$ (Ref. (7), p. 207) is

$$f_e = \sigma \sum_{\lambda, q} T(\lambda_1 \, q_1 \, \varrho) \, (l^N \, \alpha \, S \, L \, J \, J_z \, \|U_{q+\varrho}^\lambda\| \, l^N \, \alpha^1 \, S \, L^1 \, J^1 \, J_z^1)^2 \qquad (2)$$

for rare earths, $\lambda = 2, 4, 6$.

T_λ are components of a tensor operator, the explicit expression for them being

$$T_\lambda = \sum_{\varrho, q} \frac{8\pi^2 \, mc}{3 \, he^2} \frac{(n^2 + 2)^2}{9} \, Y^2 \, (\lambda, q, \varrho) \qquad (3)$$

where n is the refractive index of the medium. Other constants have their usual meaning and

$$Y(\lambda_1 \, q_1 \, \varrho) = -\frac{2}{E \, av} \sum_k (-1)^{q+\varrho} \, (2\lambda + 1)$$

$$x \begin{pmatrix} 1 & \lambda & k \\ \varrho - (\varrho + q) & q \end{pmatrix} x \begin{pmatrix} l \, l' \, \lambda \\ 1 \, k \, k'' \end{pmatrix} x <l \, \|C^{(1)}\| \, l'> \, <l' \, \|C^{(k)}\| \, l> B_q^k \, e$$

$$\int R_{nl} \, rR_{n'l'} \, dr \qquad (4)$$

Eav is the average energy difference between the configuration l^N an the perturbing configuration $l^{N-1} \, l \, (l = 5d$ or $5g)$. The quantities B_q^k are the odd parity terms in the static crystal field expansion, where $k = 1, 3, 5$ and $7, q$ are the components of k and depend on the symmetry of the medium (see Ref. (3) Table 4.2), ϱ refers to the expansion of the dipole moment and determines the polarization of the transition

$\varrho = 0$: π component of the transition
$\varrho = \pm 1$: σ component of the transition

R_{nl} and $R_{n'l'}$, are the radial parts of the wave function in the l^N and the l^{N-1} configuration, respectively.

In theoretical calculations of the oscillator strengths, the free ion-reduced matrix elements U_q^λ are used and the coefficients T_λ are treated as adjustable parameters. All attempts to derive the value B_q^k from micro-crystalline parameters of the lattice using the electrostatic point charge approximation theory have been, thus far, unsuccessful. This is probably due to the extreme sensitivity of B_q^k to the exact positions of the ions, the exact form on the radial part of the wave functions, and to the percentage of covalence of the bond. It is clear that a molecular orbital treatment should be applied in such cases. *Jørgensen et al.* (6) and *Ellis* and *Newman* (7) showed that this treatment gives a much more realistic picture of the rare earth in a crystal. The existence of the nephelauxetic effect as first recognized by *Jørgensen* (8) clearly indicates the overlapping of the charge clouds of the central ion and its ligand. The work of *Katzin* and *Barnett* (9) has also provided direct evidence for the participation of f-orbitals in the chemical bonding of rare earth complexes. The observation of hypersensitive transitions obeying the selection rule, $\Delta J = \pm 2$, by *Jørgensen* and *Judd* (10) and subsequently those in a number of works dealing with spectral characteristics of rare earth ions, emphasize the influence of the ligand on spectral levels of the rare earth ion.

In order to continue to develop the M.O. treatment for the spectra of rare earth ions, a large number of experimental data are needed for such quantities as:

1. Oscillator strength of trivalent rare earth ions in various media; from this it is possible to obtain the radiative transition probabilities and the T_λ parameters.

2. The charge transfer spectra of rare earths in various media from which it is possible to obtain the optical electronegativity of the medium as proposed by *Jørgensen* (11, 12,) and *Blasse* and *Bril* (13).

The above quantities have been measured in glass matrices and the results are presented in the first part of this paper.

The existence of vibronic transitions in rare earth crystals shows that the rare earth ions are coupled to the vibronic modes of the lattice; we shall also show that such coupling plays a dominant role in the energy transfer process between rare earth ions in glasses. We believe that this treatment of phonon-assisted energy transfer can be extended to biologically active organic molecules such as DNA, ATP, DTP and similar molecules.

II. Absorption and Fluorescence Spectra of Rare Earths in Glasses Arising from f-f Transitions

Considerable information may be obtained from even electron systems, as their degeneracy may be entirely removed in lower symmetries. We therefore first focus our attention on Eu^{3+} and Tm^{3+}. The discussion will then be extended to cover odd electron systems exemplified by Er^{3+}.

a) Europium

A systematic study of the luminescent characteristics of glasses containing rare earths was done in this laboratory (14—20). There has been considerable spectroscopic investigation involving europium activated phosphors for several reasons. The phosphors are of practical use in color television and more information about crystal levels can be obtained for even-electron systems than those with odd-electron systems. *Reisfeld et al.* (21—25) and *Rice* and *DeShazer* (26) have used the fluorescence of europium as an indicator of site symmetry of rare earth ions in glasses.

The ground level electronic configuration of trivalent europium is f^6. Transitions within the f shell are responsible for the crystal spectra. Transitions are forbidden in a free ion by the parity rule for electric dipole transitions. In a crystal or glass, forced electric transitions become allowed as a consequence of coupling of odd electronic wave functions due to the odd parity terms in the crystal field expansion. Considering the static field approximation in the theory developed by *Judd* (4) and *Ofelt* (5), the contribution of the odd parity part of the crystal field is calculated by mixing states of different parity.

The measured intensity of an absorption band is related to P by the following expression (27).

$$P = 4.318 \times 10^{-9} \int \varepsilon_i(\sigma) \, d\sigma \tag{5}$$

where ε is the molar absorptivity at the energy $\sigma (cm^{-1})$.

In cases where the transitions occur wholly or partially by a magnetic dipole mechanism following the selection rules $\Delta J = 1$, $\Delta L = 0$, $\Delta S = 0$ and $\Delta l = 0$

$$P_{expt\,1} = P_{md} + P_{ed} \tag{6}$$

where $P_{expt\,1}$ is the experimentally observed oscillator strength, P_{ed} is the induced electric dipole transition, and P_{md} is the magnetic dipole transition.

The oscillator strengths of Eu^{3+} in glasses were measured and compared to the results obtained in aqueous solution by *Carnall, et al.* (27).

The results which follow allow us to derive conclusions concerning glasses and aqueous solutions with respect to the following parameters: 1. change in line intensity, 2. shift of spectral lines, and 3. splitting and broadening of the spectral lines.

Table 1. *Oscillator strengths of europium (III) in phosphate, silicate and germanate glasses and in aqueous solution*

Transition Assignment	Wave number, cm^{-1} [a]	Oscillator strength x 10^7			
		Phosphate	Silicate	Germanate	Exptl ($HClO_4$) (27)
$^7F_2 \rightarrow {}^5D_0$	16,319	1.544	1.854		—
$^7F_1 \rightarrow {}^5D_0$	16,771	0.351	0.551		—
$^7F_0 \rightarrow {}^5D_0$	17,256	0.006	0.063	0.035	0
$^7F_1 \rightarrow {}^5D_1$	18,700	0.505	2.425	4.681	—
$^7F_0 \rightarrow {}^5D_1$	18,993	0.146	0.251	0.156	0.14
$^7F_0 \rightarrow {}^5D_2$	21,493	1.248	1.257	2.760	0.21
$^7F_0 \rightarrow {}^5D_3$	24,009	0.547	0.474	0.065	—
$^7F_0 \rightarrow {}^5L_6$	25,380	8.981	3.371	1.853	17.7
$^7F_1 \rightarrow {}^5G_5, {}^5G_6$	26,041	6.019			4.3
$^7F_0 \rightarrow {}^5G_3$	26,507	2.726		2.888	—
$^7F_1 \rightarrow {}^5L_8$	27,285	0.391		1.710	—
$^7F_0 \rightarrow {}^5D_4$	27,567	1.926		0.874	1.7
$^7F_1 \rightarrow {}^5H_3$	30,464	1.242	—	—	—
$^7F_1 \rightarrow {}^5H_5$	31,152	3.001	—	—	—
$^7F_0 \rightarrow {}^5H_6$	31,397	6.776	—	—	7.3

[a]) Maximum w.n. for phosphate.

The oscillator strength of Eu^{3+} was calculated by use of formula (5) for the spectral bands due to f—f transitions. The oscillator strength for phosphate (22), silicate (23) and germanate (24) glasses are presented in Table 1. They are compared with the oscillator strengths obtained in perchlorate solution by *Carnall et al.* (27). The wave numbers in the Table are for phosphate glasses. There is a slight change (few cm^{-1}) when changing from phosphate to silicate and germanate glasses.

As can be seen from the Table, the oscillator strengths obtained for the glasses are higher than those obtained for aqueous solutions with the exception of the $^7F_0 - {}^5L_6$ transition. We also note that the oscillator strength for the $^7F_0 - {}^5D_0$ transition in the glass is appreciable considering the zero value reported for the europium perchlorate solution. The value of $^7F_0 - {}^5D_0$ is larger in germanate and silicate glass relative to the phosphate glass. Also, the magnitude of the hypersensitive transition (10) with $\Delta J = 2$ for the $^7F_0 \rightarrow {}^5D_2$ line is significantly larger in germanate ($f = 5.113$) compared to phosphate and silicate (~ 1.25). The facts indicate that the asymmetric part of the crystal field of the glass is larger in germanate compared to silicate glass, and in turn larger in silicate glass than in phosphate.

As was pointed out in the introduction, the matrix elements for the trivalent rare earth appearing in equation (2) $(f^N \psi J \| U_\lambda \| f^N \psi' J')$ do not depend strongly on the host lattices and do not differ significantly from their free ion values. Hence, the three quantities T_λ are mainly responsible for the variation of intensity with host matrix. The explicit dependence of the T_λ of the crystal field in *Axe's* notation is (28).

$$T_\lambda = (2\lambda + 1) \sum_{t,p} \frac{|Aip|^2}{2t+1} E(t\lambda) \qquad (7)$$

where Aip are the odd parity terms in the static crystal field expansion, and the parameters $E(t\lambda)$ contain integrals involving the radial parts of the $4f$ functions and the opposite-parity $n^1 l$ wavefunctions, as well as the energy separation between the $4f^n$ and the $4f^{n-1} n^1 l$ configurations. The $5d$ orbital makes a dominant contribution to the configuration mixing responsible for the electric dipole oscillator strengths; the contribution of $5g$ is much smaller. The sensitivity of the intensity on host structure is therefore through the parameter T_λ. In principle, one should be able to calculate the T_λ's by use of the free ion model, however, that model produces ambiguities in the evaluation of the radical integrals $E(t\lambda)$ and in the odd parity lattice sums because of complicated crystal field shielding effects. We, therefore, feel that it is best to calculate the parameters T_λ from experimental oscillator strengths using a least squares technique similar to that used by *Carnall et al.* (27) to calculate these parameters in solutions. Clearly, measurement of as many oscillator strengths as possible of a given electronic configuration of rare earth ions in glasses is vital to producing statistically meaningful results from the least squares analysis.

Magnetic dipole transitions may occur only between states of the same parity. These transitions should not be as sensitive to the host matrix as electric dipole transitions, however there will be a small contribution from the host matrix arising from J mixing due to even parity terms in the crystal field expansion, and also due to the variations of the refractive indices from one host to another. In glasses, however, the refractive index varies little from glass to glass, hence this correction is negligible. This near invariance of the magnetic dipole transition strengths can be utilized as a standard to which electric dipole transitions in emission may be referred. This is needed because in emission, only relative oscillator strengths for a particular host can be measured. We can obtain an estimate as to a given electric dipole moment for a particular host by taking the ratio of the transitions of $\Delta J \leq 2$ to $\Delta J = \pm 1$, 0, the latter being assigned mainly to the magnetic moment (30). It has been hypothesized (29) that the electric dipole transitions responsible for the transitions in Tb^{3+} in Y_2O_3 are those due to T_2 mainly with a minor contribution from the quantities T_4 and T_6.

The T_2 component of the transition probabilities contains contributions from the linear and the third order terms of the crystal field expansion. The linear terms occur only when the rare earth is embedded in the following lattice sites (29): C_s, C_2, C_{2v}, C_3, C_{3v}, C_4, C_{4v}, C_6, C_{6v}.

We note that the oscillator strength for the $^7F_0 \rightarrow {}^5D_0$ transition in the glass is significantly large in view of the zero value reported for the europium perchlorate solution. This transition has been reported to exist in cases where the site symmetry allows an electric dipole process. The symmetries allowing such a process are C_s, C_n and C_{nv} (30). This is consistent with the results of *Rice* and *DeShazer* (26), who deduced that Eu^{3+} was situated in a C_s symmetry. By comparing the half-bandwidth of the $^7F_0 \rightarrow {}^5D_0$ transition in an europium sesquioxide crystal, which is about 2 cm^{-1}, to the half-bandwidths of these transitions in glasses (Table 2) which are 119 cm^{-1} in phosphate glass and 149 cm^{-1} in silicate and germanate glass, it is concluded that there are approximately 50 slightly different sites of C_s symmetry in these glasses. The slight differences in environment are caused by small variations in the crystal field parameters. The energy-level shifts for ions in different environments lead to slight changes in the transition wavelengths producing the broadening of the spectra. We believe that a similar situation exists in solution where the broadening of the $^5D_0 \rightarrow {}^7F_1$ and $^5D_0 \rightarrow {}^7F_2$ bands resulted from a variety of sites for the same symmetry due to a slight change in the distances between the ligands (water, nitrate or perchlorate) and the rare earth ions.

In glasses, the rare earth ion is surrounded by nonbridging oxygens of phosphate (21) borate (31) silicate (23) and germanate (24). The MO_4 (M = B, P, Ge, Si) tetrahedra are undistorted since their covalent bonding strongly favors preservation of the tetrahedral geometry. On the other hand, the relative positions of the tetrahedra can be changed in relation to the rare earth. *Reisfeld* and *Eckstein* (31) proposed that the rare earth in glass is coordinated by four MO_4 tetrahedra, each tetrahedron contributing two oxygens to the coordination with the rare earth. The overall coordination is 8 (which is the most common coordination of the rare earth oxides).

Table 2. *Relative intensities and half-bandwidths of $^5D_0$ emission of Europium (III)*

Transition	Borax			Phosphate			Silicate			Germanate			Tungstate		
Assign-ment	Band Max	RI	1/2 BW (61) cm⁻¹	Band Max	RI	1/2 BWª⁾ cm⁻¹	Band Max	RI	1/2 BWª⁾ cm⁻¹	Band Max	RI	1/2 BW (24) cm⁻¹	Band Max	RI	1/2 BW (25) cm⁻¹
$^5D_0 \rightarrow ^7F_0$	578.0	0.166	135	578.5	0.036	119	579.7	0.130	149	580	0.110	149	580	0.072	103
$^5D_0 \rightarrow ^7F_1$	592.7	1.000	485	591.9	1.000	313	592.2	1.000	313	590	1.000	342	593	1.000	313
$^5D_0 \rightarrow ^7F_2$	614.0	8.945	343	612.1	3.000	289	610.2	4.255	295	612.5	6.666	159	613	6.618	329
$^5D_0 \rightarrow ^7F_3$	653.2	0.194	293	654.7	0.219	234	655.7	0.314	381	655	0.260	282	655	0.251	235
$^5D_0 \rightarrow ^7F_4$	—	—	—	692.3	1.026	142	693.2	0.398	121	705.5	0.497	131	—	—	—

1/2 BW Half-bandwiths.

ª⁾ For references to this table, see Ref. (21).

61

It was shown in a series of papers by *Reisfeld et al.* (*20, 21, 31, 43, 48*) that the behavior of rare earths in glasses is similar to that of rare earths in inorganic crystals of low symmetry except for inhomogeneous broadening of the spectra, because of the multiplicity of rare earth sites in glasses. According to this theory (*31*), a rare earth ion occupies the center of a distorted cube which is made of four tetrahedra of borate, phosphate or silicate. Two oxygens belonging to such a tetrahedron produce an edge of the cube (Fig. 1). The coordination of rare earth in such an arrangement is eight oxygens. The site differences can occur from the existence of non-uniform non-identical ligand fields caused by slightly different ranges of rare earth oxygen distances. An average RE—O distance was calculated by an assumption of a R.E. situated in the center of a regular cube. The average RE—O distance so calculated in borate glass is 2.1 Å and in phosphate glass is 2.22 Å. It should be realized that in reality the symmetry is lower than cubic, and for particular cases studied, it was found to be C_8 for Eu in phosphate glass (*27*) and C_2 for Tm in phosphate and borate glasses (*31*). It is because of this lower symmetry that forced electric dipole transitions become possible and absorption and fluorescence is observed.

The fluorescence can in some cases be quenched, due to a transfer of electronic excitation energy to the vibrations of the surrounding medium, Such interactions reduce the quantum yield of the fluorescence. In general, the larger the energy difference between the two electronic states, the smaller the non-radiative transition probability between them (*22*). For rare earths, the energy difference between the electronic levels below which considerable quenching occurs is about four phonons. The non-radiative decay of the higher excited levels in the glasses is generally assisted by the phonons of the glass. The particular vibrational quanta (phonons) which are responsible for the quenching of various glasses are given below.

Bond	stretching frequency	bond length
silicate Si—O	1010—1115	1.62
metaphosphate P—O	1140—1300	1.57
borate B—O	1310—1380	1.39 (triangle)
		1.48 (tetrahedron)
germanate Ge—O	840— 930	1.70

From this we conclude that in borate glasses, quenching will occur between levels which are separated by about 5000 cm^{-1}, while in silicate by about 4000 cm^{-1}, and germanate about 3500 cm^{-1}.

We shall now show how this general theory works for specific cases.

We can picture the rare earth ion in a cube (Fig. 1) with 8 oxygens at the corners. Each edge of the cube is common to the cube and tetrahedron. The cube is not a regular one, but distorted by the relative twisting of the tetrahedra, which can be situated at angles other than 90° respective to each other. Therefore, a low symmetry C_s site is formed. From the inhomogeneous broadening we can conclude that while this low symmetry is conserved for all the sites, the position of a given site relative to the surrounding oxygens may depend on the exact situation of the rare earth in the cube. A convenient parameter for measuring the average distance from the Eu to its surrounding oxygens is the average Eu—O distance assuming that Eu is situated in the center of a regular cube. The tetrahedra in phosphate glass have P—O and O—O distances of 1.57 Å and 2.56 Å respectively. In silicate glass, the Si—O and O—O distances are 1.48 Å and 2.40 Å respectively, in germanate glass, 1.70 Å and 2.76 Å respectively. As we assume the rare earth ion to be at the center of the cube, its distance from the nearest oxygen at the cube corner is equal to the length of half the cubic diagonal. Hence we obtain 2.22 Å for phosphate glass, 2.29 Å for silicate, 2.07 Å for borate, and 2.39 Å for germanate glass.

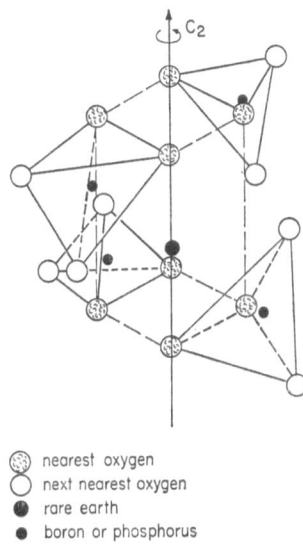

○ nearest oxygen
○ next nearest oxygen
● rare earth
• boron or phosphorus

Fig. 1. Proposed rare earth site model. ○ — represent nearest oxygen neighbors surrounding the rare earth

It is seen from experiment that the inhomogeneous broadening of the $^7F_0 \to {}^5D_1$ Eu^{3+} transition in phosphate and silicate glasses is less than it is in germanate glass. Also, the oscillator strengths of this transition are considerably less in the phosphate and silicate relative to the germanate glass. This is consistent with the rare earth-oxygen average distance for these glasses. We have seen that the size of the cube in the phosphate and silicate glasses is less than in the germanate glasses. Hence, there is more lattitude for distortion in the germanate glass than in the others (see Tables 1 and 2).

As $^5D_0$ is a nondegenerate level ($J = 0$), the splitting of the lines arising from the $^5D_0 \to {}^7F_1$ transition reflect the crystal field splitting of the terminal $^7F_1$ level. The $^5D_0 \to {}^7F_1$ line of europium is split into three components in glasses indicating the C_s symmetry. The splitting is clearly seen in the emission lines of germanate and borax glasses where three maxima are seen, and also in tungstate glass, where two maxima and a shoulder are seen. The splitting of the $^5D_0 \to {}^7F_1$ line of Eu in phosphate and silicate glasses is better seen in the absorption spectrum. The results of the splitting as obtained by Gaussian analysis are presented in Table 3.

The difference between solutions and glasses in the $^7F_0 \to {}^5D_2$ oscillator strengths and intensities can be seen in Tables 1 and 2, respectively. It should also be noted that the relative electric dipole intensities of Eu^{3+} in LaF_3 are smaller than those in glasses with C_s symmetries (Table 2).

It is known that the absorption bands of europium in solution due to $J = 2 \leftrightarrow J = 0$ transitions are intensified by a factor of about 10 when the aquo ions are replaced by other complexing ions. *Jørgensen* has named these as "hypersensitive" transitions (*10*). *Judd* pointed out that only the following symmetries will give rise to hypersensitive transitions: C_s, C_1, C_2, C_{2v}, C_{3v}, C_{4v} and C_6 (*35*). It is therefore of interest to compare the hypersensitivity of the $^5D_0 \to {}^7F_2$ and $^7F_0 \to {}^5D_2$ transitions in media other than solutions (Table 2).

By comparing the relative intensity of the $^5D_0 \to {}^7F_2$ transition from the emission spectra, we also see that although the site symmetry of Eu^{3+} is C_s in phosphate, silicate, borate, germanate and tungstate glasses, the forced electric dipole is stronger for Eu^{3+} in the germanate and tungstate glasses.

The magnetic dipole transitions ($\Delta J = \pm 1$) are, as previously stated in the introduction, insensitive to the surroundings of the ion, and may therefore be used as a standard for comparison of other transition strengths.

It is known (*10*) that electronic transitions within Eu^{3+}, in which J is changed by 2 are strongly dependent on the surrounding medium. The large ratio of the fluorescence intensities for the $^5D_0 \to {}^7F_2$ and

Table 3. Crystal field splitting of the $^5D_0$—$^7F_1$ line of Europium in borax, phosphate, silicate, germanate and tungstate glasses

Eu³⁺ in Silicate[a])		Eu³⁺ in Germanate[b])		Eu³⁺ in Tungstate[b])		Eu³⁺ in Borax[b])		Eu³⁺ in Sod. Phosphate[a])		Eu³⁺ in Magnesium Phosphate[b])	
SiO₂ 75 mol% BaO 5 mol% Na₂O 15 mol% ZnO 5 mol%		K₂O 17 mol% BaO 17 mol% GeO₂ 66 mol%		WO₃ 42 mol% P₂O₅ 19 mol% Na₂O 39 mol%		Na₂O. 2 B₂O₃		Na₂O. P₂O₅		MgO. P₂O₅	
σ_{max} cm⁻¹	$\Delta\sigma$	σ_{max} cm⁻¹	$\Delta\sigma$	σ_{max} cm⁻¹	$\Delta\sigma$	σ_{max} cm⁻¹	$\Delta\sigma$	σ_{max} cm⁻¹	$\Delta\sigma$	σ_{max} cm⁻¹	$\Delta\sigma$
16781	119	16775	112	16702	180	16740	136	16797	82	17656	117
16900	149	16887	116	16828	126	16876	125	16879	101	16873	117
17049		17003		17008		17001		16980		16990	

a) From absorption spectrum.
b) From emission spectrum.

$^5D_0 \rightarrow {}^7F_1$ transitions implies a low symmetry field at the europium site (23).

The $^5D_0 \rightarrow {}^7F_2$ transitions relative to the magnetic dipole $^5D_0 \rightarrow {}^7F_1$ transitions are $3:1$ in phosphate, $4.255:1$ in silicate and $6.666:1$ in germanate.

The trend already noticed, in which the symmetry surrounding the ion progressively decreases as we pass from phosphate to silicate to germanate glasses, is thus further borne out.

b) Thulium and Erbium

The absorption and emission spectra of thulium and erbium were studied in phosphate and borate glass by *Reisfeld* and *Eckstein* (31, 32). Their objectives were 1) to study the influence of two different glass hosts on the transition probabilities of trivalent Er and Tm ions (henceforth R. E.), 2) to compare the intensities of the spectra of these R. E. in glasses with those of liquid solutions and doped oxide lattices, and 3) to compare the broadening of the R. E. fluorescence bands in glasses with those in aqueous solutions and doped crystals.

Thulium and erbium sesquioxides have only one stable polymorph, which belongs to the cubic type. Each unit cell of cubic form has one site of eight R. E. on C_{3i} sites, one set of 24 R. E. on C_2 sites and 48 oxygen ions on C_1 sites (34). By the analogy with europium, we can expect these symmetries of erbium and thulium to be found in glasses. The C_2 and C_{3i} are distinguishable from the optical transitions in the spectrum of the rare earth, it being known that the first can give rise to a forced electric dipole, while the latter can be effective only for a magnetic dipole transition due to the presence of a center of inversion (4). The character of the absorption spectrum can thus be used to define the site symmetry of the rare earth in glasses.

The results detailed below allow us to derive conclusions about the crystal field in different glasses, and to compare glasses and aqueous solutions with respect to the following parameters:

1. change in line intensity,
2. shift of the spectral lines,
3. splitting and broadening of the spectral lines.

The experimentally determined oscillator strengths are given in Tables 4 and 5. A knowledge of these parameters would give insight into the crystal field strength of glasses.

Tables 4 and 5 give the oscillator strengths (expressed as the f numbers) and spectral assignments of Tm^{3+} and Er^{3+} respectively in borate and phosphate glasses, as compared to aqueous solutions, molten KNO_3-

Table 4. *Oscillator strengths (f) of Tm^{+3} in borate and phosphate glasses $f = 4.318 \times 10^{-9}$ $\varepsilon(\sigma) d\sigma$*

Energy level	Borate Glass			Phosphate Glass		
	Wave no. (cm^{-1})	Wave length (nm)	$P \times 10^6$	Wave no. (cm^{-1})	Wave length (nm)	$P \times 10^6$
$^3H_4$	5847 to 6134	1630.0 to 1710.0	3.94	5714	1750.0	...
$^3H_5$	8257	1211.0	2.09	8250	1212.0	1.82
$^3F_4$	12626 12903	792.0 775.0	4.39	12626	792.0	3.00
$^3F_3$	14619	684.0	4.02	14550	687.2	2.53
$^3F_2$	15100	662.0	0,70	15060	664.0	0.22
$^1G_4$	21231 21505	471.0 465.0	2.19	21052 21505	475.0 465.0	1.58
$^1D_2$	27855 28050	356.5 355.0	4.75	27855	359.0	3.06
$^1I_6$ $^3P_0$	34246 34662 35082	292.0 288.5 285.0	4.14	33057 34364 35057	302.5 291.0 286.0	31.87
$^3P_1$	36363	275.0		36166	276.5	
$^3P_2$	37950 38387	262.5 260.5	6.13	38095	262.5	14.46

$LiNO_3$, Tm_2O_3 and doped Y_2O_3 crystals. We were able to observe transitions from the $^3H_6$ ground state of Tm^{3+} to all the remaining ^3H, ^3F, ^1G, ^1D and ^3P multiplets.

We also observed almost all the absorption bands of Er^{3+} corresponding to transition from the ground state $^4I_{15/2}$ to higher multiplets. Comparison of the absorption spectra of the R. E. in the two different glass media shows that in the UV region, the relative intensities of a given transition are higher in the phosphate than in the borate glass. In the latter case, higher splitting occurs as well. In the visible and the near infrared regions, the results are opposite: higher intensities are observed in borate but the splittings are higher in phosphate.

The differences in the UV absorption spectra of Er^{3+} between the two kinds of glasses are smaller than those for Tm^{3+} in the same media. The reproducibility of the absorption spectra in all the spectral regions investigated excludes the possibility of contributions to the absorption

Table 5. *Oscillator strengths (f) of Er^{+3} in borate and phosphate glasses* $f = 4.318 X 10^{-9}$ $\varepsilon(\sigma) d\sigma$

Energy Level	Borate Glass			Phosphate Glass		
	Wave no. (cm^{-1})	Wave Length (nm)	$P \times 10^6$	Wave no. (cm^{-1})	Wave Length (nm)	$P \times 10^6$
$^4I_{13/2}$	6544	1528.0		6527	1532	
	6697	1493.0	3.19			2,31
	7235	1382.0		6688	1495	
$^4I_{11/2}$	10256	975.0	1.2	10256	975.0	1.09
$^4I_{9/2}$	12547	797.0	1.2	12500	800.0	0.5
$^4F_{9/2}$	15349	651.5	3.2	15209	657.5	2.8
				15325	652.5	
$^4S_{3/2}$	18390	543.7		18331	544.5	
$^2H_{11/2}$	19175	521.5	16.28	18921	528.5	13.75
				19230	520.0	
$^4F_{7/2}$	20470	488.5	2.76	20470	488.5	2.55
$^4F_{5/2}$	22172	451.0		21857	457.5	
			1.19	22148	451.5	1.05
$^4F_{3/2}$	22573	443.0		22547	443.5	
$^2H_{9/2}$	24539	407.5	1.5	24539	407.5	2.94
$^4G_{11/2}$	26176	382.0	30.61	26246	381.0	26.24
	26420	378.5		26420	378.5	
$^2K_{15/2}$	27322	366	4.7	27322	366.0	4.0
$^2G_{9/2}$						
$^2G_{7/2}$	28011	357	1.2	28030	356.7	1.44
$^2P_{3/2}$	31595	316.5		31207	320.5	
	32200	310.5		31527	317.2	
	32679	306.0	1.72	31986	312.7	3.24
	33361	299.7		32597	306.7	
				33147	301.7	
				33682	296.7	
$^2D_{5/2}$	34050	293.5	1.45	34071	293.5	3.03
	34722	288.0		34752	287.7	
$^4G_{9/2}$	35590	281	2.27	35345	283.0	5.22
	36363	275		36403	274.7	
$^4D_{5/2}$	38834	257.5	15.24	38989	256.5	15.55
$^4D_{7/2}$	39138	255.5				

Table 6. *Emission spectrum of Tm^{3+} at various excitation wavelengths*

Excitation	Assigned	Emission			
		Borate		Phosphate	
(nm)	transition	λ (nm)	R.A.[a]	λ (nm)	R.A.[a]
	$^1G_4 \rightarrow {}^3H_4$			651.5	1.000
	$^1G_4 \rightarrow {}^3H_5$			752.7	0.114
468.0 ($^1G_4$)	$^3F_2 \rightarrow {}^3H_6$			665.5	0.083
	$^3F_3 \rightarrow {}^3H_6$			690.0	0.708
	$^1D_2 \rightarrow {}^3H_4$	456.0	1.000	453.0	1.000
	$^1D_2 \rightarrow {}^3H_5$	517.0	0.008	513.0	0.011
358.0 ($^1D_2$)	$^1D_2 \rightarrow {}^3F_4$	665.0	0.127	663.5	0.034
	$^1G_4 \rightarrow {}^3H_6$	478.0	0.118	478.0	0.063
	$^1G_4 \rightarrow {}^3H_4$	652.5	0.180	651.5	0.074
	$^3F_2 \rightarrow {}^3H_6$	665.5	0.014	665.5	0.009
	$^1I_6 \rightarrow {}^3H_4$	355.0	0.446	350.0	1.000
	$^1I_6 \rightarrow {}^3H_5$	385.0	0.125	383.0	0.094
	$^3P_0 \rightarrow {}^3F_4$ $^1I_6 \rightarrow {}^3F_4$	465.0	0.155	463.0	0.200
	$^1I_6 \rightarrow {}^3F_3$	500.0	0.110	498.0	0.060
	$^3P_0 \rightarrow {}^3F_3$ $^1I_6 \rightarrow {}^3F_2$	530.0	0.061	521.5	0.094
288.0 ($^3P_0$)	$^3P_0 \rightarrow {}^1G_4$ $^1I_6 \rightarrow {}^1G_4$	705.0	0.754	701.0	1.000
	$^1D_2 \rightarrow {}^3H_6$	367.5	0.275	365.0	0.133
	$^1D_2 \rightarrow {}^3H_4$	456.0	1.000	453.0	0.455
	$^1D_2 \rightarrow {}^3H_5$	517.0	0.067	513.0	0.044
	$^1D_2 \rightarrow {}^3F_4$	...	...	...	...
	$^1G_4 \rightarrow {}^3H_6$	480.0	0.190	478.0	0.105
	$^1G_4 \rightarrow {}^3H_4$	...	...	...	...
	$^3F_2 \rightarrow {}^3H_6$	...	...	...	...
	$^3F_3 \rightarrow {}^3H_6$	...	...	...	...

[a] Relative areas; for each excitation the areas are given relatively to the strongest emission band which is taken as unity.

from background absorption or from the tail of the large bands with maxima 50000 cm^{-1}.

Comparison of the absorption intensities in the ultraviolet, visible and near infrared regions in glasses with those obtained in aqueous solutions ($HClO_4 - DClO_4$) shows that all absorption bands are more intense in the glass "lattices" and are comparable to the results obtained in molten $LiNO_3 - KNO_3$, Tm_2O_3 crystals and Er^{3+} doped Y_2O_3 crystals.

Table 7. *Relative intensities and half-bandwidths of* $^1D_2$ *and* $^1G_4$ *emissions of* Tm^{3+}

Transition	Borate			Phosphate			$Y_2O_3:Tm^{3+}$[7]	
	λ (Å)	R.I.	$\Delta\lambda$ (Å)	λ (Å)	R.I.	$\Delta\lambda$ (Å)	λ (Å)	$\Delta\lambda$ (Å)
$^1D_2 \rightarrow {}^3H_4$	4560	56.0	830.0	4530	68.0	670.0	4540	13.9
$^1G_4 \rightarrow {}^3H_6$	4800	1.1	1410.0	4780	0.8	1080.0	4884	19.3
$^1D_2 \rightarrow {}^3H_5$	5170	0.06	750.0	5130	0.07	750.0	5154	4.2
$^1G_4 \rightarrow {}^3H_4$	6525	5.3	175.0	6515	5.6	170.0	....	...

R.I.-Relative intensity; $\Delta\lambda$-half-bandwidth.

Table 8. *Emission spectrum of* Er^{3+} *at various excitation wavelengths*

Excitation (nm)	Assigned transition	Wavelength (nm)
	$^2P_{3/2} \rightarrow {}^4I_{15/2}$	310.0—320.0
	$^2G_{7/2}, {}^2G_{9/2}, {}^2K_{15/2} \rightarrow {}^4I_{15/2}$	360.0—370.0
	$^4G_{11/2} \rightarrow {}^4I_{15/2}$	388.0
255.0 a)	$^2P_{3/2} \rightarrow {}^4I_{13/2}$	404.0
	$^2P_{3/2} \rightarrow {}^4I_{11/2}$	473.0
	$^2P_{3/2} \rightarrow {}^4I_{9/2}$ $^2H_{11/2} \rightarrow {}^4I_{15/2}$	530.0
	$^2H_{9/2} \rightarrow {}^4I_{13/2}$ $^4S_{3/2} \rightarrow {}^4I_{15/2}$	540.0—550.0
382.0 b)	$^2H_{11/2} \rightarrow {}^4I_{15/2}$	524.0—530.0
	$^2H_{9/2} \rightarrow {}^4I_{13/2}$ $^4S_{3/2} \rightarrow {}^4I_{15/2}$	548.0—557.0

a) Excitation with Mercury Low Pressure source
b) Excitation with Xenon source.

Table 6 presents the emission spectrum of Tm^{3+} in borate and phosphate glass as obtained at various excitation wavelengths. Table 7 gives the wavelength maxima, relative intensities and half bandwidths for several fluorescence bands from the $^1D_2$ to the ^3H multiplet for Tm^{3+} in borate and phosphate glasses as compared to Y_2O_3 doped crystals.

The UV and visible fluorescence of Er^{3+} in our glasses is much weaker than that of Tm^{3+}. The fluorescence was observable only in phosphate glasses, with 382 nm excitation (from a xenon source) and 255 nm excitation (from a mercury low-pressure lamp). In Table 8, we present the wavelengths of fluorescence together with the transition assignments obtained under excitation at 382 and 255 nm.

The R.E. in the glass "lattice" will assume a site symmetry similar to that existing in oxide crystals, which is C_2 and C_{3i} for Tm^{3+} and Er^{3+}. As shown above this assumption was confirmed by *Rice* and *DeShazer* (*26*) and our experimental studies on Eu^{3+} ions (*22*) in different glasses. Consequently we assume that the predominant point symmetry of Tm^{3+} and Er^{3+} in glasses is C_2. The degeneracy of each electronic J level of Tm^{3+} (f^{12}) in the C_2 sites is completely lifted, giving rise to $2J+1$ crystalline Stark levels. The degeneracy of each J level of Er^{3+} (f^{11}) in the C_2 sites is only partially removed, giving rise to $J+1/2$ crystalline *Stark* levels, because of *Kramers'* rule. Electronic transitions are governed by different selection rules for the ions in two different sites. If Tm^{3+} or Er^{3+} are present in the C_{3i} symmetry, then only magnetic dipole transitions are allowed. For ions on the C_2 sites, magnetic as well as forced electric dipole transitions are allowed, since there is no center of inversion. From our experimental results, we see that mainly the transitions obeying the selection rule $1 < \Delta J \le 6$ are observed, from which we concluded that these are forced electric dipole transitions. They may occur in C_2 or C_s but not in C_{3i} symmetries. It was therefore concluded that C_2 is the predominant symmetry site of Tm^{3+} and Er^{3+} in the glasses studied. The C_2 symmetry should be favored as it is the symmetry of Tm^{3+} and Er^{3+} in Y_2O_3 crystals, and we have shown previously for Eu^{3+} that the Rare Earth has a tendency to preserve the same symmetry (or lower) as in the oxide crystals. It should be mentioned that *Fournier* and *Bartram* (*33*) have proposed recently a model in which Yb^{3+} is surrounded by three phosphate tetrahedra in phosphate glasses. For thulium and erbium in phosphate and borate glasses, we proposed a model (*31*) in which these ions are surrounded by four tetrahedra in a similar arrangement to Eu^{3+} as pictured in Fig. 1, preserving C_2 symmetry.

A comparison of transition intensities for Tm^{3+} and Er^{3+} in various media (Tables 4 and 5) shows that the intensities in the UV region are smaller and borate than in phosphate glasses by a factor of 2—3, and in water they are smaller by a factor of 6—10.

The assumed C_2 symmetry of Tm^{3+} and Er^{3+} in glasses is in accordance with *Judd*'s (*35*) prediction of symmetries, giving rise to hypersensitive transitions. In addition, we think that the great difference in the absorption spectra of Tm^{3+} and Er^{3+} in UV, as compared to small changes in the visible and I.R. regions in the glasses, can be explained only by the contribution of J mixing by the crystal field, which should be stronger in the higher levels where the J levels are closer together in energy. We also observed that transitions for $\Delta J = 2$ are the strongest. This is also observed in the absorption spectrum (see Tables 4 and 5) but in the fluorescence spectrum (Tables 6 and 8) this rule is not always obeyed because of the possibility of non-radiative transitions between levels, which are dominant when the energy gap is less than 3000 cm^{-1}. This energy corresponds to about four phonons in the glass lattices. A similar observation was made by *Weber* (*36*) for the fluorescence of Er^{3+} in LaF$_3$ crystals. In the case of Tm^{3+} doped glasses, the fluorescence arising from the transition $^1G_4 \rightarrow {}^3H_4$, $^1D_2 \rightarrow {}^3H_4$ and $^1I_6 \rightarrow {}^3H_4$ are the strongest and just satisfy the selection rule $\Delta J = 2$. The high contribution to radiative transition here is caused by the large energy gap between the emitting level and the next lower one. On the other hand, in Er^{3+} the high probabilities arising from the transitions $^4I_{15/2} \rightarrow {}^4I_{11/2}$, $^4I_{15/2} \rightarrow {}^2H_{11/2}$, $^4I_{15/2} \rightarrow {}^4G_{11/2}$ are observed only in the absorption spectrum, while the emission from the $^4G_{11/2}$ and $^2H_{11/2}$ is very weak due to the small energy gap between the emitting level and the next-lower one, thus causing nonradiative losses. According to the ligand field theory, an increase in the separation of the crystal field split degenerated levels, indicates an increase in the strength of the ligand field shown by the R. E. In cases of transition between two non-degenerated energy levels, an increase in the ligand field causes the two levels' displacement towards lower energy levels. Stronger crystal fields will induce energy level shifts towards longer wavelengths; this is reflected in the longer wavelength emissions in borate glasses compared with phosphate glasses.

In order to approximate the glass host crystal fields, we used a simple expression for an "effective" ionic field strength, $F = z/r^2$, where z is the valence of the network forming ions and r their ionic radius (in Å). The "effective" field strengths for the network forming ions B^{3+} and P^{5+} are 75.0 and 43.2 respectively. It can be seen therefore that the sodium borate glass host has a stronger "effective" ionic field than the phosphate glass (*18*).

In comparing the bandwidth of the $^1D_2 \rightarrow {}^3H_4$ transition of Tm^{3+} in borate glass with phosphate glass and doped yttrium oxide crystals (830 Å, 670 Å and 13.9 Å respectively (see Table 7)), we came to the conclusion that there are about 50 different sites of C$_2$ symmetry in a glass host. This result is consistent with results obtained for Eu^{3+} in phosphate, silicate and germanate glass. As previously explained, the slight differences in environment are caused by small variations in the crystal field parameters. The energy level shifts for ions in different environments lead to slight changes in the transition wavelength from one ion to another, thus producing the broadening of the spectra. We believe that a similar situation exists in aqueous solutions, as can be seen from the example of Eu^{3+} (*22*), where the broadening of the $^5D_0 \rightarrow {}^7F_1$ and $^5D_0 \rightarrow {}^7F_2$ bands resulted in a variety of sites for the same symmetry due to a slight change in the distance between the ligands and the R. E., as shown above.

The half-bandwidths in the absorption and fluorescence spectra are larger in phosphate than in borate (see Table 7). This can be attributed to the fact that the cavity in which the R.E. can be situated is larger in phosphate glasses, permitting a larger variety of sites with slightly different crystal field parameters, producing inhomogeneous broadening. From this, we can conclude that the large half-width in the emission spectrum of R.E. in a glass is produced by inhomogeneous broadening rather than by crystal field splitting, which would be expected to decrease as the cavity size increases. A similar effect was observed in Eu^{3+} (22), where the emission half-widths in silicate glass are slightly higher than those in the phosphate glasses. The crystal field strength expressed as $\frac{z}{r^2}$ is 43.2 in phosphate and 23.8 in silicate, while the respective ionic radii are 0.34 Å and 0.41 Å.

The fact that the change in the relative absorption intensities (Table 5) between phosphate and borate glasses is smaller for Er^{3+} doped glasses than for Tm^{3+} doped glasses can be epxlained by a somewhat weaker interaction of Er^{3+} with the glass network. The cation radius of Tm^{3+} (0.87 Å) is smaller than that of Er^{3+} (0.89 Å) and as a consequence the cation-oxygen nuclear distance becomes also smaller. Both the increase in the effective charge and the smaller cation-oxygen distance cause an increase in the electrostatic attraction of cation for oxygen. The increased attractive force must be balanced by an increased oxygen-oxygen repulsion, and this increase should be reflected in a shortening of the B—O or P—O bond and an increase in the frequency of vibration of the borate and phosphate ions. Thus, as the cation changes from Er^{3+} to the smaller Tm^{3+}, we shall get a slightly more dense structure which thus gives eventually a different arrangement of the structural unit. Even small differences in arrangement of the structural unit can cause the intensities of absorption bands of the two R.E. to differ.

Oscillator strengths as well as fluorescence lines in the visible and the U.V., together with the quantum efficiencies for Gd^{3+}, Tb^{3+}, Sm^{3+}, Eu^{3+}, Tm^{3+} and Er^{3+} in borate and phosphate glasses, may be found in a paper by *Reisfeld* (38).

Jørgensen (39) has proposed that the seven electronic states of the partly filled f shell differ slightly in energy. The position of the baricenters of the J levels depends slightly on the medium. This is due to the formation of molecular orbitals (MO) having the 4f central atom orbitals as the main constituents and to a slight contribution of the filled orbitals of the ligands.

We feel that also in glasses some covalent bonding exists between the f orbitals of the rare earth ions situated at the center of the cube and the linear combination of atomic orbitals of the eight surrounding oxygens.

Such covalent bonding could explain the slight shift in the baricenters of the hypersensitive lines, and also the variation of the intensity for different glasses. *Kuse* and *Jørgensen* (*40*) obtained similar results for $ErPO_4$ and $ErVO_4$.

III. Absorption Spectra of Ce^{3+} and Tb^{3+} Arising from f—d Transitions

The $4f - 5d$ transitions in the case of Ce^{3+} and Tb^{3+} fall within the limit of standard spectrophotometers. The strong ultraviolet bands belonging to the ^2D terms of Ce^{3+} may split by crystal field into maximum five levels (*Kramers'* degeneracy).

Jørgensen and *Brinen* (*41*) have found in aqueous solutions five main bands due to the $4f - 5d$ transition. *Loh* (*42*) reported four bands for Ce^{3+} in CaF_2, SrF_2 and BaF_2 and suggested that they correspond to the four $5d$ subshells of tetragonally distorted $Ce(III)F_8$. In the gaseous Ce^{3+}, the two J-levels of $5d$ are situated at 49740 and 52230 cm^{-1} respectively Ref. (*39*), p 120). The average energy of the $5d$ orbitals are 43500 in CaF_2, 44000 in SrF_2 and 44700 cm^{-1} in BaF_2, somewhat below the average energy of 51230 cm^{-1} in free Ce^{3+}. We observe here a slight nephelauxetic effect, even in fluoride where the bonding is believed to be totally electrostatic.

The average energy of the $5d$ orbitals of Ce^{3+} in water is 44700 cm^{-1}, the same as in BaF_2. The decrease of the energy 4f − 5d in Ce^{3+} is seen in octahedral complexes; $CeCl_6^{-3}$ has a strong band already at 30300 cm^{-1} and $CeBr_6^{-3}$ at 29150 cm^{-1} (*Ryan* and *Jørgensen*) (*45*). Here we can clearly see the nephelauxetic effect. *Jørgensen* (Ref. (*39*) p. 121) has proposed that the reason for the low wavenumbers in the hexahalide complexes is the absence of σ (though not π) anti-bonding effects on the lowest $5d$ subshell.

Jørgensen believes that the three adjacent bands in CaF_2, SrF_2 and BaF_2 correspond to transitions to the higher σ anti-bonding subshell (xy, xz, yz), which may be separated energetically because of the distortion of the cubic symmetry due to the differences in radii of Ce^{3+} and Ca^{2+}, Sr^{2+} and Ba^{2+}.

The present author has investigated Ce^{3+} in phosphate (*43*) and in borax (*44*) glasses. The wavenumbers of the absorption maxima for Ce^{3+} in phosphate and borax glasses are presented in Tables 9 and 10. Whereas in phosphate, the splitting into five components similar to water solution is clearly seen, in borax glasses some of the bands are overlapping and the peaks of these bands are observed only after resolution by Gaussian

Table 9. *Absorption spectra of Ce^{3+} in aqueous solution, in halides, CaF_2, SrF_2, BaF_2 and in borax and phosphate glasses*

Water (41)	CaF_2 (42)	SrF_2	BaF_2	Borax (44)	Phos- phate (43)	$CeCl_6^{-3}$	$CeBr_6^{-3}$ ((12), see Table 5)
50000	53900	53400	53500	47177	50000		
47393	51100	50300	51700	45125	47000		
45147	49500	48800	50000	39201	44000		
41754				33454	40000		
39604	32500	33600	34200	31959	33333	30300	29150

values in cm^{-1}

analysis. The ground state of $4f^1$ configuration of Ce^{3+} is $^2F_{5/2}$, the next state $^2F_{7/2}$ lies at about 2000 cm^{-1}. Therefore, the $^2F_{5/2}$ is the only populated state at room temperature. The next excited states are the crystal field components of the 5d configuration. The occurrence of 5 components in the absorption spectrum is due to the total lifting of degeneracy (with the exception of *Kramers*) by the crystal field of the glass. We neglect here the effect of the spin orbit splitting, since the spin orbit constant for 5d is about 10^3 cm^{-1}. We have proposed (44) that Ce^{3+} in borax and phosphate glasses is surrounded by eight non-bridging oxygens belonging to the corners of PO_4 tetrahedra, and each tetrahedron donates two oxygens. This behavior of Ce^{3+} is similar to that previously described of Eu^{3+} and Tm^{3+} (see Fig. 1).

The oxygens may be arranged in the corners of a tetragonally distorted cube; such an arrangement corresponds to site symmetry lower than the site symmetry D_{2d} in crystals (42, 46).

The average energy of the 5d orbitals in phosphate glass is 42866 cm^{-1} and in borax 39383 cm^{-1}, compared to 44700 cm^{-1} in water. The decrease of energy difference between 4f and 5d of Ce^{3+} in borax relative to phosphate and water solution can be explained in our opinion as follows:

The shift of the absorption band of Ce^{3+} to longer wavelengths in the order of water, phosphate, borax indicates that the covalency of the Ce — O bonds increases in the same order. Similar phenomena were observed by *Reisfeld et al.* for Tl^+ doped glasses (38, 47) and Eu^{3+} doped glasses (48, 49). In borax, the electrons of the oxygens participating in the Ce — O bond are shifted more to the Ce^{3+} than in phosphates. This means that the interaction between the rare earth and the oxygen ligands is greater in borax glasses than in phosphate glasses. Hence the covalency of the R.E. — O bond is greater in the former than the latter, thus the B — O bond must be less covalent than the P — O bond. We may say,

Table 10. *Absorption spectra of Ce^{3+} in sodium phosphate and borax glass*

		x^2-y^2	$2z^2-x^2-y^2$	xy	y^2	x^2	eg	t_{2g}	spherical symmetry	10 Dq
				Assignment of d orbitals						
sodium phosphate glass	peak wavenumber cm^{-1}	33500	40733	43956	47169	51282	36666	47000	42866	11444
	half-bandwidth cm^{-1}	816.8	1407.9	836.1	1133.7	986.55				
	oscillator strength $\times 10^3$	1.647	1.807	0.7358	0.8107	0.6512				
borax glass	peak wavenumber cm^{-1}	31952	33454	39201	45125	47177	32706	43834	39383	11128
	half-bandwidth cm^{-1}	2744	5184	6902	4572	3231				
	oscillator strength $\times 10^3$	2.48	4.70	5.19	2.02	1.97				

therefore, that the electronegativity of the BO_4 glass-formed tetrahedra is less than the PO_4 tetrahedra.

According to *Jørgensen* (*39*), in most chromophors of low symmetry all 5d orbitals are σ anti-bonding to some extent. *Jørgensen* (*6*) also emphasizes that most inter-shell transitions have wavenumbers varying with the ligands in much the same way as the nephelauxetic series. The particular d orbitals responsible for the various splittings are shown in Table 10. The assignments are similar to those proposed by *Loh* for Ce^{3+} in CaF_2 (*42*).

From the sharper splitting of the d orbitals in phosphate, we conclude that the asymmetric part of the crystal field is higher in phosphate, in agreement with the larger cavity in the phosphate available for Ce^{3+}. We therefore conclude:

1. The energy shift from the free ion to the glass indicates the lowering of the d orbitals due to the electrostatic perturbation of the surroundind oxygens. The perturbation is stronger the more the electrons are localized on the oxygens (higher polarity of $B - O$ than $P - O$ bond).

2. Crystal field splitting in glass indicates the influence of non-symmetrical crystal field terms.

3. The microsymmetry of Ce^{3+} in a glass is either C_s or C_2.

4. The optically active electron does not take part in the bond between Ce^{3+} and the oxygens.

5. The partly covalent σ bond is formed by overlaps of linear combination of σ orbitals of the eight oxygens and linear combination of a_{2u} (f_{xyz}) and the set of three t_{lu} $f_{(x^3-3/5xr^2)}$, $f_{(z^3-3/5\ 2r^2)}$, $f_{(y^3-3/5yr^2)}$.

The last conclusion is based on the shape of the f functions; the function a_{2u} indicated above can point to the corners of the cube (see also Ref. (*50*), p. **838** and Ref. (*51*)), and the t_{lu} forms a suitable linear combination for σ bonding.

The π bond can be formed between the linear combination of π orbitals of the oxygens and linear combination of t_{2u} f orbitals of Ce^{3+} which are $f_{z(x^2-y^2)}$, $f_{x(z^2-y^2)}$ and $f_{y(z^2-x^2)}$ (*6, 52*). These functions point to the center of edges of the cube. The π bonding here will be directed from the oxygens to the central Ce^{3+} atom. Here the effect of the mixing is to increase the energy of the t_{2u} orbitals relative to the d orbitals thus decreasing the Δ.

Jørgensen and *Rittershaus* (*53*) studied the strong nephelauxetic effect in mixed oxides of the rare earth and proposed a relation

$$\sigma - \sigma_{aquo} = d\sigma - (d\beta)\ \sigma_{aquo} \qquad (8)$$

where σ is the wavenumber in the absorption maximum of the compound studied and σ_{aquo} is the maximum of the absorption spectrum of the ion in the aqueous solution, $d\sigma$ accounts for the difference between the average sub-level population of the ground level of the compound studied at a definite temperature, and the energy of the lowest sub-level of the aquo ion relative to the baricenter of the ground J level.

The present author calculated the nephelauxetic parameter β from the Eq. (9)

$$\frac{\sigma_f - \sigma}{\sigma_f} = \beta \tag{9}$$

where σ_f is the wavelength of the appropriate absorption for the free ion. Following *Sinha* (55), we feel that σ_f, the wavenumber of the free ion, should be used whenever possible, as the aquo complex produces a slight nephelauxetic effect (as does even the fluoride).

Table 11 presents such calculations for mercury-like ions, Tl^+ and Pb^{2+}, in which the first absorption band arises from spin-forbidden transition $^1S_0 \rightarrow {}^3P_1$ in the free ion and $^1A_{1g} \rightarrow {}^3T_{1u}$ ($^1\Gamma_4 - {}^3\Gamma_4°$) in octahedral symmetry, and Ce^{3+} and Tb^{3+} ions in which the UV band arises from $4f - 5d$ transitions. Energy level schemes of $4f^7 5d$ for the Tb^{3+} and $4f 5d$ for the Ce^{3+} configuration have been recently discussed by *Hoshina* and *Kuboniwa* (50). The electrostatic interaction between $4f^7$ (^8S) and $5d$ produces two excited levels, the lower lying ^9D and the higher, ^7D. The transition $^7F \rightarrow {}^9D$, responsible for the 36.8 band obtained by *Ryan* and *Jørgensen* (45) in the hexachloride complex of Tb^{3+}, is very weak because it is spin forbidden. The $^7F \rightarrow {}^7D$ lying at 42.75 is a spin-allowed strong transition. The D terms are split by the crystal field into ^9E, ^7E, $^7T_2$ and $^9T_2$, but only the first one is observed in glasses The stability of the half-filled $4f$ shell in the $4f^7 5d$ configuration explains why these transitions are at relatively low energies. In the case of Ce^{3+}, the ^2D term is at lower energy when Ce^{3+} is embedded into a complex, crystal or glass.

The nephelauxetic effect, as seen from Table 11, is similar for mercury-like ions and rare earth ions. This effect expresses the relative decrease in interelectronic (*Racah*) parameters by mixing $6p$ (for mercury-like ions) and $5d$ and ligand orbitals. The reduction of the s-p and f-d distance causing the absorption at longer wavelength will be more pronounced if the electronegativity of the ligand decreases and the mean cation — ligand distance decreases. The comparison of β in different hosts is most reliable when the hosts are changed and the central atom kept constant. Actually, the nephelauxetic series obtained in this way is analogous to the tendency of covalent bonding. Thus, fluoride and water forming the least covalent bonding have smaller β than chloride or bromide, which form more covalent bonding. In this respect, the phosphate glasses have similar properties to water, because of the low tendency of oxygen bound

Table 11. Absorption maxima σ_0 and parameter β for „Rydberg" transitions in Tl^+, Pb^{2+} ($^1S_0$—$^3P_1$) and Ce^{3+}, Tb^{3+} ($4f$—$5d$) in various hosts

	Gas	Water	Chloride	Bromide	Borax	Phosphate	Germanate	Silicate	Tungstate (75%)
$^1S_0$—$^3P_1$									
Tl^+ σ_0 cm^{-1}	52390 [a]	46700 [a]	40500 [b]	38100 [b]	43450 [b]	47000 [b]			
β		0.1086	0.2270	0.7272	0.1706	0.1029			
Pb^{2+} σ_0 cm^{-1}	64390 [a]	48000 [a]	36800 [c]	33200 [c]	38100 [b]	41600 [b]	33900 [e]		
β		0.2545	0.4285	0.4844	0.4083	0.3539	0.4735		
$4f$—$5d$									
Ce^{3+} σ_0 cm^{-1}	51000 [f]	39600 [g]	26000 [g]	19200 [g]	31959 [g]	33333 [g]		31500 [h]	
β		0.2235	0.4902	0.6235	0.3734	0.3464		0.3824	
Tb^{3+} σ_0 cm^{-1}	46500 [i]	45900 [j]	42750 [j]	36000 [j]	44500 [b]	46000 [b]	35700 [k]		43500 [k]
β		0.0129	0.0806	0.2258	0.0430	0.0108	0.2322		0.0645

a) Reference (39) — Table 6.3.
b) Reference (38) — Table 10.
c) Reference (56).
d) Reference (57).
e) Reference (24) (from excitation spectrum).
f) Reference (58).
g) Reference see Table 9 (the wavenumbers are of the first intense transition).
h) Reference (38) from excitation spectrum.
i) Reference (39) Table 6.1 in CaF$_2$.
j) Reference (45).
k) Reference (59) from excitation spectrum.

to phosphorus to donate electrons to the central atom (situated at the center of a cube between eight oxygens). We can write a nephelauxetic order of glasses analogous to the halide. The nephelauxetic parameter is increased in the following order: aquo ion $\geq$ phosphate glass $\geq$ borate glass $\geq$ silicate glass $\geq$ tungstate glass $\geq$ germanate glass.

Correlation between s-p spectra and the nephelauxetic ratio in glass chemistry was also indicated recently by *Duffy* and *Ingram* (62, 63). This effect is parallel with the smaller electronegativity of the glass-forming tetrahedra in the order above, or with the increase of the reducing character of the tetrahedra. We may also use here the *Fajans* concept of polarizability. Considering partly covalent bonding as a second order perturbation, we expect it to become more pronounced as the energy difference between the orbitals of the partly filled shell and of the ligand's lone pair decreases. The nephelauxetic effect is explained by *Jørgensen* as an expansion of the partly filled shell due to the transfer of the ligands to the core of the central atom; it is a measure of covalency effect. The values of β presented in Table 11 express the amount of covalency in each glass with the central rare earth or mercury-like ion.

IV. Charge Transfer Bands of Eu^{3+} in Various Glasses and Other Hosts

The electron transfer (or charge transfer spectra) occurs by electron jumping from one of the highest filled molecular orbitals of the ligand to the partly filled shell of the central atom. *Jørgensen* was the first to report electron transfer spectra for the rare earths in 1962 (11, 12). In particular, the hexahalide complexes have electron transfer spectra arising from the filled M.O. mainly delocalized on the ligands to the empty or partly filled f shell. The filled p shell of each of the six halide ligands splits to a first approximation into two subshells when the local symmetry is lowered: one σ orbital at low energy and two π orbitals of somewhat higher energy. The radial functions are slightly contracted in the bonding M.O. whereas the nephelauxetic effect is connected with an expansion of the partly filled shell participating in the formation of antibonding M.O. A related effect of charge separation occurs in electron transfer spectra where the wavenumbers of maximum absorption observed are due to the difference between the ionization energy of the filled M.O., and the electron affinity of the empty or partly filled shell, corrected for the attraction energy between the electron which was transferred and the "hole" left in the filled M.O. Professor *Jørgensen* is of the opinion that the identification of the M.O. involved in each case is more a question of induction, by comparing the rich experimental data, than deductive calculation. In his paper on electron transfer spectra (12), he correlated the uncorrected optical electronegativity of the metal χ_{uncorr} (M) with the electronegativity of the ligand χ_{opt} (X) and the σ_{obs} wavenumber of the absorption peak by the relation (10)

$$\sigma_{obs} = [30\,000\ (\chi_{opt}\ (X) - \chi_{uncorr}\ (M))] \qquad (10)$$

The chemical criteria for the oxidation ability of a central atom depend on χ_{uncorr} (M). The values of χ (M) as taken from Ref. (12) are

$\chi_{uncorr.}$	Ce(IV)	Sm(III)	Eu(III)	Tm(III)	Yb(III)
	2.1	1.6	1.9	1.5	1.8

Clearly, Eu(III), in which only one electron is missing from the half-filled stable shell, is the most oxidizing of the trivalent rare earth ions and the charge transfers of Eu(III) are at energies low enough to be measured by standard spectrophotometers. Therefore, Eu(III) was chosen in our work in order to calculate the electronegativity of glasses. In the cases of germanate, tungstate, $Mg_3(PO_4)_2$ and $Ca_3(PO_4)_2$, we have to use excitation spectra, as the host glass has self-absorption in the region of the charge transfer band.

Table 12. *Charge transfer transitions of Eu^{3+} in various hosts*

Host		Wave-number (cm^{-1})	χ_L	Reference
Eu^{3+} in aquo		53190	3.67	(55) (page 91)
	Cl_6^{-3}	42600	3.32	(45)
	Br_6^{-3}	24500	2.72	(45)
	I_6^{-3}	14800	2.39	(12) (Table 5)
	YPO_4	45000	3.40	
	$LaPO_4$	37000	3.13	
	$LaBO_3$	37000	3.13	(60)
	YBO_3	42700	3.32	
	Y_2O_3	41700	3.29	
glasses	Na_3PO_4	49020	3.53	(59)
	Borax	45660	3.42	
	Borate	43103	3.34	
	Germanate	38462	3.18	(24)
	Tungstate (45%)	39216	3.21	(25)

The results of the calculations are presented in Table 12. From the Table we can see that the electronegativity of the various glasses increases in the opposite order as the nephelauxetic parameter. The slope β vs χ_L is a measure of the interaction of the cation orbitals with the surrounding medium. The larger the covalent bonding between the rare earth or mercury-like cationthe and smaller the electronegativity of the glass, the longer is the absorption wavelength. Assuming that the absorption occurs from an electron in the nonbonding orbitals of the cation to the antibonding molecular orbital, this implies that the energy gap between nonbonding orbitals and antibonding orbitals is lowered in glasses which have a stronger tendency to form a partially covalent bond.

V. Energy Transfer

This process is observed when there are two different ions in a matrix. We may excite one ion, the donor, and observe fluorescence from another ion, the acceptor. We shall discuss here only the energy transfer between rare earth ions in a glass matrix.

Rare earths are especially suitable for energy transfer studies because of their well-defined and narrow electronic levels, to which absorption occurs and from which fluorescence is observed. Symbolically, energy transfer can be written

$$2(D) \rightarrow 1(D) \rightsquigarrow 1(A) \rightarrow 2(A) \tag{11}$$

or

$$D^* + A \rightarrow D + A^* \tag{12}$$

The donor system returns from the excited state $2(D)$ to the ground state $1(D)$ and the energy released is used to bring the activator system from the ground state $1(A)$ to its excited state $2(A)$. In case of the rare earth, the transfer is a nonradiative one, *i.e.* no photon will appear in the system during the transfer. *Förster (64, 65)* predicted that for the organic system, the rate of energy transfer is proportional to the overlap of the donor emission and the acceptor absorption spectra and to R^{-6}, where R is the distance between the donor and the acceptor. The discussion of *Förster* was extended by *Dexter (66)* for ions in inorganic crystals.

The probability of energy transfer by dipole-dipole *(dd)* interaction in a simple case where the Born-Oppenheimer approximation holds is given by *Dexter (66)* as

$$\tag{13}$$

$$P_{da(dd)} = (3h^4 c^4 Q_a/4\pi R^6 n^4 \tau_d) (\varepsilon/k^{1/2} \varepsilon_c)^4 x \int [f_d(E) F_a(E)/E^4] dE$$

where h is Planck's constant, c is the velocity of light, R is the separation of the nuclei of D and A, ε_c is the electric field within the crystal, ε is the electric field in vacuum, n is the index of refraction of the medium, and $f_d(E)$ is the observed shape of the emission band normalized to unity; $\int f_d(E)\,dE = 1$. E is the energy; the subscripts d and a refer to donor and acceptor, respectively; $F_a(E)$ is the normalized function of the acceptor absorption such that $\sigma(E) = F_a(E)$, $Q_a = \int \sigma(E)\,dE$ the measured area under the absorption band, and $\int F_a(E)\,dE = 1$. τ_d is the decay constant of the pure donor; σ is the absorption cross section.

The equation may also be written as follows:

$$P_{da(dd)} = \frac{3hc^2\,Q_a\,Q_d\,g_d}{4\pi^3\,nR^6\,g_{d'}} \left(\frac{\varepsilon}{k^{1/2}\,\varepsilon_c}\right)^4 \frac{f_d(E)\,F_a(E)}{E^2}\,dE, \tag{14}$$

where g_d is the degeneracy of the ground state of the donor, $g_{d'}$ is the degeneracy of the excited state fo the donor, and Q_d is the area under the donor absorption curve.

Formula (13) may be written for the borate glasses used in our experiments as follows:

$$P_{da(dd)} = \frac{1.7 \times 10^{18}}{C_d\,C_a\,l_d\,l_a} \frac{\int A_d(E)\,dE\,A_a(E)\,dE}{n^2\,R^6} \times \frac{f_d(E)\,F_a(E)\,dE}{E^2} \tag{15}$$

(for the phosphate glass the number 1.7 is replaced by 1.14) where C_d and C_a are the donor and acceptor concentrations in weight per cent, l_d and l_a are the thicknesses of the donor and acceptor glass samples in millimeters, $\int A_d(E)\,dE$ and $\int A_a(E)\,dE$ are the areas under the donor and acceptor absorption curves on a wavenumber scale (cm^{-1}), the quantity R is the interionic distance in angstroms, and n is the refractive index.

For a dipole-quadrupole mechanism the transfer probability is given by *Dexter* as (66)

$$P_{da(dq)} = \frac{135\,ah^6\,c^6\,Q_d}{4\pi\,R^8\,n^4\,\tau_a} \times \frac{g_{a'}\,g_d}{g_a\,g_{d'}} \frac{\varepsilon}{k^{1/2}\,\varepsilon_c}^4 \times \frac{f_d(E)\,F_a(E)}{E^6}\,dE \tag{16}$$

The ratio between dipole-quadrupole and dipole-dipole transition probabilities is given in Ref. (66) as

$$P_{da}/P_{dd} \simeq (a/R)^2, \tag{17}$$

where a is the atomic radius of the rare earth and R the interionic distance.

For glasses, an additional factor K should be put in the numerator (67). The factor K is provided to account for varying orientation of the dipole moments of D and A. $K = 1$ if all D and A ions have parallel dipoles and $K = 2/3$ if the orientation is totally random as it should be in the glass.

In addition to the multipolar transfer mechanism, a quantum mechanical exchange process can be effective at very short interionic distances.

The probability $p(\text{exc})$ of these can be written according to *Dexter* (66)

$$p(\text{exc}) = \frac{2\pi}{h} Z^2 f_d(E) F_a(E) dE \qquad (18)$$

In this formula, Z is not accessible to direct optical measurement and Z^2 varies with distance R as $\exp(-R/L)$, where L is an effective average Bohr radius of donor and acceptor ions for the excited and unexcited states.

While the former two interactions are electrostatic in origin, the exchange interaction arises from the antisymmetry requirements of the electronic wavefunction for a system consisting of a donor and an acceptor. Exchange does not seem to be a likely mechanism for energy transfer between rare earths in glasses. *Nakazawa* and *Shionoya* (68), and the present author (37, 70, 71, 72) studied the mechanism of energy transfer between rare earths in glasses.

The purpose of the paper of *Nakazawa* and *Shionoya* was to clarify the mechanism of excitation energy transfer between unlike trivalent rare earth ions in inorganic solids. They measured the manner in which the intensities of the luminescences of Tb^{3+} ($^5D_4 \rightarrow {}^7F_5$) and Eu^{3+} ($^5D_0 \rightarrow {}^7F_2$) were quenched and the donor decay times increased by the presence of other kinds of trivalent rare earth ions as a result of energy transfer. As the host material, calcium metaphosphate glass was used. The combination of Tb^{3+} donor and Nd^{3+} acceptor was studied most in detail. The results were discussed in terms of the resonance theory of energy transfer, and the experimental data are analysed using the numerical calculations of the theory recently made by *Inokuti* and *Hirayama* (69). The overlap integrals of donor emission spectra with acceptor absorption spectra were estimated from the data of spectral measurements.

Among the resonance mechanisms due to various electrostatic multipole interactions, the dipole-quadrupole interaction gave the best fit between theory and experiment. The transfer due to the exchange interaction was inferred not to be operative. It was finally concluded that the mechanism of the energy transfer between unlike trivalent rare earth ions in ionorganic solids is predominantly governed by the dipole-quadrupole interaction.

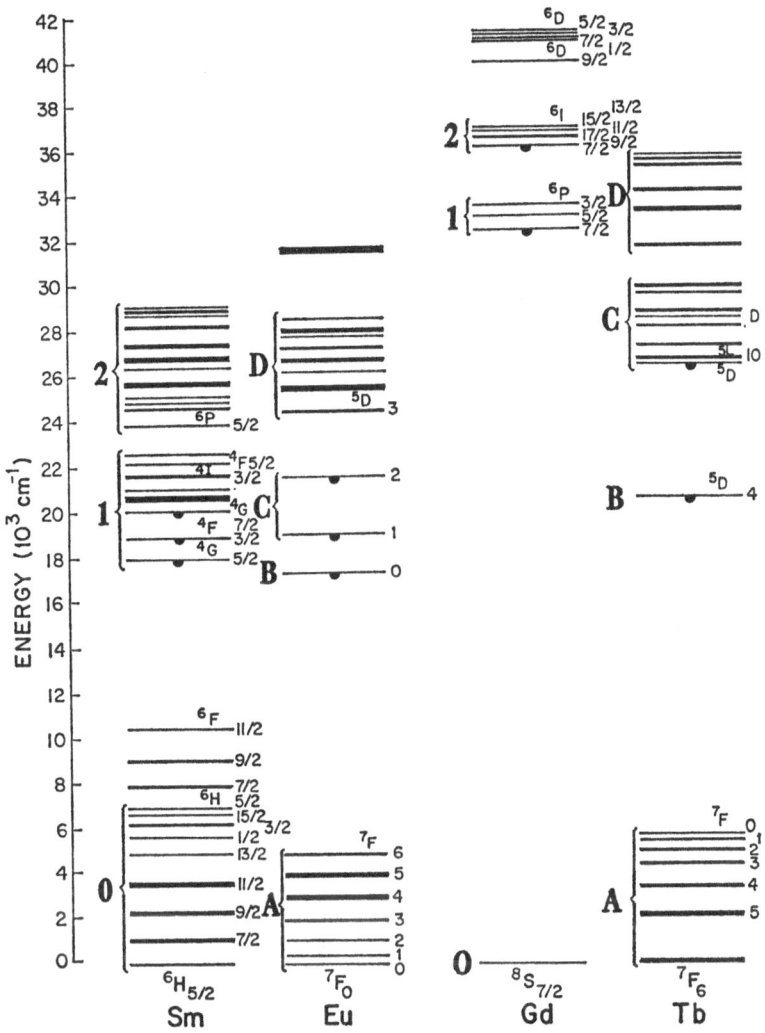

Fig. 2. Schematic energy level diagram for the donor acceptor system

A method for calculating the probability and efficiency of energy transfer between inorganic ions, which have well-defined electronic levels from donor and acceptor luminescence intensities, and for luminescent lifetime was proposed by *Reisfeld et al.* (71). Formulae were derived from rate equations applicable to a system consisting of a pair of unlike rare earth ions in a glass medium.

The system is shown schematically in Fig. 2 and a specific example of two pairs in which the energy was studied is shown in Figs. 3 and 4.

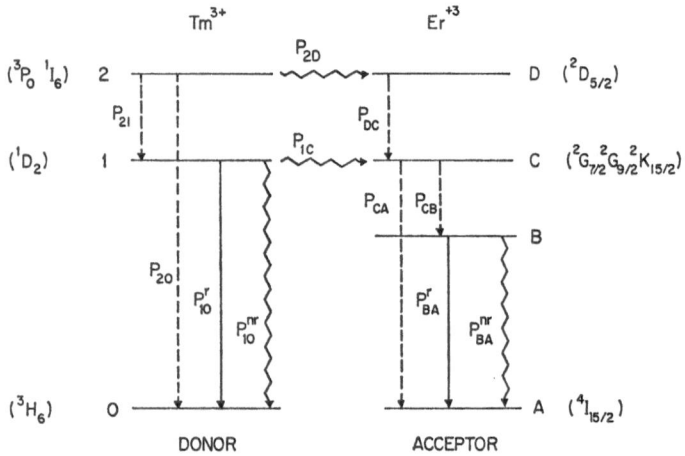

Fig. 3. Electronic energy levels of Tm[3+] and Er[3+]

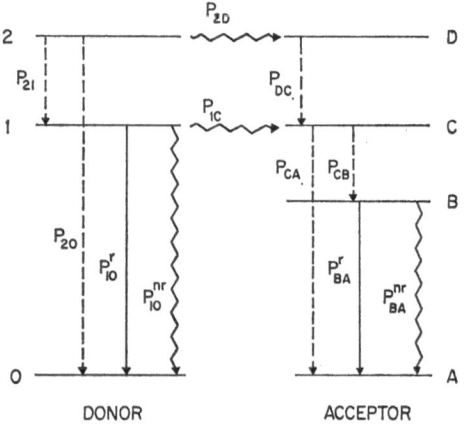

Fig. 4. Electronic levels of Sm[3+] — Eu[3+] and Gd[3+] — Tb[3+]

Here the numbers specify the levels of the donor ion and the letters, the levels of the acceptor ions. The superscripts "r" and "nr" refer to radiative and nonradiative transitions respectively. The p's are the transition probabilities between various levels designated in the figures. The φ are the quantum efficiencies defined as

$$\varphi_1 = \frac{P^r_{10}}{P^{nr}_{10} + P^r_{10}} \quad (8)V \qquad \varphi_2 = \frac{P_{21}}{P_{21} + P_{20}} \tag{19}$$

P_{1C} is the probability of energy transfer between lower levels and P_{2D} is the probability of energy transfer between the higher levels (see Fig. 1).

The donor emission quantum yield in the presence of c concentration of the acceptor η_d from level $1 >$ to $0 >$ is

$$\frac{\eta_d^0}{\eta_d} = (1 + \tau_d\, P_{1C}) \left(1 + \varphi_2 \frac{P_{2D}}{P_{21}}\right) \tag{20}$$

η_d^0 is the donor emission quantum efficiency when no acceptor is present.

From this equation it is possible to determine whether one or more energy transfer channels are operative. This is because the energy transfer probability is proportional to some power of concentration and a single channel transfer process would give a linear dependence $\dfrac{\eta_d^0}{\eta_d}$ on this power. Where a single channel is operative, as in most practical cases (70, 71, 37), and $P_{2D} \ll P_{21}$ then

$$\frac{\eta_d^0}{\eta_d} = 1 + P_{1C}\,\tau_d \tag{21}$$

and

$$P_{1C} = \frac{1}{\tau_d} \left(\frac{\eta_d^0}{\eta_d} - 1\right) \tag{22}$$

When $P_{2D} \ll P_{21}$, the efficiency of energy transfer is given by

$$\eta_t = \frac{P_{1C}\,\tau_d}{1 + P_{1C}\,\tau_d} = 1 - \frac{\eta_d}{\eta_d^0} \tag{23}$$

The last two equations enable calculation of P and η from the donor emission intensity and fluorescence lifetime data.

An alternative calculation of P_{1C} was proposed in Ref. (71) from the increase of the acceptor fluorescence $\Delta\eta_a$ in the presence of the donor, from the relations for the quantum efficiencies of acceptor fluorescence

$$\varphi_B = \frac{P_{BA}^r}{P_{BA}^r + P_{BA}^{nr}} \tag{24}$$

and the efficiency of transfer from level $|C>$ to level $|B>$

$$\varphi_C = \frac{P_{CB}}{P_{CB} + P_{CA}} \tag{25}$$

Since $P'_{10} = \varphi_1(P'_{10} + P^{nr}_{10})$ and $\tau_{meas} = 1/(P'_{10} + P^{nr}_{10})$. The energy transfer probability was obtained as

$$P_{1C} = \frac{\varphi_1}{\varphi_B \, \varphi_C \, \tau_{meas}} \frac{\Delta \eta_a}{\eta_d} \tag{26}$$

The product $\varphi_B \, \varphi_C$ appearing in Eq. (26) is the quantum yield of level $|B>$ on excitation to level $|C>$, denoted by $\varphi_B^{(C)}$. Finally,

$$P_{1C} = \frac{\varphi_1}{\varphi_B^{(C)}} \frac{1}{\tau_{meas}} \frac{\Delta \eta_a}{\eta_d} \tag{27}$$

The Eq. (27) gives the energy transfer probability in terms of experimentally measurable quantities.

Reisfeld et al. (70) have studied energy transfer between Gd^{3+} and Tb^{3+} in borate glasses and have used the above relations for calculating the energy transfer probability from the quantum efficiency of pure donor η_0 and from η, the quantum efficiency of the donor in the presence of a given concentration of the acceptor. They have shown the effect of Gd addition on the excitation spectrum of Tb. It is seen that on addition of Gd two new peaks appear in the Tb excitation spectrum at 312 nm and 273 nm. These are identical with the Gd peaks due to the transitions $^8S_{7/2} \to {}^6P_{7/2}$ and $^8S_{7/2}$ and $^8S_{7/2} \to {}^6I$. The addition of Tb to a borate glass containing Gd causes a decrease in the fluorescence of the $^6P_{7/2}$ level of Gd while simultaneously there is an increase in the fluorescence of the $^5D_4$ terbium level. This effect is seen under excitation at 273 nm. It should be noted that the Tb absorption is not increased by addition of Gd.

Two series of measurements were made for the decrease of Gd^{3+} fluorescence and lifetime for varying concentrations of Tb^{3+}. The results of these measurements are presented in Tables 13 and 14.

Two series of Gd fluorescence and lifetime measurements were made for varying Tb concentrations. In all measurements of the time dependence of the decay of fluorescence from the $^6P_{7/2}$ level of Gd, it was seen that the decay was a simple exponential with a lifetime dependency on the Tb concentration. The lifetime is 2.84 msec (compared with 4.10 msec for 3% Gd alone). The dependence of η/η_0 on c (where η is the Gd fluorescence in the presence of a concentration c% of Tb and η_0 is the fluorescence for $c = 0$) does not vary with the concentration in the same way as τ/τ_0 (τ is the Gd lifetime for a concentration c% of Tb and τ_0 is the lifetime for $c = 0$).

In order to determine the form of the energy transfer rate constant as a function of donor-acceptor distance, the ratio was plotted of η_0/η versus

Table 13. *Efficiency and probabilities of energy transfer from Gd to Tb*

Concentration of acceptor Tb^{3+} wt%	η/η_0	τ_{meas} (msec)	τ_0/τ	R-donor acceptor distance (Å)	η_t	$P_{1c} = \dfrac{1}{\tau d}$ $\left(\dfrac{\eta^\circ d}{\eta d} - 1\right)$
Gd 2 wt %						
0	1	4.1	1	16.56	0	0
1	0.67	3.51	1.17	14.47	0.33	120
2	0.45	3.21	1.28	13.14	0.55	298
3	0.31	3.09	1.33	12.27	0.69	543
4	0.22	2.69	1.54	11.54	0.78	865
Gd 3 wt %						
0	1	4.1	1	14.47	0	0
1	0.56	3.64	1.13	13.14	0.44	192
2	0.35	3.44	1.19	12.27	0.65	453
3	0.23	2.86	1.43	11.54	0.77	817
4	0.18	2.54	1.61	11.00	0.82	1111

Table 14. *Increase of the fluorescence of Tb (3 wt %) and decrease of fluorescence of Gd (3 wt %) due to energy transfer*

Ion	Wavelength (nm)		Fluorescence Intensity	Increase of Tb fluorescence	Decrease of Gd fluorescence
	Excitation	Emission	(in arbitrary units)		
Gd	273	312	34		
Gd + Tb	273	312	8		26
Gd + Tb	273	486, 543, 587	87		
Tb	273	486, 543, 587	55	32	
Gd + Tb	312	489, 543, 587	15		
Tb	312	489, 543, 587	6	9	

the concentration of both the donor and the acceptor for various concentrations. It was found that η_0/η is linear in $(C_{Tb} + C_{Gd})^2$ for all concentrations measured. Such a dependence suggests that the rate constant is proportional to the inverse sixth power of the distance.

The number of channels by which the energy takes place was obtained from the relation η_0/η from Ref. (71). This relation is

$$\eta_0/\eta = (1 + \varphi_1 P_{1c} \tau_d)(1 + \varphi_2 P_{2d} \tau_{21}{}^r) \tag{28}$$

where P_{1c} and P_{2d} are the energy transfer probabilities from the ^6P and ^6I Gd multiplets, respectively, τ_d is the pure Gd $^6P_{7/2}$ lifetime, $\tau_{21}{}^r$ is the radiative lifetime for the $^6I \rightarrow {}^6P$ transition, and φ_1 and φ_2 are the quantum yields of the $^6P_{7/2}$ level excited at ^6P and ^6I, respectively.

Since both P_{1c} and P_{2d} are concentration dependent, it would seem that only one of them can be operative because of the linearity of η_0/η with $(C_{Gd} + C_{Tb})^2$. We see from experiment that on excitation to the ^6P multiplet energy transfer takes place. On excitation to the ^6I multiplet the ^6P level fluoresces. Hence, if there were an energy transfer channel via the ^6I multiplet, this would mean that two channels are operative; this is inconsistent with our previous results. Hence, we concluded that in the main it is the ^6P channel which is operative for energy transfer.

The theoretical dipole-dipole transition probabilities calculated from *Dexter's* formula are 0.6 sec^{-1} for 3% Gd and 3% Tb. Comparing this result with the experimentally determined energy transfer probability (817 sec^{-1}) given in Table 13 we see that the experimental results do not agree with the theory of the resonant energy transfer process.

The large difference (4 orders of magnitude) between the probability of energy transfer calculated by use of formulae (23) and (27) to that calculated by the use of *Dexter's* (2) formulae (15) for resonant energy transfer, shows that the resonant mechanism is inadequate to explain the experimental results for energy transfer between rare earth ions in glasses. We would like to suggest therefore that the energy transfer is assisted by the vibrational modes of the glass matrix.

The stretching frequency of the P—O bond is 1140—1300 cm^{-1} and that of the B—O bond 1310—1380 cm^{-1}. The energy of these phonons is quite sufficient to match the gap between the levels of the donor and the acceptor, without the necessity of the cooperation of more than two or three phonons.

Based on the above results, we suggest that a phonon-assisted energy transfer mechanism is operative in glasses. In this respect, glasses behave similarly to crystals, for which phonon-assisted energy transfer was proposed (*73, 74, 75*) and verified experimentally by *Yamada, Shionoya* and *Kushida* (*76*).

The energy transfer study between Sm^{3+} and Eu^{3+} in phosphate glass is described by *Reisfeld* and *Boehm* (*72*).

In the case of the Sm^{3+}—Eu^{3+} pair, the Sm^{3+} was excited from the $^6H_{5/2}$ ground state to $^4L_{13/2}$, $^4F_{7/2}$, $^6P_{3/2}$ grouping at 402 nm having a relatively high oscillator strength of 4×10^{-6} (compared to the oscillator strength of 4.8×10^{-8} for the $^4G_{5/2}$ level, which is the fluorescing level from which the energy transfer occurs).

Fluorescence was observed from the $^4G_{5/2}$ level of Sm^{3+} to the $^6H_{5/2}$, $^6H_{7/2}$, $^6H_{9/2}$, $^6H_{11/2}$ ground multiplet. It was shown (*37, 71, 72*) that the energy transfer takes place from the metastable long living levels from which also fluorescence occurs. The Eu^{3+} fluorescence was measured for the transition from the $^5D_0$ level to the ^7F ground multiplet.

It is seen that on excitation at 402 nm, the Sm^{3+} fluorescence decreases in the presence of Eu^{3+} and that of the Eu^{3+} increases.

The lifetime of the pure $^4G_{5/2}$ level of Sm^{3+} is 1.91 msec. The quantum efficiency of the Sm $^4G_{5/2}$ level, excited to this level φ_1, was found to be 0.95 (38), the quantum efficiency of Eu $^5D_0$ excited to $^5D_1$ was $\varphi_B^c = 0.91$.

Using these data the probabilities of transfer were calculated from formulas (23) and (27). The results are presented in Tables 13, 14 and 15.

Table 15. *Efficiency and probability of energy transfer from Sm to Eu*[a])

Concentration of acceptor Eu^{3+} wt%	R-donor-acceptor distance (Å⁰)	$\dfrac{\eta}{\eta_0}$	η_t	Probability of transfer cal'd. by formula $P_{1c} = \dfrac{1}{\tau_d}\left(\dfrac{\eta^\circ d}{\eta_d} - 1\right)$ (sec⁻¹)	Probability of transfer calc'd. by formula P_{1c}[b]) $= \dfrac{\varphi_1}{\varphi_B^{(c)}}\dfrac{1}{\tau_{meas}}\dfrac{\Delta\eta_a}{\eta_d}$ (sec⁻¹)
0.5	24.0	0.90	0.10	59.5	—
1.0	19.2	0.85	0.15	73.0	70.6
1.5	16.8	0.83	0.17	109.9	—
2.0	15.2	0.75	0.25	178.6	183.3
2.5	14.4	0.65	0.35	285.7	278.4
3.0	13.0	0.57	0.43	400.0	389.7

[a]) Donor concentration constant 1 *wt* %.
[b]) The quantum yield of $^5D_0$ fluorescence of Eu^{3+} is 0.95.

Additional evidence for the phonon-assisted energy process in glass is presented in the work of *Reisfeld* and *Eckstein* (37), who have studied energy transfer between Tm^{3+} and Er^{3+} in borate and phosphate glass.

As can be seen from Tables 16 and 17, the following levels of thulium are close to the levels of erbium:

Tm^{3+} levels		Er^{3+} levels
$^3P_0$, $^1I_6$	is close to	$^2D_{5/2}$
$^1D_2$	is close to	$^2G_{7/2}$, $^2G_{9/2}$, $^2K_{15/2}$
$^1G_4$	is close to	$^4F_{7/2}$

These pairs of levels can be operative in direct energy transfer between them.

Table 16. Efficiency and probability of energy transfer from Tm $^1D_2^*$ to Er ($^2G_{7/2}$, $^2G_{9/2}$, $^2K_{15/2}$)

Concentration of acceptor Er[43] (wt %)	Borate Glass			Phosphate Glass		
	R — donor-acceptor dist. (Å)	Quantum efficiency of transfer $\eta_t = 1 - \dfrac{\eta}{\eta_0}$	$P_{da} = \dfrac{1}{\tau_d}\left(\dfrac{\eta^\circ_d}{\eta_d} - 1\right)$ sec^{-1} (10^3)	R — donor-acceptor dist. (Å)	Quantum efficiency of transfer $\eta_t = 1 - \dfrac{\eta}{\eta_0}$	$P_{da} = \dfrac{1}{\tau_d}\left(\dfrac{\eta^\circ_d}{\eta_d} - 1\right)$ (sec^{-1}) (10^3)
0.25	23.9	0.065	4.8	22.0	0.195	12.6
0.50	21.1	0.125	17.2	19.6	0.296	31.1
0.75	20.5	0.265	26.2	18.5	0.430	56.3
1.00	19.5	0.398	45.5	17.7	0.556	101.5

Tm^{3+} — donor, concentration constant 1.0 wt %.

Table 17. *Ratio of decrease of maximum fluorescence of thulium (η_0) to its fluorescence in the presence of erbium (η)*

Excited level	Emission transitions									
	Phosphate					Borate				
	$^3P_0 \to {}^3H_6,\ ^1I_6 \to {}^3H_4$	$^3P_0,\ ^1I_6 \to {}^3H_4$	$^1D_2 \to {}^3H_4$	$^1G_4 \to {}^3H_4$	$^3P_0,\ ^1I_6 \to {}^1G_4$	$^3P_0,\ ^1I_6 \to {}^3H_6$	$^3P_0,\ ^3I_6 \to {}^3H_4$	$^1D_2 \to {}^3H_4$	$^1G_4 \to {}^3H_4$	$^3P_0,\ ^1I_6 \to {}^1G_4$
	292 nm	350 nm	453 nm	651.5 nm	701 nm	294 nm	355 nm	456 nm	651.5 nm	705 nm
	η_0/η									
$^3P_2$ (262 nm)	1.8	2.17	2.45		1.8	1.6	1.6	1.7		1.6
$^3P_0$ (288 nm)		2.15	2.50		1.8		1.4	1.6		1.6
$^1D_2$ (358 nm)		2.3	2.2					1.6	1.5	
$^1G_4$ (469 nm)			2.5					1.8	1.8	

η_0 — intensity of thulium fluorescence in glass containing 1.0 *wt* % of thulium only

η — intensity of thulium fluorescence in presence of erbium, when the concentration of both ions in a glass is 1.0 *wt* %

The theoretical dipole-dipole transition probabilities for energy transfer from Tm $^1D_2$ to Er ($^2G_{7/2}$, $^2G_{9/2}$, $^2K_{15/2}$) calculated from *Dexter* (formulae (15) are 13.6 sec^{-1} and 27.42 sec^{-1} in phosphate and borate glasses respectively, containing both thulium and erbium at concentrations of 1.0 wt %. By comparison of these results with experimentally-determined energy transfer probabilities, 101.5×10^3 sec^{-1} in phosphate and 45.5×10^3 sec^{-1} in borate glasses (see Table 17), we see that the experimental results are higher by three orders of magnitude than the theoretical results for dipole-dipole and by four to five orders for dipole-quadrupole interactions. Similar results were obtained for the Gd—Tb systems and Sm—Eu systems as shown above.

According to the theory of *Miyakawa* and *Dexter* (77) on phonon-assisted energy transfer in solids, the probability of this process depends on the energy gap, ΔE, between the levels of the donor and the acceptor in the form:

$$W = W(0) \exp(-\beta \Delta E) \tag{29}$$

β being a constant determined by the phonon nature of the host lattice and by the strength of electron-lattice coupling, and $W(0)$ the energy transfer probability when zero phonon lines overlap, which is the rate at the temperature of absolute zero.

The phonon-assisted energy transfer is governed by factors similar to those of the multipolar relaxation process. *Moos* (78) and *Weber* (79) found that $W(0)$ for the relaxation process is nearly independent of the individual nature of the rare earth ions and depends strongly on the host. This last conclusion has also been verified experimentally for energy transfer processes between pairs of rare earth ions in solids, by *N. Yamada et al.* and *Shionoya* (76). In resonant energy transfer, $W = W(0)$; therefore if the energy is transferred between various levels which are in resonance, the probability of phonon-assisted energy transfer should be independent of the specific levels involved. These expectations were confirmed by the similar factor of decrease of emission intensities from different levels of thulium in the presence of erbium (see Table 16).

Phonon-assisted energy transfer will have the effect of increasing the overlap integral of the absorption spectra of donor and acceptor as well as increasing the coupling between them. This follows from the fact that probability of a phonon-assisted transfer depends on the difference between the matrix element of the dynamic part of the orbit-lattice interaction between the excited states of the acceptor ions and between the ground and excited states of the donor ion, in addition to the matrix elements of the multipole interaction, which are determined by Eq. (15). The magnitude of these elements depends strongly on the extent to which the lattice is deformed.

The rate of energy transfer is faster in phosphate than in borate glasses by a factor of about 2 (see Table 16). In a paper by *Reisfeld et al.* *(21)*, it was found that the absorption and emission half-bandwidths of rare earths are larger in phosphate than in borate glasses. This was attributed to the fact that the cavity in which the rare earth ion can be situated is larger in phosphate glasses, thus permitting a larger variety of sites with slightly different crystal field parameters to exist, producing inhomogeneous broadening rather than by crystal field-splitting (the glass host crystal field is stronger in borate than in phosphate glasses).

The higher inhomogeneous broadening of the electronic levels of the rare earths, as a consequence of the stronger possibility of lattice deformation in phosphate than in borate glasses, in addition to broadening of the host vibrational bands, will increase the overlap integral of the absorption spectra of both donor and acceptor and will increase coupling between them. As a result, the probability of energy transfer from thulium to erbium should be higher in phosphate than in borate glasses, which was confirmed by the experimental results.

In the Tm—Er system, a mutual migration of excitation energy occurs. The energy transfer form thulium to erbium is a multichannel process in which the energy is transferred from all the metastable levels of thulium to the matching energy levels of erbium. In addition, back-transfer of energy from erbium to thulium occurs by cross-relaxation of respective erbium transitions. The efficiency of energy transfer from the $^1P_2$ level of erbium ranges between 0.14—0.60 in phosphate and 0.065—0.40 in borate glasses, for concentrations of 1.0 wt % thulium and 0.25—1.0 wt % erbium.

Energy transfer rates from the $^1D_2$ level to the respective level of erbium lie between 12.6×10^3—101.5×10^3 sec^{-1} for phosphate and between 4.8×10^3—45.5×10^3 sec^{-1} for borate glasses, for the ranges of concentration given above.

The efficiency of energy transfer from thulium to erbium is independent of the levels between which the transfer occurs, but is dependent on the matrix. Hence, it is concluded that the energy is transferred via the phonons of the host glass.

VI. Conclusions

1. From the absorption spectra arising from the $f \rightarrow f$ transition, we can see that the radiative probabilities for the hypersensitive transition $^7F_0 \rightarrow {}^5D_2$ of Eu^{3+} increase in the order of phosphate, silicate, germanate due to the stronger mixing of the $4f$ with $5d$ configuration in that order.

An alternative explanation of this phenomenon is the increase in the covalency of the Eu—O in the glasses in the order of phosphate, silicate and germanate. The increase of covalency (or interconfigurational mixing) can also be detected from the ratio of the $^5D_0 \rightarrow {}^7F_2$ (electric dipole) transition of Eu to the $^5D_0 \rightarrow {}^7F_1$ (magnetic dipole) transition. This ratio increases in the order of phosphate $<$ silicate $<$ tungstate $=$ germanate $<$ borax. The increase in the relative intensity of the $^5D_0 \rightarrow {}^7F_2$ transition is followed by a shift towards longer wavelengths of this transition in borax relative to phosphate, resulting from a lowering of the $^5D_0$ level in the former relative to the free ion.

2. The nonradiative relaxation of the highest level to the lower metastable level, as exemplified by relaxation to $^5D_0$ of Eu, is highest in borate glasses (because of its high phonon energy) and lowest in germanate glasses. This means that fewer phonons are needed in order to match the various energy gaps in borate. As a result, germanate glasses should be used whenever fluorescence of higher levels is desired.

3. Splitting of the $5d$ orbitals of Ce^{3+} in glasses into five components of different energies indicates that Ce^{3+} is ituated in the C_2 or C_s site symmetry in glass. The electrostatic perturbation exerted on Ce^{3+} levels is larger in borate than in phosphate glass, as is the amount of covalency of the Ce—O bond. This is seen from the shift of the center of gravity to longer wavelengths in borax relative to phosphate.

4. It is proposed that molecular orbitals are formed from the f orbitals of the rare earth and the linear combination of atomic orbitals of its surrounding oxygens.

5. The nephelauxetic effect in glass is observed from the rare earth absorption spectra arising from f—d (and f—f) and for mercury-like ions from the $^1S_0 \rightarrow {}^3P_1$ transitions.

6. The electronegativity of different glasses is calculated from the position of the maxima of charge transfer bands of Eu in various glasses, and it has been shown that there exists a linear relation between the nephelauxetic parameters and optical electronegativity of glasses.

7. Energy transfer occurs between pairs of rare earths and glasses. The probability of transfer indicates that the phonons of the glass are effective in assisting the energy transfer process. This phenomenon can be used as an effective tool for studying the energy process in both inorganic and biological systems.

Acknowledgements. The author is deeply grateful to Mrs. *Esther Greenberg* for her generous help in the preparation of this paper, to Dr. *B. Barnett* for many fruitful discussions, and to my students *L. Boehm, Y. Eckstein, N. Lieblich, J. Hormodaly* and *H. Mack*, who provided me with many important experimental data during the writing of this paper.

VII. References

1. *Ziman, J. M.:* J. Phys. Chem. (Proc. Phys. Soc.) *1*, 1532 (1968).
2. *Wybourne, B. G.:* Spectroscopic Properties of Rare Earths. New York: Interscience Pub. 1965.
3. *Patek, K.:* Glass Lasers. London: Butterworth & Co., Ltd. 1968.
4. *Judd, B. R.:* Phys. Rev. *127*, 750 (1962).
5. *Ofelt, G. S.:* J. Chem. Phys. *37*, 511 (1962).
6. *Jørgensen, C. K., Pappalardo, R., Schmidtke, H. H.:* J. Chem. Phys. *39*, 1422 (1963).
7. *Ellis, M. M., Newman, D. J.:* Phys. Letters *21*, (5) 508 (1966). — *Newman, D. J., Ellis, M. M.:* Phys. Letters *23* (1), 46 (1966).
8. *Jørgensen, C. K.:* Progr. Inorg. Chem. *4*, 73 (1962).
9. *Katzin, L. I., Barnett, M. L.:* J. Phys. Chem. *68*, 3779 (1964).
10. *Jørgensen, C. K., Judd, B. R.:* Mol. Phys. *8*, 281 (1964).
11. — Mol. Phys. *5*, 271 (1962).
12. — Progr. Inorg. Chem. *12*, 101 (1970). Modern Aspects of Ligand Field Theory, Amsterdam: North-Holland Publishing Co. 1971.
13. *Blasse, G., Bril, A.:* J. Chem. Phys. *45*, 3327 (1966).
14. *Reisfeld, R., Greenberg, E.:* Anal. Chim. Acta *47*, 155 (1969).
15. — *Gur-Arieh, Z., Greenberg, E.:* Anal. Chim. Acta *50*, 249 (1970).
16. — *Biron, E.:* Talanta *17*, 105 (1970).
17. — *Greenberg, E., Kraus, S.:* Anal. Chim. Acta *51*, 133 (1970).
18. — — *Kirschenbaum, L., Michaeli, G.:* 8th Rare Earth Conference *2*, 743 (1970).
19. — *Boehm-Kirschenbaum, L.:* Israel J. Chem. *8*, 103 (1970).
20. *Velapoldi, R. A., Reisfeld, R., Boehm, L.:* 9th Rare Earth Conf. *2*, 488 (1971).
21. *Reisfeld, R., Velapoldi, R. A., Boehm, L., Ish-Shalom, M.:* J. Phys. Chem. *75*, 3980 (1971).
22. — — — J. Phys. Chem. *76*, 1293 (1972).
23. *Velapoldi, R. A., Reisfeld, R., Boehm, L.:* unpublished results (1972).
24. *Reisfeld, R., Lieblich, N.:* Phys. Chem. of Solids (1973).
25. — *Mack, H.:* unpublished results (1972).
26. *Rice, D. K., DeShazer, L. G.:* Phys. Rev. *186*, 387 (1969).
27. *Carnall, W. T., Fields, P. R., Rajnak, K.:* J. Chem. Phys. *49*, 4412 (1968).
28. *Axe, J. D., Jr.:* J. Chem. Phys. *39*, 1154 (1963).
29. *Hoshina, T.:* Japan. J. Appl. Phys. *6*, 1203 (1967).
30. *Nieuwpoort, W. C., Blasse, G., Bril. A.:* Optical Properties of Ions in Crystals; H. M. Crosswithe and H. W. Moos, Ed. (1967), p. 161.
31. *Reisfeld, R., Eckstein, Y.:* J. Solid State Chem. (1972).
32. — — Anal. Chim. Acta *56*, 461 (1971).
33. *Fournier, J. T., Bartram, R. H.:* J. Phys. Chem. Solids *31*, 2615 (1970).
34. *Roth, R. S., Schneider, S. J.:* J. Res. Natl. Bur. Stand. *A64*, 309 (1960).
35. *Judd, B. R.:* J. Chem. Phys. *44*, 839 (1966).
36. *Weber, M. J.:* Phys. Rev. *157*, 262 (1967).
37. *Reisfeld, R., Eckstein, Y.:* J. Noncrystalline Solids *5*, 174 (1972).
38. — Inorganic Ions in Glasses and Polycrystalline Pellets as Fluorescence Standard Reference Materials, Chapter in Accuracy in Spectrophotometry and Spectrofluorimetry, Ed. by N. B. S. (1972).
39. *Jørgensen, C. K.:* Oxidation Numbers and Oxidation States. Berlin–Heidelberg–New York: Springer 1969.
40. *Kuse, K., Jørgensen, C. K.:* Chem. Phys. Letters *1*, 314 (1967).

41. *Jørgensen, C. K., Brinen, S.:* Mol. Phys. *6*, 629 (1963).
42. *Loh, E.:* Phys. Rev. *147*, 332 (1966).
43. *Reisfeld, R., Hormodaly, J.:* 9th Rare Earth Conference *1*, 123 (1971).
44. — — *Barnett, B.:* Chem. Phys. Letters (1972).
45. *Ryan, J. L., Jørgensen, C. K.:* J. Phys. Chem. *70*, 2845 (1966).
46. *Blasse, G., Bril, A.:* J. Chem. Phys. *47*, 5139 (1967).
47. *Reisfeld, R., Morag, S.:* Appl. Phys. Letters *21*, (2), 57 (1972).
48. — *Honigbaum, A., Michaeli, G., Harel, L., Ish-Shalom, M.:* Israel J. Chem. *7*, 613 (1970).
49. *Sinha, S. P.:* Europium. Berlin–Heidelberg–New York: Springer 1967.
50. *Hoshina, T., Kuboniwa, S.:* J. Phys. Soc. Japan, *31*, 828 (1971).
51. *Bril, A., Blasse, G., de Poorter, J. A.:* J. Electrochem. Soc., Solid State Sci. *117*, 346 (1970).
52. *Eisenstein, J. C.:* J. Chem. Phys. *25*, 142 (1956).
53. *Jørgensen, C. K., Rittershaus, E.:* Mat. Fys. Medd. Dan. Vid. Selskab. *35*, No. 15 (1967).
54. *Reisfeld, R.:* unpublished results (1972).
55. *Sinha, S. P.:* Complexes of the Rare Earths. Oxford: Pergamon Press, 1966.
56. *Reisfeld, R., Glasner, A.:* J. Chem. Phys. *43*, 2923 (1965).
57. — — J. Chem. Phys. *42*, 2983 (1965).
58. *Lang, R. L.:* Can. J. Res. *14A*, 127 (1930).
59. *Reisfeld, R., Greenberg, E.:* unpublished results (1972).
60. *Blasse, G.:* J. Solid State Chem. *4*, 52 (1972).
61. *Reisfeld, R., Boehm, L., Greenberg, E., Ish-Shalom, M.:* unpublished results (1972).
62. *Duffy, J. A., Ingram, M. D.:* J. Chem. Phys. *52*, 3752 (1970).
63. — — J. Chem. Phys. *54*, 443 (1971).
64. *Förster, Th.:* Discussions Faraday Soc. *27*, 7 (1959).
65. — Ann. Physik *2*, 55 (1948).
66. *Dexter, D. L.:* J. Chem. Phys. *21*, 836 (1953).
67. *Patek, K.:* p. 94, See Ref. 3.
68. *Nakazawa, E., Shionoya, S.:* J. Chem. Phys. *47*, 3211 (1967).
69. *Inokuti, M., Hirayama, F.:* J. Chem. Phys. *43*, 1978 (1965).
70. *Reisfeld, R., Greenberg, E., Velapoldi, R. A., Barnett, B.:* J. Chem. Phys. *56*, 1698 (1971).
71. — *Boehm, L.:* J. Solid State Chem. *4*, 417 (1972).
72. — *Barnett, B.:* unpublished results (1972).
73. *Soules, T. F., Duke, C. B.:* Phys. Rev. B. *3*, 262 (1971).
74. *Fong, F. V., Miller, M. M.:* Chem. Phys. Letters *10*, 408 (1971).
75. *Riseberg, L. A., Gandrud, W. B., Moos, H. W.:* Phys. Rev. *159*, 262 (1967).
76. *Yamada, N., Shionoya, S., Kushida, T.:* J. Phys. Soc. Japan *30*, 1507 (1971).
77. *Miyakawa, T., Dexter, D. L.:* Phys. Rev. B. *1*, 2961 (1970).
78. *Moos, M. W.:* J. Luminescence *1*, 2, 406 (1970).
79. *Weber, M. J.:* Phys. Rev. *157*, 262 (1967); Phys. Rev. *171*, 283 (1968).

Received July 7, 1972

The Crystal Chemistry of the Rare-Earth Silicates

J. Felsche

Institut für Kristallographie der ETH, Zürich, Switzerland

Table of Contents

J. Felsche

1. Abstract

Rare-earth silicate compounds form a chapter in the chemistry of large-cation silicates. With RE^{3+} ionic radii ranging from 0.85 to 1.28 Å, the crystal structures consist of isolated (SiO_4) tetrahedra or groups of (Si_2O_7) and (Si_3O_{10}). Rare-earth cations show six- to tenfold oxygen coordination.

In the binary systems RE_2O_3–SiO_2, compounds are known of composition 1:1, 7:9, and 1:2. The 1:1 compounds display two different structural types for $RE_2(SiO_4)O$; their range of stability along the lanthanide series is determined by a critical value of the ionic radius between Tb^{3+} and Dy^{3+}. Compounds $7\,RE_2O_3 \cdot 9\,SiO_2$ were found for all the rare earths. The crystals represent a cation-deficient type of apatite structure which is best described by the formula $(RE_{3.33} \square_{0.67}) RE_6(SiO_4)_6O_2$. Compounds $1\,RE_2O_3 \cdot 2\,SiO_2$ show extensive polymorphism. Seven different structure types have been observed. The fields of stability are given by units at the positions Eu^{3+} and Ho^{3+}. Transition temperatures are in the range 1300–1500 °C. All disilicate structures $RE_2(Si_2O_7)$ contain double-tetrahedra groups either in the staggered or in the eclipsed configuration. An exception is the group of compounds $(Eu, \ldots Er)_4(Si_3O_{10})\,(SiO_4)$, which show linear (Si_3O_{10}) tetrahedra groups plus isolated (SiO_4) tetrahedra instead of (Si_2O_7). All phase transitions are of the reconstructive type. In the systems RE O–SiO_2 the existence of four types of compounds, 3:1, 2:1, 3:2, and 1:1, was suggested for (Sm^{2+}), Eu^{2+} and Yb^{2+} by powder crystal data (x-ray, infrared spectroscopy). So far, single-crystal investigations on dimorphic $Eu_2(SiO_4)$ and $Eu_3(SiO_4)O$ have confirmed their isostructural relationship to polymorphic $Ca_2(SiO_4)$ and $Sr_3(SiO_4)O$, respectively.

Ternary compounds of type $Alk.RE(SiO_4)$ show three different structures closely related to alkaline-earth silicates. Other orthosilicate compounds $RE_9Alk.(SiO_4)_6O_2$ and $RE_8M_2(SiO_4)_6O_2$ have the apatite structure. Stillwellite compounds $RE\,B(SiO_4)O$ are known for La, ...Nd, whereas garnets $RE_2Mg_3(SiO_4)_3$ exist only with the small rare earths Y, Ho, ...Lu. Ternary compounds with (Si_2O_7) double tetrahedra are known for $Na_3(Y, Er, \ldots Lu)\,(Si_2O_7)$. The structure of $K_3Eu(Si_2O_7)$ is not yet known in detail.

Effective ionic radii of the trivalent rare earths with different coordination numbers were obtained from plots of $r^3(RE^{3+})$ vs. cell volume by evaluating all experimental data on rare-earth silicate structures available so far in the literature, and from our own investigations.

A considerable variation of Si–O bond lengths in the (Si_2O_7)- and (Si_3O_{10})-tetrahedra groups, with extreme values 1.56 and 1.76 Å for individual bonds, was found to be strongly correlated to the variance in charge balance on the individual oxygens.

Introduction

Modern solid-state technology has recently made use of rare-earth oxide compounds in microwave devices, semiconductors, ferromagnetics, ferroelectrics, lasers and phosphors. The unique properties of this group of compounds, which also includes rare-earth silicates and vitreous silica doped with rare earths, are due to their special $4f^n$ electronic states and the general shielding of these orbitals by higher orbitals in the rare-earth ion. The term rare earths is commonly assigned to the 14 lanthanides which represent the $4f$-transition elements. Yttrium and scandium are sometimes included if they are considered to be congeners of the element lanthanum because of their equivalent electronic configurations. In this study of rare earths (RE) we will refer exclusively to the 14 lanthanide elements, unless otherwise stated.

The $4f$ elements form the longest continuous series of chemically similar elements in the periodic table. Starting with La^{3+}, which has the electronic configuration of the closed Xe-shell, the following ions in the series Ce, ... Lu successively add an electron to the $4f$ subshell up to a total of 14. The shrinkage of the atomic vulume as the $4f$ subshell is filled is commonly referred to as the 'lanthanide contraction'. The contraction is due to an especially imperfect shielding from the nuclear charge of one f electron followed by another f electron. As the orbitals are shielded by $5s^2$ and $5p^6$ electrons, they do not contribute significantly to chemical bonding. The rare earths are strongly electropositive, so that bonding to ligands, such as O^{2-} and F^-, is essentially electrostatic. The poor screening of the nuclear charge by the f electrons is also responsible for the high polarizing power of the rare-earth cations. These strong positive fields produce greater bonding strength than cations of the same group of elements having only s, p and d electrons.

Some evidence of periodicity within the series is apparent in the oxidation state, magnetic properties and thermodynamic stability. All these data reflect preferences for the $4f^0$, $4f^7$ and $4f^{14}$ electronic configurations. With regard to the oxidation state, the normal valence of the rare-earth ions is known to be 3+. However, as a result of the distinct $4f^0$, $4f^7$ and $4f^{14}$ stability, there are known in solid-state chemistry isoelectronic or closely isoelectronic groups of ions which also possess ions with charges other than 3 +: La^{3+}, $Ce^{4+}(4f^0)$, $Pr^{4+}(4f^1)$; Sm^{2+} $(4f^6)$, Eu^{2+}, Gd^{3+}, $Tb^{4+}(4f^7)$; Yb^{2+}, $Lu^{3+}(4f^{14})$.

Unlike the d-transition elements, the lanthanides show a gradual, monotonic variation in the ionic radii which amounts to about 20% from La^{3+} (1.06 Å) to Lu^{3+} (0.85 Å) for sixfold coordination. The influence of this varying ionic size on coordination number and coordination geometry is a unifying concept in chemistry. It is within the scope of this article to discuss this effect on silicate structures with special attention to possible discontinuities along the series of rare-earth elements. Compounds of the binary systems RE_2O_3–SiO_2 are of special interest because their properties should show a closer relationship with the lanthanide periodicity than is the case with more complex compounds. Therefore, some charateristics and the history of the investigation of these systems will be given first, followed by brief comments on ternary rare-earth silicate compounds.

Pure rare-earth compounds are unknown in nature; minerals (see also appendix I) usually contain groups of rare earths because of their nearly identical chemical character. It is not surprising, therefore, that a systematic investigation of the individual binary silicate systems RE_2O_3–SiO_2 revealed a large number of new phases and unknown structural types. These experiments were started about 15 years ago, when modern methods of ion separation were developed and provided very pure rare-earth elements.

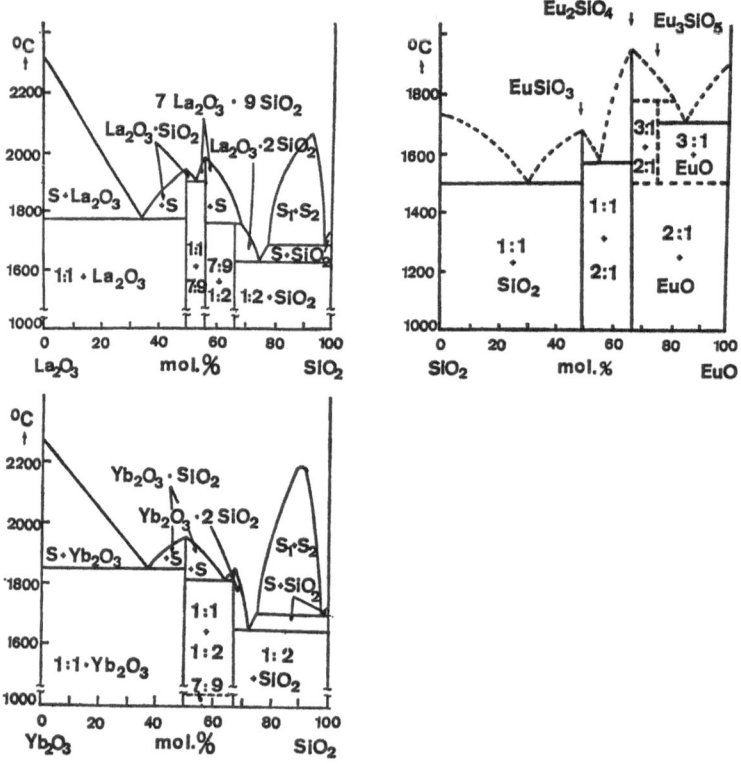

Fig. 1. The Systems of La_2O_3–SiO_2 and Yb_2O_3–SiO_2 after Ref. (81) and of EuO–SiO_2 after Ref. (36).

First results on the complete binray systems RE_2O_3–SiO_2 were published about ten years ago by Russian authors, essentially (1—9). The phase diagrams of La_2O_3–SiO_2 and Yb_2O_3–SiO_2, as shown in Fig. 1, are from a recent review made by this group of authors of their own work in the rare-earth silicate field (81). The phase diagrams have been slightly revised as compared to the data from the early period of investigation. Originally, silicates of composition $1 \, RE_2O_3 \cdot 1 \, SiO_2$, $2 \, RE_2O_3 \cdot 3 \, SiO_2$ and $1 \, RE_2O_3 \cdot 2 \, SiO_2$ were described. The existence of these compounds was suggested by various X-ray powder patterns of a rather complicated character, vibration spectra or optical data, but with hardly any single-crystal information. Data of this quality were published in a first survey of the rare-earth silicate compounds (10). The introduction of single-crystal X-ray methods into this field led to some progress in understanding the highly polymorphic character of the compounds.

Analysis of the crystal structure of compounds $2 \, RE_2O_3 \cdot 3 \, SiO_2$ revealed the true composition to be $7 \, RE_2O_3 \cdot 9 \, SiO_2$ (11). The crystals represent a cation-deficient type of oxyapatite structure $RE_{9.33} \square_{0.67} (SiO_4)_6 O_2$. Subsequently the crystal structures of 1:1 and 1:2 rare-earth silicate compounds were studied. The single-crystal information led to reinterpretation of the powder diffraction patterns of the samples prepared during the first stage of phase diagram investigation. The X-ray powder diffraction patterns of the single phases are shown in Appendix II. Two different structure types were identified in 1:1 compounds of type $RE_2(SiO_4)O$ (12, 13). Seven polymorphic forms were found of the compounds $1 \, RE_2O_3 \cdot 2 \, SiO_2$, largely of the type $RE_2(Si_2O_7)$ (14—21). Consequently, some major changes had to be introduced into the original version of the RE_2O_3–SiO_2 phase diagrams; Fig. 1 shows the revised data.

Many of the binary compounds were prepared by solid-state reaction of corresponding oxide mixtures, to some extent by the employment of flux methods. Some ternary compounds were synthesized under hydrothermal conditions. These compounds are polymorphic $NaRE(SiO_4)$, $Na_3RE(Si_2O_7)$ (29—31) and a garnet-type $Mg_3RE_2(SiO_4)_3$ (32). The synthesis of borosilicates $RE\,B(SiO_4)O$ showing the stillwellite structure (58) was accomplished by solid-state reaction (80). Various compounds containing divalent rare earths were reported in the systems $RE\,O$–SiO_2 for Sm, Eu and Yb, largely based on X-ray powder data (33—37). The phase diagram EuO–SiO_2 after (36) is shown in Fig. 1. Dimorphic $Eu_2(SiO_4)$ (59, 60) and $Eu_3(SiO_4)O$ (62) were shown by structural analysis to be isostructural with polymorphic $Ca_2(SiO_4)$ and $Sr_3(SiO_4)O$, respectively.

It is proposed to present in this article a detailed structural description of all rare-earth silicate compounds known so far. Compounds more complex than ternary and compounds containing OH groups or molecular H_2O are not included. The data from recent investigations cited in the literature, and from our own unpublished work, will be correlated in order to show periodicity or discontinuities along the rare-earth series. As outlined above, these is to be expected from the preferred $4f^0$, $4f^7$, $4f^{14}$ electronic configurations and the increasingly ineffective screening of the nuclear charge with increasing number of f electrons. This should affect the bonding strength of the rare earth cations against the coordinating oxygens. For this discussion, polymorphism of the rare-earth silicates and variation of RE—O bond lengths and coordination polyhedra will be taken into special consideration. Secondly, it will be of interest to study the configuration and the geometrical distortion of the individual SiO_4 tetrahedra, or groups of tetrahedra, relative to the rare-earth periodicity.

2. Structures Containing Isolated (SiO_4)

The various orthosilicate structures containing isolated (SiO_4) tetrahedra can be divided into two different groups: the first group includes the so-called oxyorthosilicates $RE_2(SiO_4)O$ which contain two different types of anions: the well-known complex ion $(SiO_4)^{-4}$ and additional, not silicon-bonded oxygen ions in the structure. All binary compounds of composition $1\ RE_2O_3 \cdot 1\ SiO_2$ belong to this group which displays two different structure types. One is stable for the large rare earths La to Tb, and the other embraces the smaller rare earths Dy to Lu. Binary compounds of composition $7\ RE_2O_3 \cdot 9\ SiO_2$ may also be called oxyorthosilicates. They crystallize with a cation-deficient type of apatite structure and the extra, not silicon-bonded oxygen replaces the anions F, OH on the main channel position in the common apatite structure. The structure of stillwellite containing mixed trivalent cations $CeB(SiO_4)O$ also belongs to the group of oxyorthosilicates.

The second group of rare-earth orthosilicates consists of structures containing isolated (SiO_4) tetrahedra only. However, apparently for stoichiometric reasons, binary compounds exist with divalent rare earths only. Compounds are known of Eu^{2+}; analogues for Sm^{2+} and Yb^{2+} have also been suggested by X-ray powder patterns (37). A charge-coupled substitution $2\ M^{2+} \longleftrightarrow 1\ M^{1+} + RE^{3+}$ leads to the polymorphic series of compounds $NaRE(SiO_4)$. For rare earths smaller than Ho the garnet structure type seems to exist with rare earths on the eight-fold coordinated (c)sites. The powder diffraction patterns of all the known orthosilicate structures are shown in Appendix II.

2.1. 1 $RE_2O_3 \cdot 1$ SiO_2 Compounds

The existence of 1:1 compounds in the binary rare-earth silicate system was suggested by several authors during the early stage of experimental work on the systems RE_2O_3–SiO_2 by X-ray powder diffraction or IR vibration spectroscopy. The first single-crystal investigation was carried out a few years later on the yttrium silicate (*12*). Subsequently the crystal structure of this compound (*38*) and of isotypic $Yb_2(SiO_4)O$ (*41*) was refined. A series of isostructural compounds of the smaller rare earths ranging from Tb to Lu has been found (*13, 39*). The crystals prepared with a flux of Bi_2O_3 contained Bi_2O_3 up to 7 Mol. % (*39*).

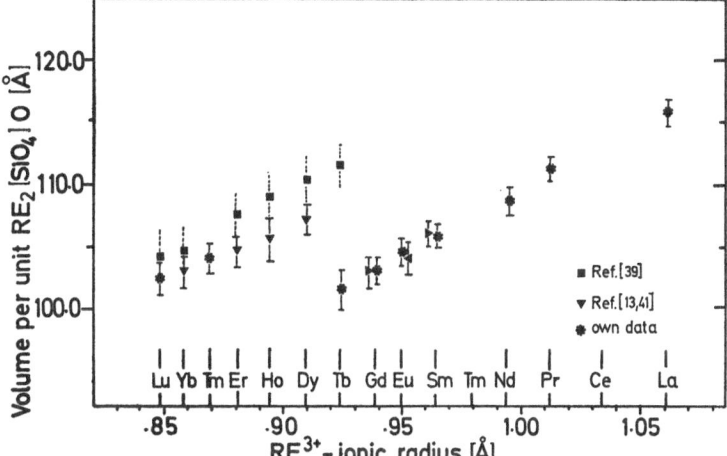

Fig. 2. Plot of mol. volume ($RE_2[SiO_4]O$) vs. Ionic radius (RE^{3+}). The different degree of space filling in the two structure types of compounds $RE_2[SiO_4]O$ is demonstrated

Compounds of the larger rare earths La to Tb crystallize with a different structure, as was shown for $Gd_2(SiO_4)O$ (*13*). This structure also has monoclinic symmetry but shows higher oxygen coordination of the rare-earth cations. A plot of the mol. volume $V(RE_2(SiO_4)O)$ vs. ionic radius (RE^{3+}) in Fig. 2 reveals the higher degree of space filling which appears to exist in this structure type (*40*). The smaller volume per $RE(SiO_4)O$ unit of the larger rare earths can be understood by the different linkage of (O–RE_4) tetrahedra in the two structure types. This will be discussed in detail in the following chapters.

Table 1. *Cell dimensions of compounds* $RE_2[SiO_4]O$. *Structure type A of monoclinic symmetry, space group* $P2_1/c$ *and* $Z = 4$ *for compounds La, ... Tb. Structure type B of monoclinic symmetry space group* $B2/b$ *and* $Z = 8$ *for compounds of Dy, ... Lu*

	$a_0[Å]$	$b_0[Å]$	$c_0[Å]$	$\beta[Å]$	$V[Å^3]$
$La_2[SiO_4]O$	9.420 (9)	7.398 (7)	7.028 (7)	108.21 (6)	465.2 (9)
$Pr_2[SiO_4]O$	9.253 (9)	7.301 (8)	6.934 (8)	108.15 (9)	445.1 (8)
$Nd_2[SiO_4]O$	9.250(11)	7.258(10)	6.886 (9)	108.30(11)	439.3 (6)
$Sm_2[SiO_4]O$	9.161 (9)	7.112 (9)	6.821 (7)	107.51 (9)	424.4 (7)
$Eu_2[SiO_4]O$	9.142 (8)	7.054 (6)	6.790 (6)	107.53 (9)	417.9 (8)
$Gd_2[SiO_4]O$	9.131 (7)	7.045 (6)	6.749 (5)	107.52 (7)	414.0 (9)
$Tb_2[SiO_4]O$	9.083(22)	6.990(11)	6.714(10)	107.31(21)	406.1(42)
	$a[Å]$	$b[Å]$	$c[Å]$	$\gamma[Å]$	$V[Å^3]$
$Dy_2[SiO_4]O$	14.38 (2)	10.42 (2)	6.74 (1)	122.0 (3)	856.5(72)*
$Ho_2[SiO_4]O$	14.35 (2)	10.37 (2)	6.71 (1)	122.2 (3)	843.0(38)*
$Er_2[SiO_4]O$	14.32 (2)	10.35 (2)	6.69 (1)	122.3 (3)	836.7(41)*
$Tm_2[SiO_4]O$	14.302(9)	10.313(9)	6.662 (6)	122.21 (9)	828.5 (9)
$Yb_2[SiO_4]O$	14.28 (1)	10.28 (1)	6.653 (5)	122.2 (1)	824.0 (7)*
$Lu_2[SiO_4]O$	14.254(9)	10.241(8)	6.641 (7)	122.20 (8)	819.3(10)

* Data from Ref. (*13*).

Table 2. *Atomic parameters of* $Gd_2(SiO_4)O$

Atom	x	y	z	$B[Å^2]$
Gd(1)	0.11453(5)	0.14600(6)	0.41628(1)	0.48
Gd(2)	0.52458(5)	0.62451(6)	0.23428(1)	0.40
Si	0.2020 (3)	0.5876 (3)	0.4598 (6)	0.30
O(1)	0.2032 (9)	0.4302(10)	0.6453(18)	0.58
O(2)	0.1317 (8)	0.4587 (9)	0.2520(17)	0.48
O(3)	0.3839 (8)	0.6361 (9)	0.5059(16)	0.45
O(4)	0.0941 (9)	0.7681(11)	0.4507(18)	0.73
O(5)	0.3837 (8)	0.3782 (9)	0.0487(16)	0.49

After Ref. (*13*).

2.1.1. Structure Type A, $(La, ... Tb)_2(SiO_4)O$, '(RE Oxy (A)'

The structure of the larger rare-earth compounds consists of isolated (SiO_4) tetrahedra, extra, not silicon-bonded oxygens, and two crystallographically independent RE atoms (Fig. 3). The extra oxygens are located at the centre of RE^{3+}-cation tetrahedra. These $(O-RE_4)$ tetrahedra form a two-dimensional network parallel to the (*100*) plane, as shown in the lower illustration in Fig. 3. Sheet-like packing of the structure is

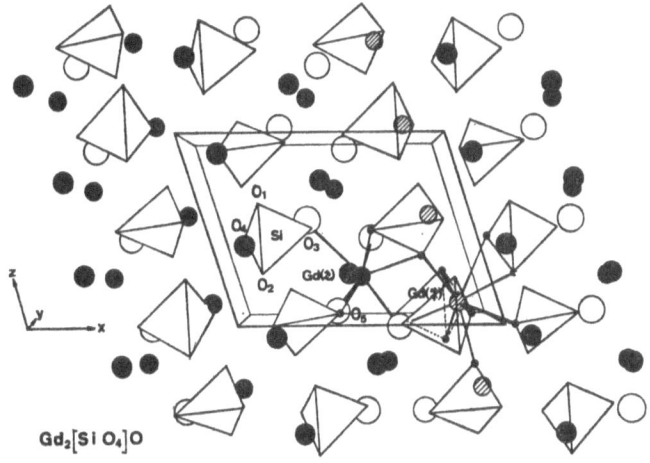

$Gd_2[SiO_4]O$

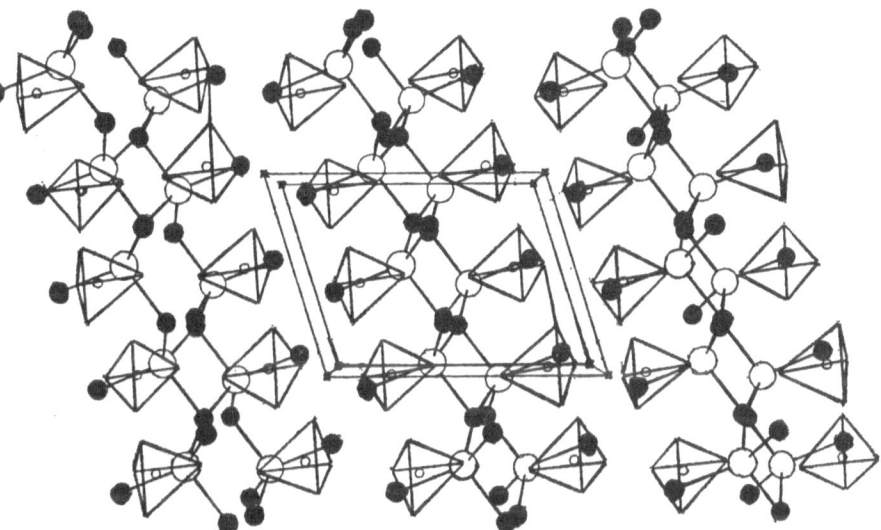

Fig. 3. Crystal structure of $Gd_2[SiO_4]O$

achieved by the introduction of (SiO_4) tetrahedra into the wide meshes of this net. This ensures charge balance and connection to the parallel-running units which are all just one a_0 translation wide. The atomic parameters for $Gd_2(SiO_4)O$ are given in Table 2.

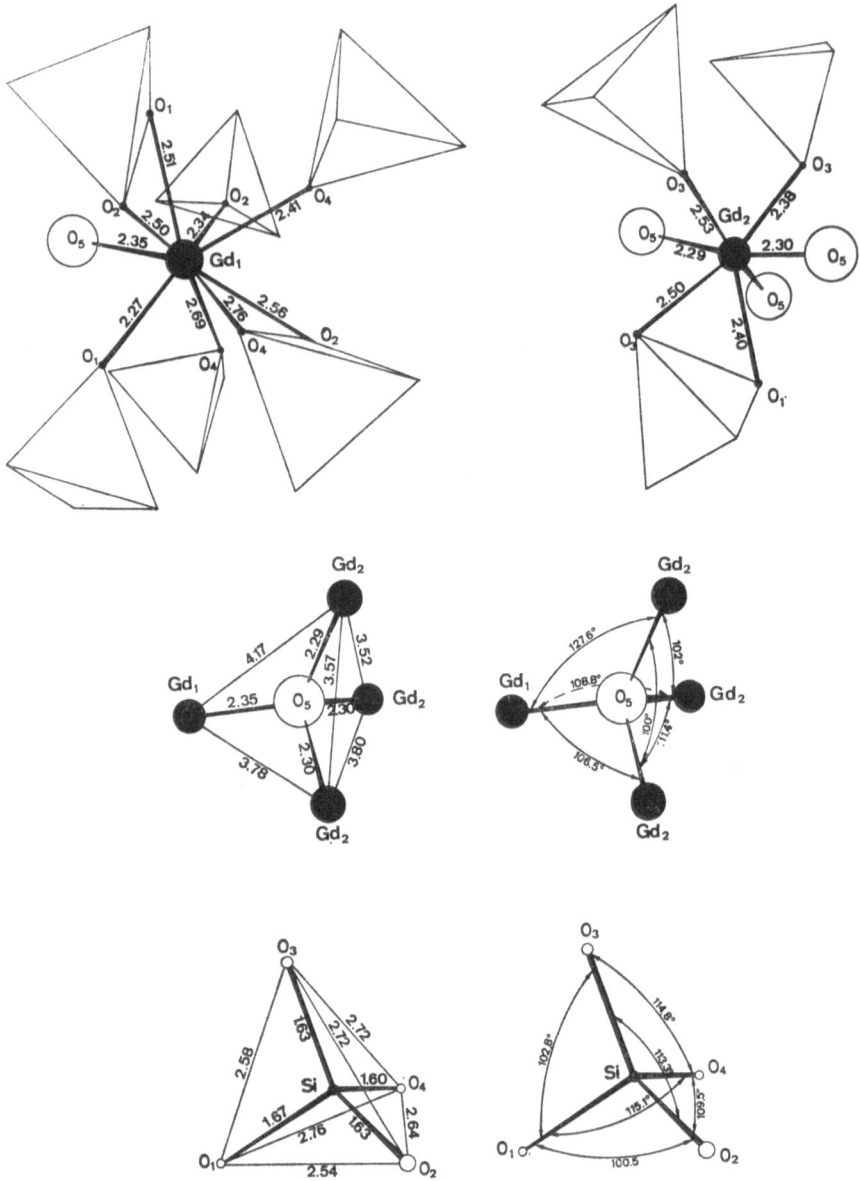

Fig. 4. Interatomic distances and angles in $Gd_2[SiO_4]O$

The monoclinic structure of space group $P2_1/c$ was determined and refined with three-dimensional intensity data to a final residual $R = 7.7\%$, which corresponds to a mean standard deviation of the oxygen-cation distances of 0.009 Å. Gd(1), which is located at the vertices of the network-forming $(O-Gd_4)$tetrahedra, is surrounded by eight further oxygens which all belong to (SiO_4) tetrahedra (Fig. 4). The mean of the nine Gd—O distances is 2.49 Å. Gd(2), on the other hand, shows sevenfold oxygen coordination. There are three 'extra' oxygens O(5) in the first shell and the remaining four belong to three different (SiO_4) tetrahedra. This Gd(2), which actually forms the individual meshes of the $(O-Gd_4)$ network, is located at the bottom face of the tetrahedron and shows a mean distance of 2.39 Å to the seven oxygen ligands. In both the oxygen coordination polyhedra of Gd(1) the Gd—O distances to the 'extra' oxygens O(5) are considerably shorter than the mean Gd—O values for the individual polyhedron. These short bonds of 2.31 Å, a length significantly below the sum of the ionic radii (2.39 Å), indicate high bonding strength corresponding to the strong polarizing forces of the rare-earth cation against the isolated (not-silicon-bonded) oxygen.

The mean value of Si—O distances in the (SiO_4) tetrahedron is 1.63 Å. As pointed out in (13), the shortest Si—O distances correspond to the relatively longest Gd—O interatomic distances: e.g. the short Gd(2)—O(1) distance of 2.40 Å corresponds to the long Si—O(1) distance of 1.67 Å. On the other hand, the long Gd(1)—O(4) distance of 2.62 Å is compensated for by the shortest distance within the (SiO_4)tetrahedron, which is Si—O(4) of 1.60 Å. These data suggested that there is an appreciable covalent component in the RE—O bonding (13).

2.1.2. Structure Type B, (Dy, ... Lu)$_2$(SiO$_4$)O, (RE Oxy B)

The crystal structure of the smaller rare-earths with $Z = 8$ contains twice as many formula units $RE_2(SiO_4)O$ in the B-centred monoclinic unit cell as are present in the RE Oxy A type of structure. The crystal structure of space group $B2/b$ has been refined by three-dimensional data for two compounds, $Y_2(SiO_4)O$ (38) and $Yb_2(SiO_4)O$ (41). $Y_2(SiO_4)O$ has been refined in space group $I2/c$, a setting which had originally been reported from single-crystal diffraction patterns for $Er_2(SiO_4)O$ (cell dimensions $a = 10.33(3)$ Å, $b = 6.69(2)$ Å, $c = 12.37(2)$ Å, $\beta = 103°$) and also for $Y_2(SiO_4)O$ (42). Both refinements confirm the proposal made initially for $Y_2(SiO_4)O$ (12). For the present discussion we will refer to the data for $Yb_2(SiO_4)O$ because of its lower R value of 5.4% and the correspondingly lower standard deviation (≤ 0.008 Å) for the interatomic cation-oxygen distances. The atomic parameters for $Yb_2(SiO_4)O$ are given in Table 3.

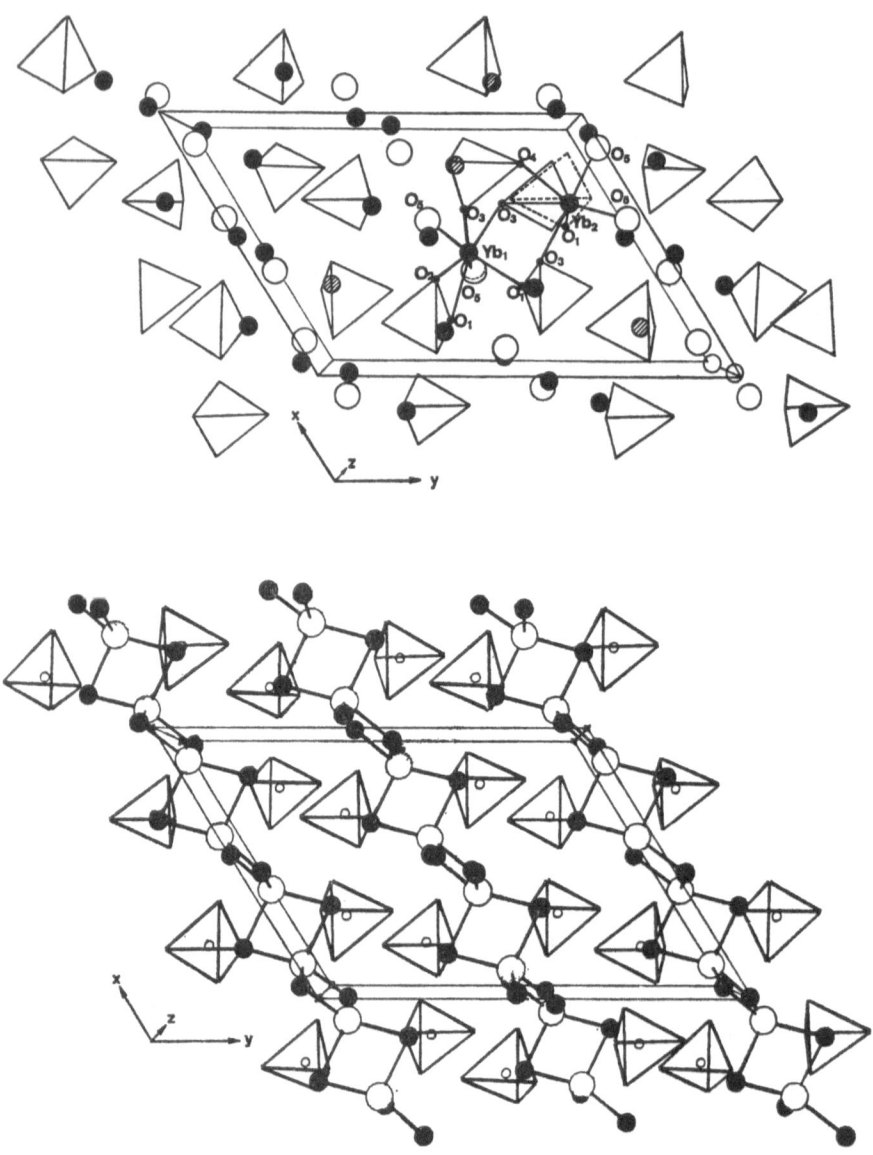

Fig. 5. Crystal structure of Yb$_2$[SiO$_4$]O

As illustrated in Fig. 5, this structure type also has ionic units consisting of isolated (SiO$_4$)tetrahedra and the other type of anion, the 'extra', not silicon-bonded oxygens. These oxygens are also surrounded by four rare-earth cations in the shape of a slightly distorted tetrahedron,

Table 3. *Atomic parameters of* $Yb_2(SiO_4)O$

Atom	x	y	z	$B[Å^2]$
Yb(1)	0.46638(4)	0.53709(3)	0.75564(8)	0.338
Yb(2)	0.66416(4)	0.35892(3)	0.87736(8)	0.340
Si	0.6928 (3)	0.3182 (2)	0.4085 (6)	0.237
O(1)	0.6747 (7)	0.3787 (5)	0.2103(16)	0.586
O(2)	0.8618 (7)	0.4122 (5)	0.4941(16)	0.596
O(3)	0.6769 (6)	0.2029 (5)	0.3535(14)	0.406
O(4)	0.5627 (9)	0.2987 (6)	0.5710(17)	0.886
O(5)	0.8965 (6)	0.5177 (5)	0.9052(16)	0.412

After Ref. (*14*).

and thus, in this case too, the main structural motifs might very well be described in terms of the arrangement of (SiO_4) and $(O-RE_4)$tetrahedra. In contrast to the structure of the larger rare-earth compounds, here the $(O-RE_4)$tetrahedra form, not a two-dimensional network, but chains and (O_2-RE_6)double tetrahedra running in the a_0 direction. The infinite chains of edge-sharing $(O-RE_4)$tetrahedra are connected to the (O_2-RE_6) groups by isolated (SiO_4)tetrahedra. This arrangement is, however, less intensively -filling than the other type found in the 'RE Oxy A' structure. The effect of these two types of packing on the cell volume has already been shown in Fig. 2. Some further details will illustrate the difference between the two structures.

As can be seen from Fig. 6, the coordination around the heavy atoms is somewhat different with respect to the degree of bond-length variation from that observed in the larger rare earth type of structure. Actually, both of the independent rare-earth cations suggest oxygen coordination number 6 instead of 7 for Yb(1). Four silicon-bonded oxygens and two not silicon-bonded oxygens form the shells. Assuming coordination number 6 for both the heavy atoms, a model of a fairly ideal closest packing of the oxygens has been suggested (*38*) with two of the five available octahedral sites occupied by the two heavy atoms and three of the octahedral holes left vacant. However, this interpretation of the structure neglects the fact that one of the rare-earth cations, Yb(1), actually shows CN 7, with the additional oxygen present at a fairly close distance of 2.63 Å. A discussion of polyhedra packing in terms of distorted octahedra, heptahedra and (SiO_4) tetrahedra is, however, much more difficult. Moreover, it fails to explain the different degree of space-filling in the two analogous structure types, as shown in Fig. 2.

The mean Yb—O distance in the oxygen polyhedron of the sevenfold coordinated Yb(1) is 2.33 Å, as compared to 2.23 Å of the sixfold coor-

J. Felsche

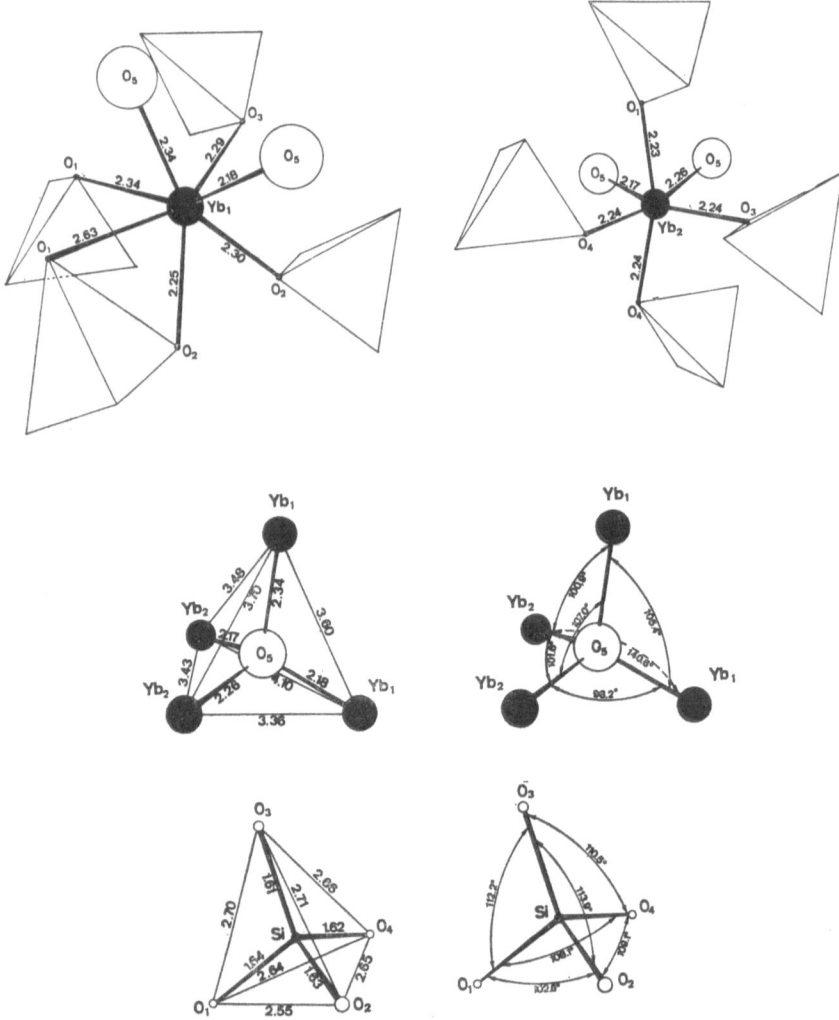

Fig. 6. Interatomic distances and angles in Yb$_2$[SiO$_4$[O

dinated Yb(2) cation. In this structure type, too, a compensating relationship can be observed between bond lengths in the (Yb—O$_{6,7}$) and the
(SiO$_4$) coordination polyhedra: the shortest Yb—O distances correspond
to the longest Si—O bonds present in the structure. The degree of
distortion in the (SiO$_4$)tetrahedron with an average Si—O distance of
1.63 Å is, however, appreciably less than that observed in the structure
of Gd$_2$(SiO$_4$)O (see Figs. 4 and 6).

2.2. Oxyapatite Compounds

Complete series of isostructural apatite-like compounds $RE_{9.33} \square_{0.67}$ $(SiO_4)_6O_2$ and Alk $RE_9(SiO_4)_6O_2$ are known, with RE ranging from La to Lu (43) (Table 4). Some other mixed cation apatites, $M_2RE_8(SiO_4)_6O_2$ with M:Ba,Sr,Ca,Mg,Pb,Cd,Mn, have been found to be restricted to certain ranges of rare earths and to depend on the size of the substituted divalent cation M (44—46). With the apatite structure we have the only set of isostructural rare earth silicate compounds which comprises the complete lanthanide series. For this reason, the oxyapatites will be discussed in more detail than most of the other rare-earth silicate structures.

Table 4. *Unit cell dimensions of compounds* $RE_{9.33} \square_{0.67}(SiO_4)_6O_2$, $LiRE_9(SiO_4)_6O_2$ *and* $NaRE_9(SiO_4)_6O_2$ *with RE:La* → *Lu showing the apatite structure (space group* $P6_3/m, Z = 1$). *Refined values from Guinier powder data have e.s.d.'s of* $\sigma(a_0) = 0.002$ Å *and* $\sigma(c_0) = 0.001$ Å

RE	$RE_{9.33} \square_{0.67}[SiO_4]_6O_2$			$LiRE_9[SiO_4]_6O_2$			$NaRE_9[SiO_4]_6O_2$		
	a_0[Å]	c_0[Å]	V[Å³]	a_0[Å]	c_0[Å]	V[Å³]	a_0[Å]	c_0[Å]	V[Å³]
La	9.713	7.194	587.8	9.681	7.160	581.2	9.687	7.180	583.5
Ce	9.657	7.121	575.1	9.623	7.091	568.7	9.628	7.117	571.3
Pr	9.607	7.073	565.3	9.575	7.040	558.8	9.580	7.080	562.6
Nd	9.563	7.029	556.9	9.529	6.994	550.1	9.535	7.027	553.4
Sm	9.493	6.946	542.0	9.464	6.918	536.6	9.472	6.943	539.6
Eu	9.472	6.905	536.6	9.437	6.876	530.3	9.456	6.912	535.2
Gd	9.431	6.873	529.4	9.413	6.852	525.8	9.419	6.878	528.4
Tb	9.401	6.825	522.4	9.381	6.803	522.0	9.390	6.840	522.2
Dy	9.373	6.784	516.2	9.362	6.769	513.9	9.362	6.800	516.2
Ho	9.346	6.744	510.3	9.337	6.736	508.6	9.337	6.760	510.4
Er	9.324	6.686	503.4	9.316	6.696	503.3	9.321	6.728	506.2
Tm	9.300	6.666	499.3	9.301	6.672	499.8	9.310	6.688	502.0
Yb	9.275	6.636	494.4	9.270	6.637	493.8	9.300	6.661	498.8
Lu	9.260	6.621	491.6	9.265	6.615	490.9	9.290	6.635	495.9

The following discussion of oxyapatite-like compounds is based mainly on two recent studies of oxyapatite compounds (43, 44), both of which incorporate the results of structural investigations carried out with La, Sm (11) and Gd crystals (47). The structural analysis confirmed a cation-deficient type of silicate apatite $RE_{9.33} \square_{0.67}(SiO_4)_6O_2$. A comprehensive survey of the relevant literature on apatite is presented in (44) and will therefore be omitted here; however, it will be referred to in cases of particular interest during the discussion of the individual

113

data. It seems worthwhile, however, to give the general impression from the numerous papers published recently on multivalent charge-couple substitution in apatites. It is tempting during synthesis experiments to establish the chemical formula of a new apatite-like compound from the overall chemical analysis, or even from the composition of the starting material as soon as the X-ray examination provides the apatite diffraction pattern. The temptation is especially strong because the apatite structure is known to be extremely tolerant to any type of charge-coupled cation and anion substitution, as well as to cation and probably also anion deficiency. It should be realized, however, that the stoichiometry of apatite is extremely complicated and that, because solid solutions and mixed phases of different degrees of crystallinity may form, reliable data can only be obtained from the analysis of single crystals.

Data on cation substitution will be presented in the following chapter, but only if they are likely to be in agreement with the given requirement. The structural interpretation of the crystal data on mixed cation apatites will be of the preliminary state. The rather anisotropic response of the unit cell dimensions suggests that cation distribution on the $(4f)$ and $(6h)$lattice sites may vary in many cases. However, most of these conclusions have to be confirmed by structural analysis.

2.2.1. Binary Compounds $RE_{9.33} \square_{0.67}(SiO_4)_6O_2$

Compounds $7\,RE_2O_3 \cdot 9\,SiO_2$ were shown to crystallize with the apatite structure of space group $P6_3/m$. The cation-deficient oxyapatite structure was established by structural analysis for La, Sm (11) and confirmed also for the Gd analogue (47). Taking into account all atoms per unit cell, the nature of this structure might best be described in terms of the formula $^{IX}(RE_{3.33} \square_{0.67})\,^{VII}RE_6 (^{IV}Si\,^{IV}O_4)_6\,^{III}O_2$, which makes special allowance for the coordination around each atom (Roman numbers above the elements give the coordination number, CN). The

Table 5. *Atomic parameters of $Gd_{9.33} \square_{0.67}(SiO_4)_6O_2$*

Atom	x	y	z	$B(\text{Å}^2)$
Gd(1)	0.24001(7)	0.23321(7)	0.75	0.67
Gd(2)	0.66666	0.33333	0.0	0.76
Si	0.4001 (3)	0.3721 (3)	0.25	0.71
O(1)	0.3178 (7)	0.4872 (7)	0.25	1.00
O(2)	0.6002 (7)	0.4740 (7)	0.25	1.10
O(3)	0.3418 (5)	0.2497 (5)	0.0575(1)	1.40
O(4)	0.0	0.0	0.25	1.09

After Ref. (47).

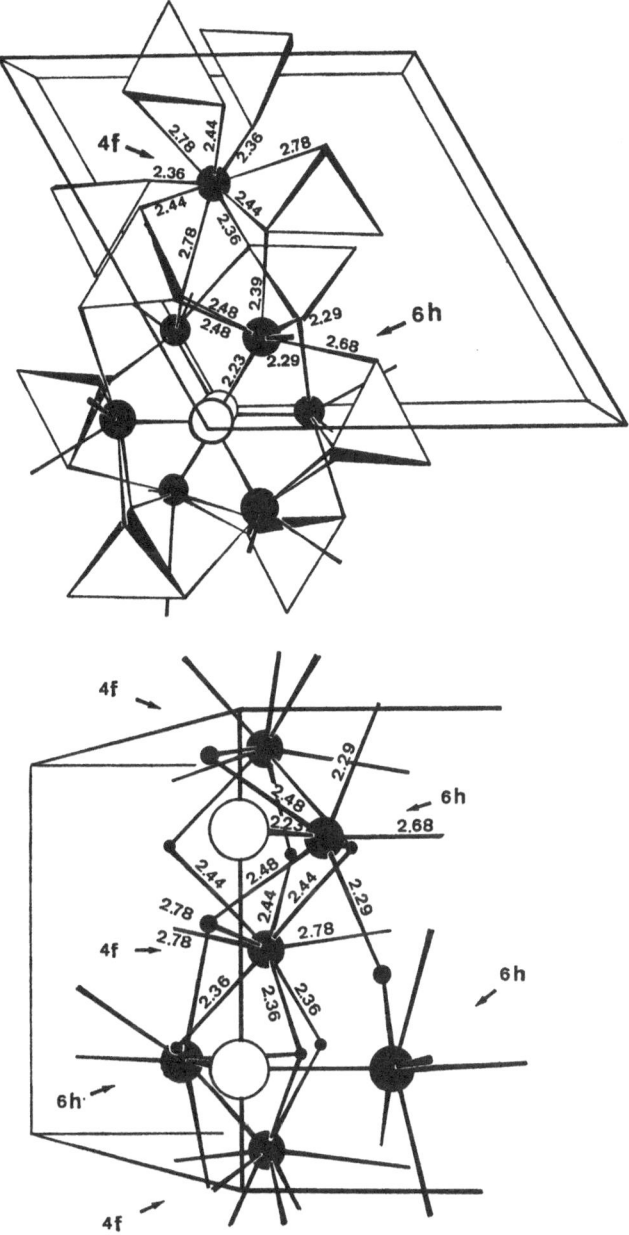

Fig. 7. Perspective view of the structure of rare-earth silicate oxyapatites $(RE_{3.33}\,\square_{0.67})RE_6(SiO_4)_6O_2$ along [OOl] and [lOO]. Interatomic distances from data on the Gd analogue (47) with e.s.d.'s ranging from 0.002 Å to 0.008 Å. Black balls: rare earths in the special positions $(4f)$ and $(6h)$, white balls: 'free', not silicon-bonded oxygen, solid (SiO_4) tetrahedra

main structural features are shown in Figs. 7 and 8, which are based on data for the Gd analogue. This structure was refined by means of three-dimensional intensity data to a residual $R = 6.5\%$ (47). The atomic parameters, corresponding to cell dimensions $a = 9.431(2)$ Å and $c = 6.873(1)$ Å, are listed in Table 5.

The refinement of the multiplicity parameter of Gd(2) has proved that the ($4f$)positions are occupied by $3\frac{1}{3}$ rare earth cations (RE^{3+}) only. Thus, $2/3$ 'cation holes' per cell might be considered to be statistically distributed on the ($4f$)lattice sites. This special position is surrounded by 9 oxygens, all silicon-bonded (Fig. 8). With a mean Gd—O distance of 2.53 Å it presents the larger polyhedral volume as compared to the seven-fold coordinated ($6h$)site with the average Gd—O distance of 2.41 Å. This relation will be of interest when we come to discuss the distribution of cations different sizes in mixed-cation apatites. Another interesting detail concerns the environment of the cation on the ($6h$) position in the main channel of the structure. This cation shows an extremely short distance of 2.23 Å to the 'free' oxygen in the special position $0,0,\frac{1}{4}$; $0,0,\frac{3}{4}$ on the 6_3 axis. This distance, which is considerably less than the sum of the ionic radii (2.39 Å), shows even more clearly than in the structures of $RE_2(SiO_4)O$ the strong polarizing forces exerted by the rare-earth cation on the oxygen which is not silicon-bonded.

The apatite-like compounds $7\ RE_2O_3 \cdot 9\ SiO_2$ are unique amongst all the RE silicates because they constitute the only isostructural group to include the complete series of rare earths. They are therefore an excellent set of compounds for testing lanthanide periodicity in terms of variation of the unit cell dimensions. To obtain high-precision crystal data, as shown in Table 4, compounds $RE_{9.33} \square_{0.67}(SiO_4)_6O_2$ were prepared by solid-state reaction from the oxides (43). Without the employment of any flux, step-by-step heating experiments (1000–1900 °C) were carried out in an attempt to obtain the purest phases on the basis of their powder diffraction patterns. Tb, Dy, and Ho specimens were always found to be accompanied by 10–20 Mol% of the disilicate polymorphs type B,D,E (27). The $7\ RE_2O_3 : 9\ SiO_2$ mixtures of the smaller rare earths (Ho), Er,Tm,Yb,Lu (Ho:T > 1550 °C), on the other hand, showed almost less than 20 Mol% of the apatite phase but corresponding $RE_2(SiO_4)O$ (39) and polymorphic disilicate phases type C and D (27), according to the reaction $5\{7\ RE_2O_3 + 9\ SiO_2\} \rightarrow 3\{5\ RE_2 (SiO_4)O + 2\ RE_2(Si_2O_7)\} + RE_{9.33} \square_{0.67}(SiO_4)_6O_2$. This indicates the metastable character of the apatite structure for the smallest rare earths. The large overlapping in the powder diffraction patterns of these mixtures of individual phases might be considered to be responsible for the lack of data on apatite compounds of the smaller rare earths reported in (81).

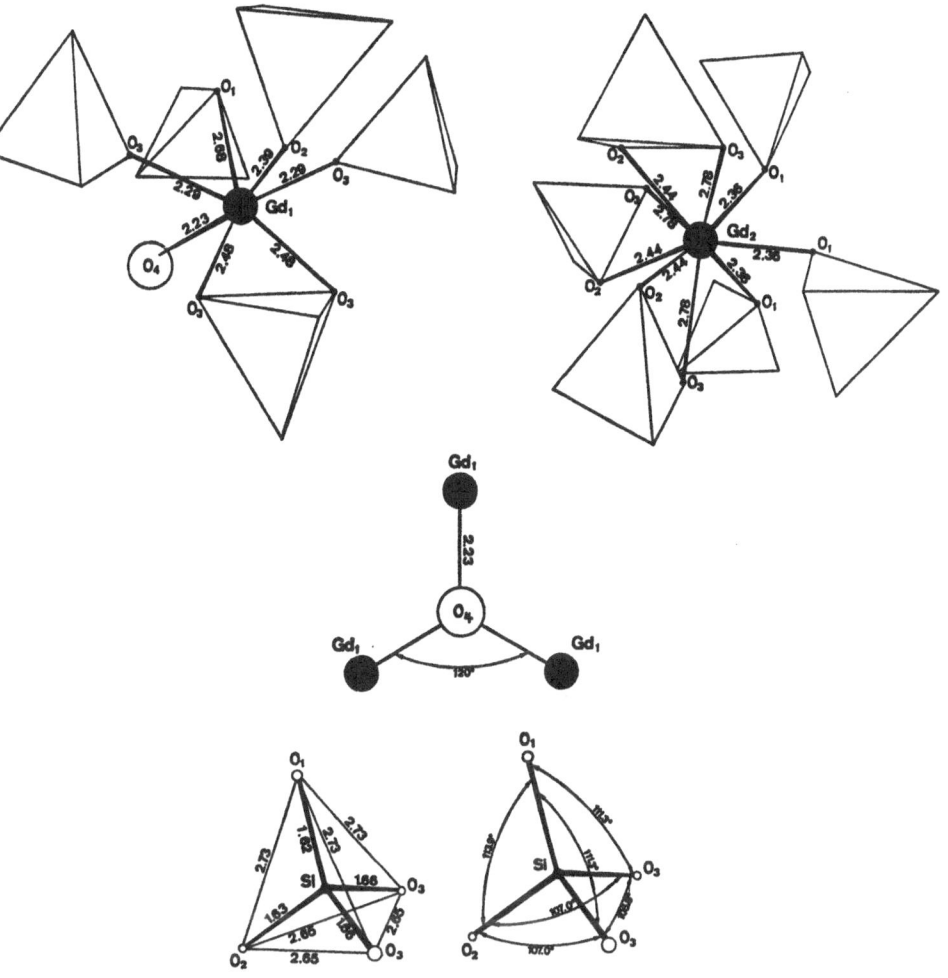

Fig. 8. Cation-oxygen coordination in the silicate-oxyapatite $(Gd_{3.33} \square_{0.67})Gd_6$-$(SiO_4)_6O_2$

The microprobe data for single crystals taken from $10\mu - 200\mu$ particles of all the 7:9 compounds showed variations of 0.2% about the theoretical RE:Si atomic ratio of 1.555. One exception was the Eu apatite phase. From the ratio Eu:Si = 1.570 ± 0.002 and the corresponding distinct value of the c_0 unit cell dimension, an appreciable amount of Eu^{2+} must be present in the structure. Probably it represents solid solution between the pure trivalent Eu apatite and the 'mixed-valency' compound $Eu_2^{2+} Eu_8^{3+}(SiO_4)_6O_2$. The latter was also synthesized from a 5:6 oxide mixture under reducing atmosphere (formiergas H_2/N_2).

117

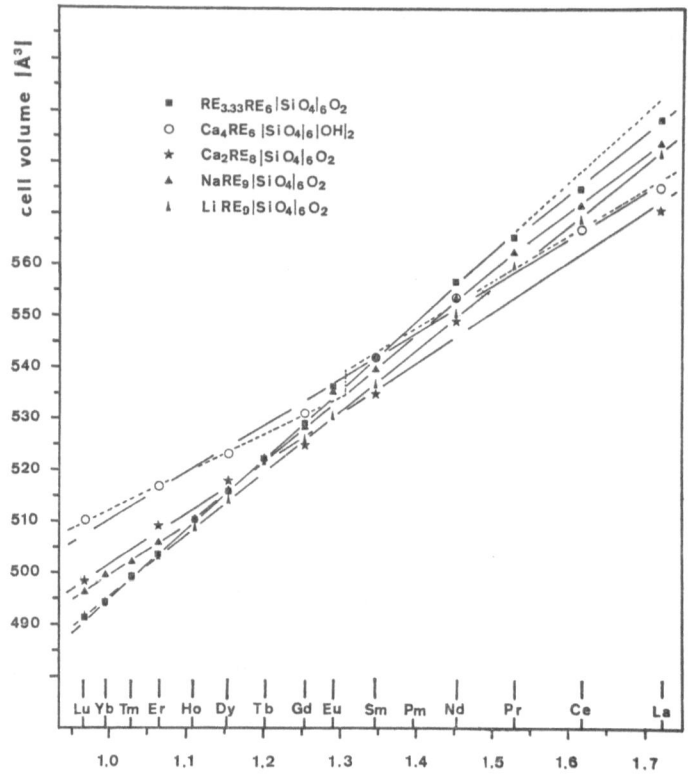

Fig. 9. Cell volume V vs. r^3 (RE^{3+} ionic radii) of rare-earth silicate apatites showing different mixed cation ratios 1:9, 2:8, 4:6 as compared to the binary cation-deficient compound $(RE_{3.33} \square_{0.67}) RE_6 (SiO_4)_6 O_2$. Data of $Ca_2RE_8(SiO_4)_6O_2$ and $Ca_4RE_6(SiO_4)_6O_2$ from (44). E.s.d.'s of the cell volume are = height of the symbols

The experiment was controlled in a simultaneous TGA/DTA run. The partial reduction of $Eu^{3+} \rightarrow Eu^{2+}$, corresponding to -0.78% loss of the total starting weight, occurred at 1050 °C:

$$5 \, Eu_3O_3 + 6 \, SiO_2 \longrightarrow Eu_2^{2+} Eu_8^{3+}(SiO_4)_6O_2 + 0.5 \, O_2 \uparrow.$$

The microprobe examination carried out on crystals ~ 0.05 mm in diameter gave $RE:Si = 1.661 \pm 0.002$, which agrees quite well with the theoretical value of 1.666 for the given formula.

To study a possible effect of the lanthanide periodicity on the apatite structure, plots were made of unit-cell volumes vs. ionic radii, $r^3(RE^{3+})$, and plots of cell dimensions a_0 and c_0 vs. ionic radii, $r(RE^{3+})$. To obtain these plots, the experimental values of RE—O bond lengths of the Gd apatite structure served to determine the effective ionic radii of the RE^{3+} cations in this structure type. Since the effective radius of any ion depends on the coordination numbers of both cation and anion, the mean RE^{3+} radii in the apatite structure were evaluated from the seven- and ninefold oxygen-coordinated Gd cations in the following way. The refined values of the Gd—O distances show e.s.d.'s of the order of 0.008 Å, corresponding to the overall R value for the structure of 6.5% (47). The average Gd—O distances in the '($4f$)polyhedron' with CN 9 and in the '($6h$)polyhedron' with CN 7 are 2.53 Å and 2.40 Å, respectively. The 24 silicon-bonded oxygens per unit cell are surrounded by four cations, whereas the two 'free' oxygens have three rare-earth neighbours. Thus, with the average oxygen radius of 1.38 Å (54), the effective Gd^{3+}—($4f$) and Gd^{3+}—($6h$) ionic radii are 1.15 Å and 1.02 Å. From these data the weighted mean value of the Gd^{3+} ionic radius in the apatite structure was determined as 1.08 Å.

Fig. 9 is the plot of apatite cell volumes vs. $r^3(RE^{3+})$. The relative scale of the trivalent rare-earth radii was maintained from the known RE^{3+} series with CN 6 (48). The complete set was shifted to the values $r(Gd^{3+}) = 1.08$ Å and $r^3(Gd^{3+}) = 1.26$ Å, respectively. The correlation is quite linear with small deviations from $r^3(La^{3+})$ and $r^3(Ce^{3+})$. These should be corrected to slightly smaller values corresponding to ionic radii of 1.19 Å and 1.17 Å. The special situation of the Eu analogue, which is also off the straight line, probably arises from mixed valence Eu^{2+}—Eu^{3+}, as already suggested from the chemical analysis on this compound. The plots of the individual cell dimensions of the apatite compounds vs. ionic radii (RE^{3+}) revealed some interesting details about the response of the structure to rare earth substitution. As shown in Fig. 10, the shrinkage of the c_0 dimension of the apatite cell is much more pronounced than that of the a_0 dimension. As compared with the La analogue, the structure of the Lu apatite shows a shrinkage of 9.6% along the c direction as compared to 4.7% along the a_0 axis. This is best understood in terms of the Gd—O interatomic distances given in Fig. 7. The shortest Gd—O bonds are directed essentially along [001] from both the cation positions ($4f$) and ($6h$).

Another interesting feature of Fig. 10 is that the changes in slope and intersections occur at different positions along a_0 and c_0. The c_0 axis indicates the three groups Lu—Er, Ho—Nd, and Nd—La, as commonly known in rare-earth chemistry, whereas a_0 shows only one intersection between Gd and Eu. This feature, for which either a change in occu-

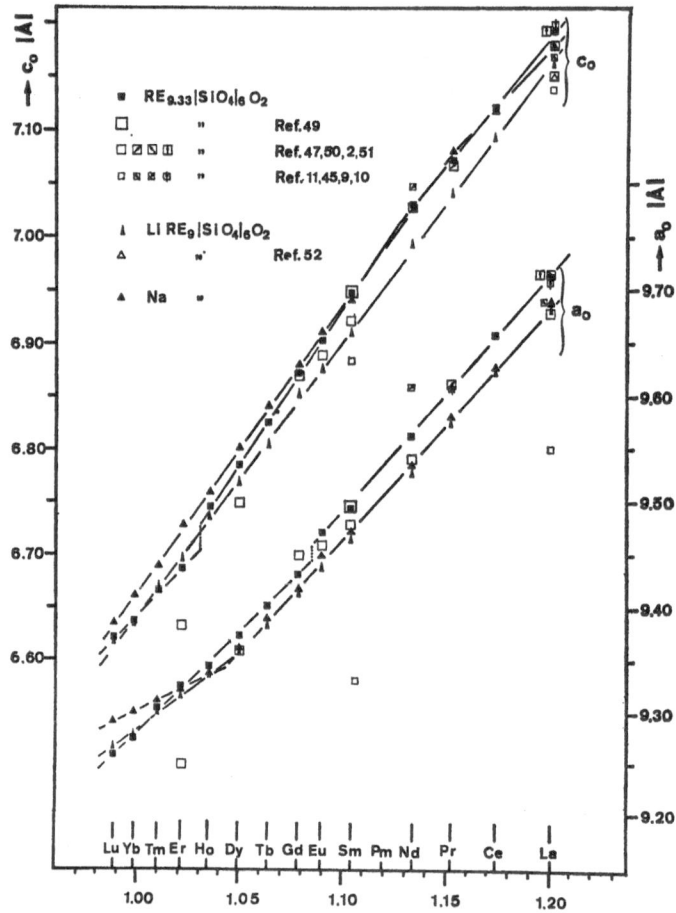

Fig. 10. Cell dimensions c_0 and a_0 vs. r (RE^{3+}) of binary rare-earth and alkali rare-earth silicate oxyapatites. Data from different sources (Ref. n) and own experiments (Table 4). E.s.d.'s are unknown for values from Refs. (*2, 9, 10, 11, 45, 52*), e.s.d.'s of 0.003 Å to 0.002 Å are given in Refs. (*9, 49, 51*), of ~0.01 Å in Ref. (*47*)

pancy of the (4*f*)cation position or a change in the electronic structure of the rare earth ions could be responsible, needs further clarification by structural analysis.

Fig. 10 includes some additional data from the literature concerning the binary compounds $7\,RE_3O_3 \cdot 9\,SiO_2$ and the isostructural alkali rare-earth apatites. Most of the crystal data reported show poor agreement of the cell dimensions for the same compound. The differences might depend on the X-ray diffraction technique used, and/or the method of evaluating of the experimental data. However, in most cases e.s.d.'s are not even given. On the other hand, traces of the flux employed (e.g. alkali fluorides, bismuth oxide) could be accommodated into the structure during the synthesis and thus be responsible for the deviation of the cell dimensions; there could also be admixtures of other rare-earth ions in the particular compound. The largest deviations from our own data in Fig. 10 concern the La and Sm analogues from Ref. (11). The values for a_0 and c_0 are smaller by 0.17 Å and 0.07 Å, which is far beyond the probable e.s.d. (not given). There is no reasonable explanation for this, especially since the structural analysis carried out with these crystals gave the same result of a cation-deficient ($4f$)position as described for the Gd apatite (47). Since the data from (47) show a better general agreement with our own data, the structural data for La and Sm were not used in the evaluation of the effective ionic radii. The differences in agreement with the various data from the same source as the Gd analogue (47) might be due to the different methods employed for the synthesis of the individual crystals.

The wide range of stability of the apatite phase in the binary systems RE_2O_3–SiO_2 is well illustrated by the fact that values from Refs. (2) and (10) were actually given for compositions $1\ RE_2O_3 \cdot 1\ SiO_2$ and from Refs. (9), (18), (22), (45), (49) for $2\ RE_2O_3 \cdot 3\ SiO_2$. Furthermore, there are no reliable experimental results to establish the cation-anion-deficient apatite type $La_{8.67}(SiO_4)_6O$, as described in Ref. (51); the cell dimensions given for this are almost identical with those of the common $7\ La_2O_3 \cdot 9\ SiO_2$ type (Fig. 10). In the ternary systems $Li_2O \cdot RE_2O_3 \cdot SiO_2$ the alkali oxyapatite also seems to exhibit a large area of stability. This is suggested, at least, by the crystal data $a = 9.69$ Å, $c = 7.15$ Å given for $LiLaSiO_4$ (52), which indicate the known apatite phase $LiLa_9(SiO_4)_6O_2$ rather than 'a new family of lanthanide compounds' with $RE = La \rightarrow Dy$. Some attempts to identify the apatite cell dimensions from the unindexed powder data given in (3—7) were not successful because of the poor resolution and the obvious mixed-phase character of most of the diffraction patterns listed.

2.2.2. Mixed-Cation Oxyapatites

Mixed-cation silicate apatites $Alk\ RE_9(SiO_4)_6O_2$ and $M_2RE_8(SiO_4)_6O_2$ with Alk: Li,Na, and M: Mg,Ca,Sr,Ba,Pb,Cd could be prepared also by solid-state reaction starting from the oxides or corresponding salts of the individual elements (42—46). It has been suggested that Alk RE apatites exist for all the rare earths (44), and they should present an excellent set of compounds for comparison with the binary rare-earth apatites. Two complete sets of compounds $LiRE_9(SiO_4)_6O_2$ with RE: La, ... Lu were synthesized in order to provide the high-precision crystal data listed in Table 4. The composition of these crystals and of some samples of compounds $M_2RE_8(SiO_4)_6O_2$ with $M:Mg,Ca,Sr,Ba$, which will be introduced later, was confirmed by microprobe analysis. Singel-crystal X-ray examination indicated space group $P6_3/m(P6_3)$ for these ternary rare-earth silicate oxyapatites, too. Reflections $[00l]$ and $[hkl]$ with $l \neq 2n$ were found to be absent on $MoK\alpha$-precession photographs in the case of $[hkl]$, if $h - k = 3n$.

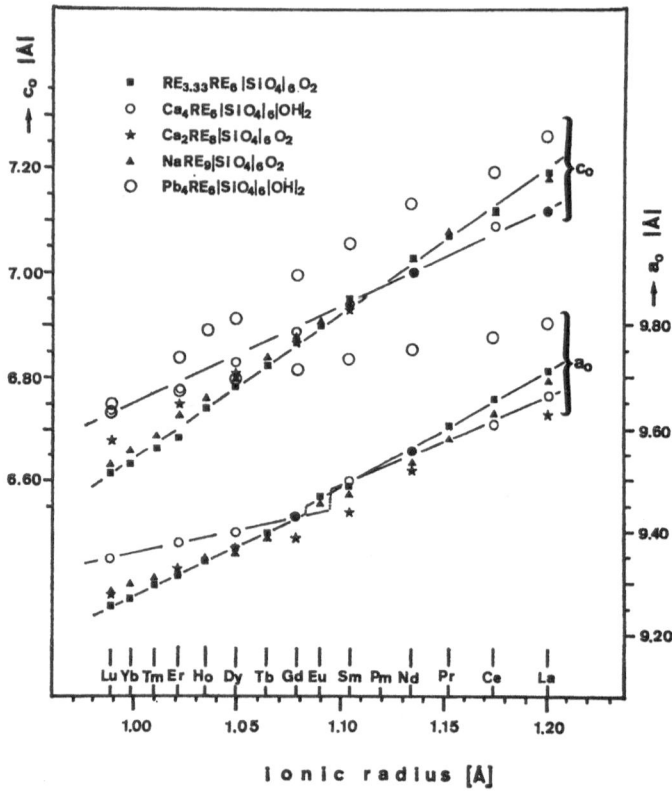

Fig. 11. Survey of cell dimensions a_0 and c_0 vs. r (RE^{3+}) of some representative RE silicate apatites showing different mixed cation ratios $1:9$, $2:8$, $4:6$ as compared to the binary cation-deficient apatite $(RE_{3.33} \square_{0.67})RE_6(SiO_4)_6O_2$. Data for Ca and Pb analogues from Ref. (44) with e.s.d.'s of ~0.01 Å (private communication)

In the following chapters the crystal data of mixed-cation oxyapatites of different stoichiometry will be discussed in relation to the cation distribution on the $(4f)$ and $(6h)$ lattice sites and the lanthanide periodicity. The general impression which emerges from the data in Figs. 9 to 11 (which also include compounds $M_4RE_6(SiO_4)_6(OH)_2$ in order to emphasize the correlation) is this: The declining slope in the ratio apatite cell volume vs. $r^3(RE^{3+})$ (Fig. 9) of compounds with an increasing proportion of foreign (not RE) cations relative to the pure binary rare-earth apatites, signifies the decreasing influence of the rare-earth periodicity on the structure. The distinct anisotropic response of apatite cell dimensions, as observed in the binary apatites (Fig. 10), is modified or vanishes completely with mixed-cation substitution. This impression will be substantiated in chapters 2.2.2.1 and 2.2.2.2.

2.2.2.1. Ternary Compounds $RE_9Alk(SiO_4)_6O_2$. Compounds $LiRE_9$-$(SiO_4)_6O_2$ show an even more ideal linear relationship between cell volume and cubic RE^{3+} radius (see Fig. 9) than the pure rare-earth apatites. The sodium analogues, however, exhibit a change in slope beyond Ho. It is interesting to see from Fig. 10 that the pronounced breaks in the curves a_0 and c_0 vs. r of the pure rare-earth apatites vanish on the introduction of a single small, but different, cation per cell. It appears to be due to this moderator function of the alkali atom that the c_0 dimension of the $LiRE_9$ apatites varies linearly with r along the complete rare-earth series; no subdivision shows up along c_0. A slight change in the slope of a_0, however, is indicated at Ho. This change in the slope at Ho of a_0 is even more pronounced with the compounds $NaRE_9(SiO_4)_6O_2$, whereas their absolute a_0 values with the larger rare earths beyond Dy are about the same. For La to Gd, c_0 for the sodium analogues is about the same as for the pure rare-earth apatite, whereas the c_0 values of the smaller $NaRE_9$ apatites are significantly larger. Since the single alkali atom per cell formally substitutes for the combination $(^1/_3\,RE + ^2/_3\,\square)^{1+}$, it seems likely that the small alkali prefers the smaller space of the $(6h)$ position in a statistical distribution. This is suggested by the significant change in slope of a_0 vs. r in Fig. 10. It occurs at a place where the Na^{1+} radius becomes competitive with the rare earths around Ho—Dy. It is apparent from Fig. 7 that the variation of the a_0 axis should depend mostly on a corresponding substitution into the $(6h)$lattice sites rather than into the $(4f)$ ones. This is because the $(6h)$position has the tighter bonds in directions $[hkO]$ (2.23 Å, 2.39 Å, 2.68 Å) than the $(4f)$ cation with three 2.78 Å Gd—O distances.

2.2.2.2. Ternary Compounds $RE_8M_2(SiO_4)_6O_2$. In order to follow the correlation between cation substitution and variation of cell dimensions in 2:8 mixed-cation apatites, we bring forward crystal data for 4:8 hydroxy apatites from *(44)*. It is especially atractive to introduce the 4:6 apatite compounds into the discussion of crystal chemistry because their mixed-cation ratio exactly corresponds to the ratio of the symmetry-equivalent positions $(4f):(6h)$ per cell. Order-disorder on these lattice sites should thus show up more readily with these than with any of the other mixed-cation compounds. We would expect to see distinct effects from the variation of the cell dimensions on substitution of cations of different sizes into the $(4f)$ and $(6h)$ positions. The results as shown for the Ca,Sr,Ba,Pb analogues in Fig. 12 are less clearcut than would be anticipated from what is stated in Ref. (45). The authors concluded from a comparison of four relative X-ray powder intensities that the distribution of the cations in $(4f)$ and $(6h)$ lattice sites, was random.

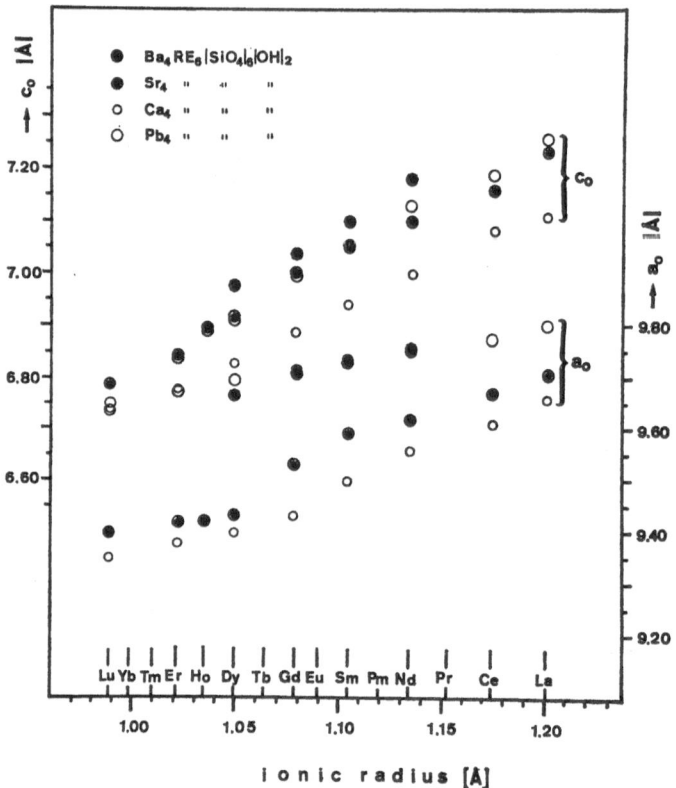

Fig. 12. Cell dimensions a_0 and c_0 vs r (RE^{3+}) of RE silicate hydroxyapatites with mixed-cation ratio 4:9. Data from Ref. (44), with e.s.d.'s. 0.01 Å (private communication)

In compounds $M_4RE_6(SiO_4)_6(OH)_2$ with M: Mg,Ca,Sr,Ba,Pb,Cd,Mn (44) of mixed-cation ratio 4:6, the two 'free' oxygens located in the main channels of the oxyapatite structure (Fig. 7) are replaced by OH groups. However, the presence of OH groups apparently only slightly changes the interatomic distances within the $(6h)$ coordination polyhedron as compared to the values known from the Gd oxyapatite (46). This is suggested by the corresponding data on the structure of synthetic Ca hydroxyapatite (53).

The standard deviation of the cell dimensions given in (44) (~ 0.01 Å: private communication) is larger than that for our own data (≤ 0.002 Å). Therefore, the scale of the a_0 and c_0 values in Figs. 11—13 was changed as compared with Fig. 10 in order to give a more comprehensive representation. The following points concerning Fig. 12 seem worth mentioning. The largest type of M cation substituted is Ba^{2+} with $r = 1.39$ Å (CN 7) and $r = 1.47$ Å (CN 9), followed by Pb^{2+} with $r = 1.25$ Å and $r = 1.33$ Å, according to Ref. (54). The almost identical values of a_0 in the two analogous apatite structures and the significantly larger c_0 cell dimensions suggest that cations Ba^{2+} and Pb^{2+} are both in the $(4f)$-position, since $(4f)$ is known to be most sensitive in the c_0 direction. With the next smaller cation Sr^{2+}, the c_0 dimension of the unit cells is almost the same as for Pb^{2+}, whereas the a_0 values vary by 0.10 Å to 0.25 Å, the difference increasing with rare-earth atomic numbers. This indicates a change in cation distribution, with Sr^{2+} also present on the 'smaller' $(6h)$sites, considered mainly responsible for the response of the apatite structure along a_0. The random distribution of Sr^{2+} on $(6h)$ and $(4f)$lattice sites yields a similar slope for both the cell dimensions along the rare-earth series up to a critical point at $\sim$Tb. There the ionic radius of Sr^{2+} exceeds in size the progressively smaller RE^{3+} ions and thus blocks any further shrinkage of the '$(6h)$-sensitive' a_0 direction. The Ca analogue behaves similarly. The blocking of the variation in the a_0 axis thus appears to be responsible for the break in both the corresponding curves for V vs. r^3 in Fig. 9.

Oxyapatite compounds of mixed-cation ratio 2:8 show some analogy with those just discussed. Data for $M_2RE_8(SiO_4)_6O_2$ with M: Mg,Ca,Sr,Ba from (44) are presented in Fig. 13. The Mg and Ca analogues show a similar change in slope of a_0 at Gd, whereas their c_0 dimension varies smoothly. This means that Mg and Ca ions are probably also accommodated on $(6h)$sites for the reasons stated above. Substitution of the larger Sr^{2+} affects the structure differently. a_0 and c_0 values vary similarly over the range of rare earths La → Er, the slope slightly increasing with the rare-earth atomic numbers. Thus, a change in cation distribution seems indicated with $(2 Sr^{2+} + 2 RE^{3+})$ in the larger coordination of $(4f)$. This should also be true for the 2:8 analogues of Ba which exist, however, only for the larger rare earths La → Sm. This limit extends from La to Dy for the 4:6 analogues of the Ba hydroxyapatites. Fig. 13 includes some cell dimensions of various mixed (Sr,La) oxyapatites from Ref. (51) which describes two different types of deficient apatite structures, in addition to the known cation-deficient apatite $RE_{9.33}$ $\square_{0.67}(SiO_4)_6O_2$, by the formulas $Sr_2La_{6.67}(SiO_4)_6$, $Sr_3La_6(SiO_4)_6$, $Sr_4La_{5.33}(SiO_4)_6$ (↑) and $Sr_4La_6(SiO_4)_6O$ (↗). However, no reliable experimental data exist, so far to confirm these types of oxygen deficiency in the silicate apatite structure. The criticism concerning $La_{8.67}(SiO_4)_6O$ from the same source (51) applies to $La_{9.33}\square_{0.67}$-$(SiO_4)_6O_2$ too, since the cell dimensions given for the various Sr—La apatites indicate that only one type of compound is present, namely $Sr_2La_8(SiO_4)O_2$ (↖). The arguments are restricted by the limits of accuracy indicated for the data given.

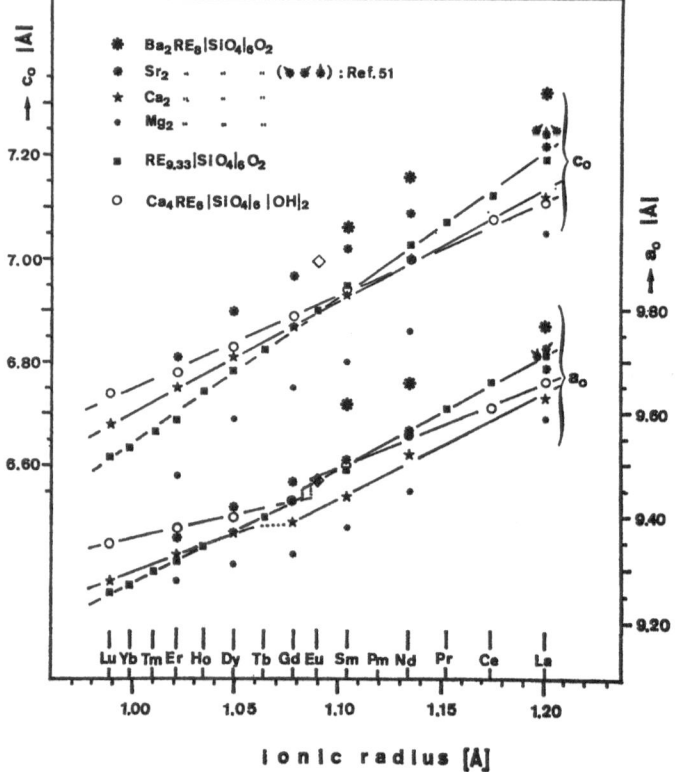

Fig. 13. Cell dimensions a_0 and c_0 vs. r (RE^{3+}) of rare-earth silicate oxyapatite with mixed-cation ratio 2:8 from Ref. (44) with e.s.d.'s ~0.01 Å (private communication) compared with the binary rare-earth silicates (own data) and the Ca hydroxyapatites with mixed-cation ratio 4:6, also from (44). The rhombus-like symbol represents cell dimensions of $Eu_2^{2+} Eu_8^{3+} (SiO_4)_6O_2$

2.3. Mixed-Cation Compounds RE $M^{1+}(SiO_4)$

Three different types of structure are known for ternary compounds RE Na(SiO_4). They were obtained by hydrothermal synthesis in the systems Na_2O–RE_2O_3–SiO_2–H_2O largely at a pressure of 1000 atm. at 450 °C (29–31). The order of their occurrence along the rare-earth series is given in Table 6 together with some crystal data.

Table 6. *Distribution of structure types in RENa(SiO₄) compounds*

RE	Structure Type	Cell Dimensions	Space Group	Z	$\varrho_{exp.}$ (g/cm³)
(La) Ce Pr Nd	Na RE A	$a = 20.00$ Å, b $= 9.28$ Å, $c = 5.45$ Å	Pna2₁	12	5.1
Nd Sm (Eu) Gd (Tb) Dy Ho	Na RE B	$a = 11.84$ Å, $c = 5.45$ Å	I4/m	8	4.7
Er Yb Tm Lu	Na RE C	$a = 5.09$ Å, b $= 10.96$ Å, $c = 6.35$ Å	Pbn2₁	4	

Phase formation during the hydrothermal experiments, which also yielded compounds $Na_3RE(Si_2O_7)$, is apparently controlled mainly by the Na(OH) reduction in each run. Analogous potassium-containing compounds have not been obtained during the corresponding experiments. However, an isostructural group of Li-containing compounds seems to exist for the smaller rare earths with the orthorhombic structure of type C NaRE(SiO_4) (52).

Structurally all compounds of type NaRE(SiO_4) are related to polymorphic $Ca_2(SiO_4)$. This supports the general feature, often described in geochemistry, of a $Na^{1+}RE^{3+} \leftrightarrow 2\,Ca^{2+}$ substitution. This is found in many minerals (see also Appendix I).

J. Felsche

2.3.1. Compounds (Ce, ... Nd)Na(SiO₄)

The crystals of space group $Pna2_1$ have a relatively large unit cell, containing $Z = 12$ units RE Na(SiO₄) (Table 6). The structure of NaNd(SiO₄) was determined from three-dimensional X-ray diffraction film data and refined to an R value of 13% (55). Individual temperatures were not reported in (55).

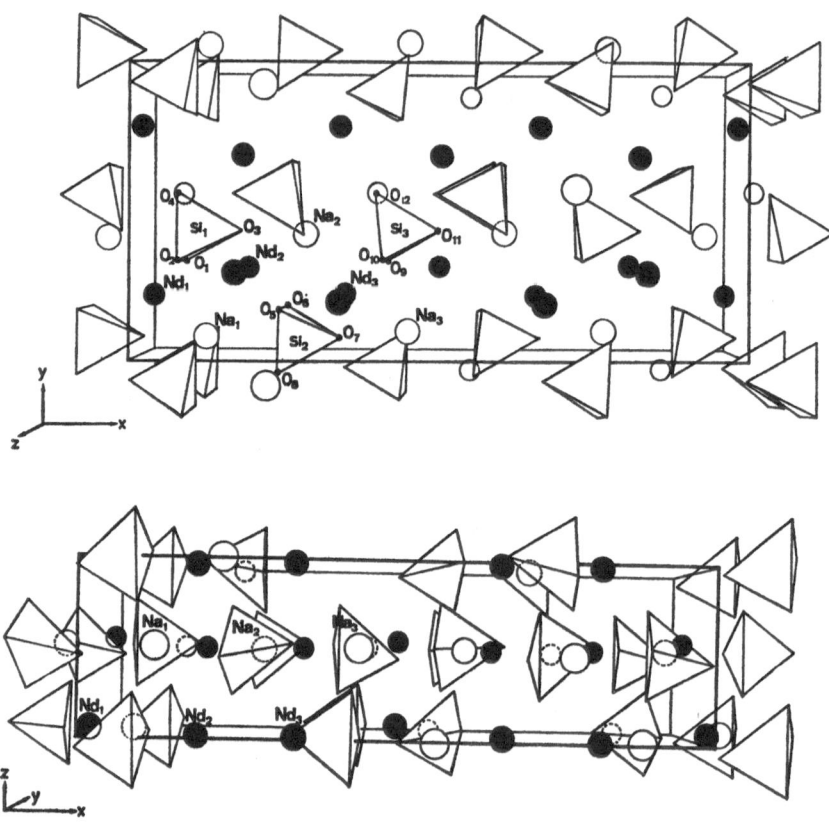

Fig. 14. Crystal structure of NaNd(SiO₄)

$\rightarrow$

Fig. 15. Cation-oxygen coordination in NaNd(SiO₄)

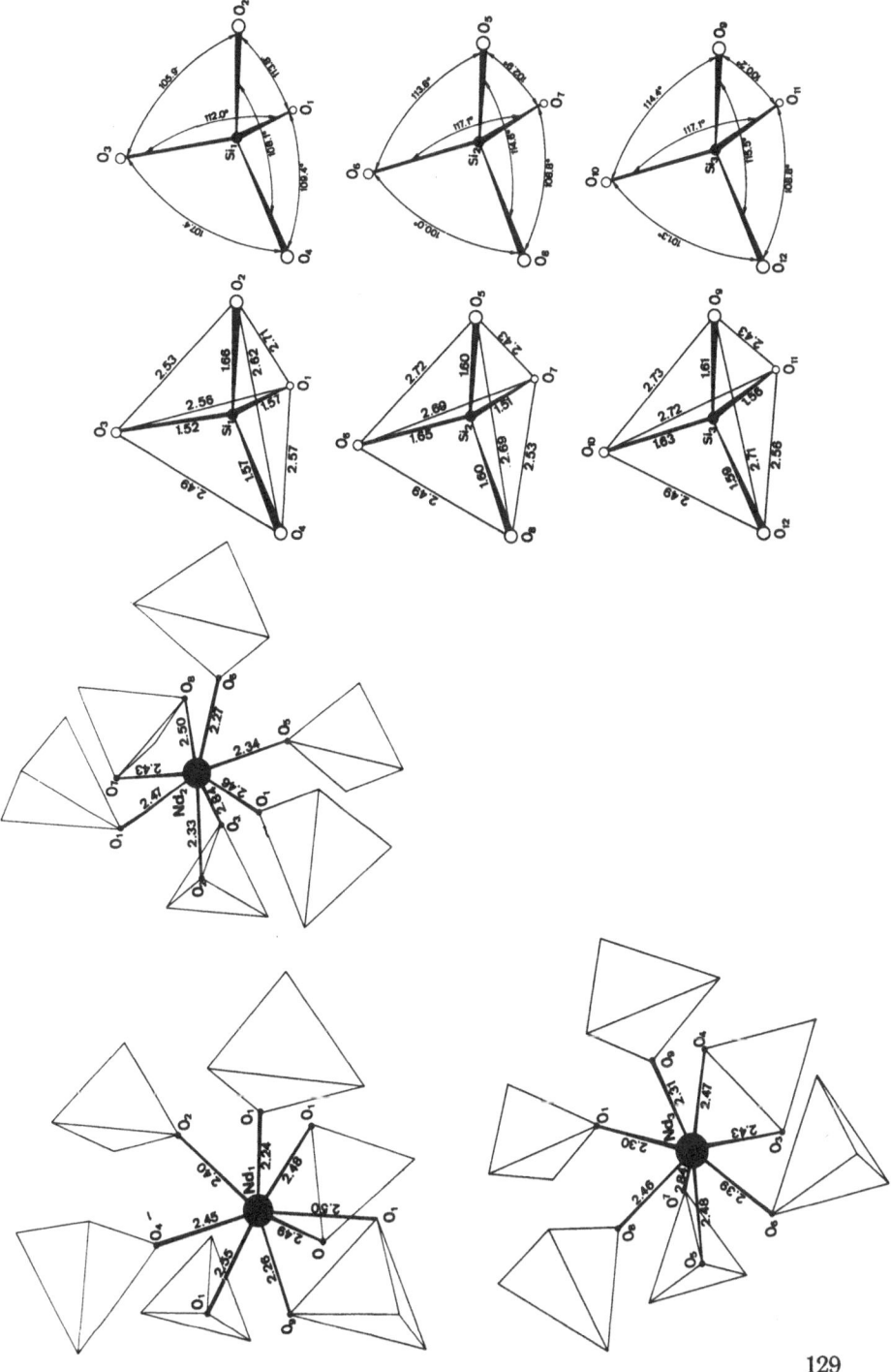

Table 7. *Atomic parameters of NaNd(SiO$_4$)*

Atom	x	y	z	Atom	x	y	z
Nd(1)	0.0	0.198	0.052	O (3)	0.166	0.437	0.533
Nd(2)	0.167	0.302	0.0	O (4)	0.442	0.070	0.017
Nd(3)	0.334	0.198	0.0	O (5)	0.235	0.166	0.737
Si(1)	0.091	0.416	0.517	O (6)	0.235	0.165	0.238
Si(2)	0.258	0.084	0.493	O (7)	0.332	0.063	0.533
Si(3)	0.424	0.416	0.500	O (8)	0.276	0.430	0.953
Na(1)	0.111	0.072	0.506	O (9)	0.402	0.333	0.253
Na(2)	0.278	0.428	0.483	O(10)	0.403	0.334	0.753
Na(3)	0.445	0.075	0.517	O(11)	0.0	0.064	0.450
O(1)	0.070	0.333	0.277	O(12)	0.109	0.070	0.040
O(2)	0.069	0.333	0.774				

After Ref. (*55*).

The structure is illustrated in Fig. 14: it contains olivine-like ribbons extending to [*100*]. These ribbons have a core of Nd polyhedra and edges of Na polyhedra and are linked by (SiO$_4$) tetrahedra and by shared corners. The three crystallographically independent Nd cations show an eightfold coordination in the shape of trigonal prisms with additional oxygens outside the midpoints of the three prism faces (Fig. 15). The structure of NaNd(SiO$_4$) is apparently not very stable since it shows extensive morphologic twinning. Another structural feature indicating considerable instability is the joining of Nd polyhedra into a strip by face-to-face contact. NaNd(SiO$_4$) forms a second modification of tetragonal symmetry. This structure is described in 2.3.2.

2.3.2. Compounds (Nd, ... Ho)Na(SiO$_4$)

The crystal structure of NaSm(SiO$_4$) represents the structural type of the medium-sized rare earths in compounds of composition NaRE(SiO$_4$). The crystal structure of space group I4/m has $Z = 8$ formula units of NaSm(SiO$_4$) in the unit cell with $a = 11.89$ Å and $c = 5.45$ Å. The structure was determined by 3-dimensional X-ray intensity film data and refined to an R value of 14%, which corresponds to standard deviations in the oxygen cation distances of about 0.03 Å. The atomic parameters without individual temperature factors are given in Table 8 after (*56*).

The main structural features are shown in Fig. 16. The structure appears to be determined by fourfold rings of Sm—O polyhedra of CN 6 arranged on two levels in the centered cell along [*001*]. The individual polyhedra are linked by their vertical edges and have the shape of fairly undistorted trigonal prisms. However, these fourfold rings of Sm—O prisms do not form a continuous three-dimensional spatial linkage.

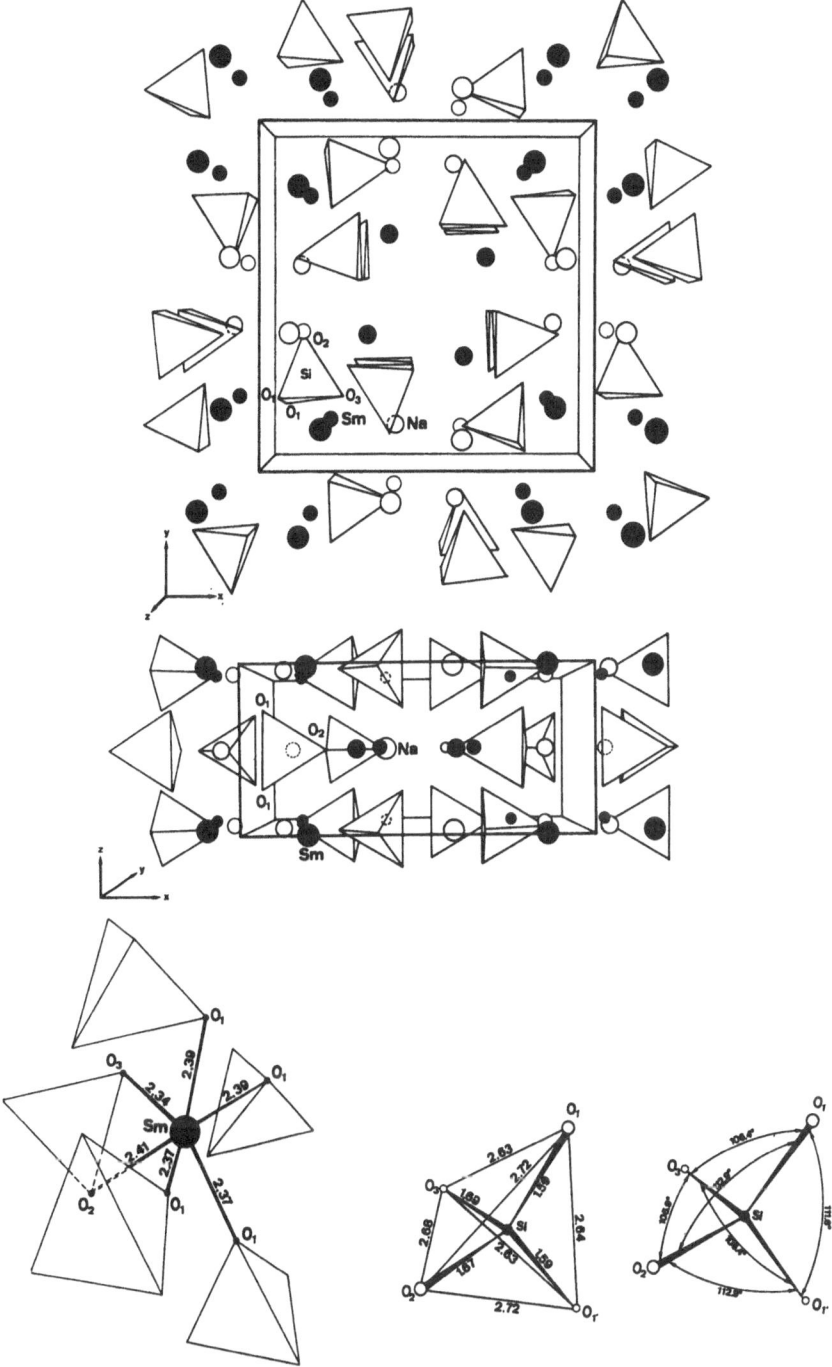

Fig. 16. Crystal structure of NaSm(SiO₄)

They are joined together by the single (SiO_4)tetrahedra. The (SiO_4) tetrahedra are closely associated with the Sm-O polyhedra, sharing three edges with three of them. The (SiO_4)tetrahedra also have a pair of common edges with two Na polyhedra which show sevenfold oxygen coordination and are located at two levels around the 4_2 axis. The presence of shaft-like channels in the structure prompted the suggestion that H_2O molecules might be found there (56). In fact, an electronic density peak was found during structural analysis on the three-dimensional Fourier synthesis at the origin of the cell, i.e. at the centre of the ring of Sm—O coordination polyhedra. However, the chemical analysis did not definitely confirm this suggestion.

Table 8. *Atomic parameters of NaSm(SiO4)*

Atom	x	y	z
Na	0.410	0.106	0.500
Sm	0.185	0.117	0.0
Si	0.102	0.249	0.500
O(1)	0.044	0.200	0.258
O(2)	0.389	0.110	0.0
O(3)	0.237	0.201	0.500

After Ref. (56).

2.3.3. Compounds (Er, ... Lu)Na(SiO4)

This structure type is represented by the data given for the analogous compound of NaY(SiO_4) (57). The orthorhombic crystal structure shows space group $Pbn2_1$ with $Z = 4$ formula units NaY(SiO_4) per unit cell. The structure was resolved by two-dimensional (*hkl*) and (*0kl*) X-ray film intensity data. The atomic parameters with individual B's, not given in (57), are listed in Table 9.

Table 9. *Atomic parameters of NaY(SiO4)*

Atom	x	y	z
Y	0.002	0.220	0.0
Na	0.500	0.010	0.254
Si	0.066	0.105	0.489
O(1)	0.225	0.162	0.705
O(2)	0.185	0.184	0.320
O(3)	0.349	0.457	0.478
O(4)	0.261	0.400	0.017

After Ref. (57).

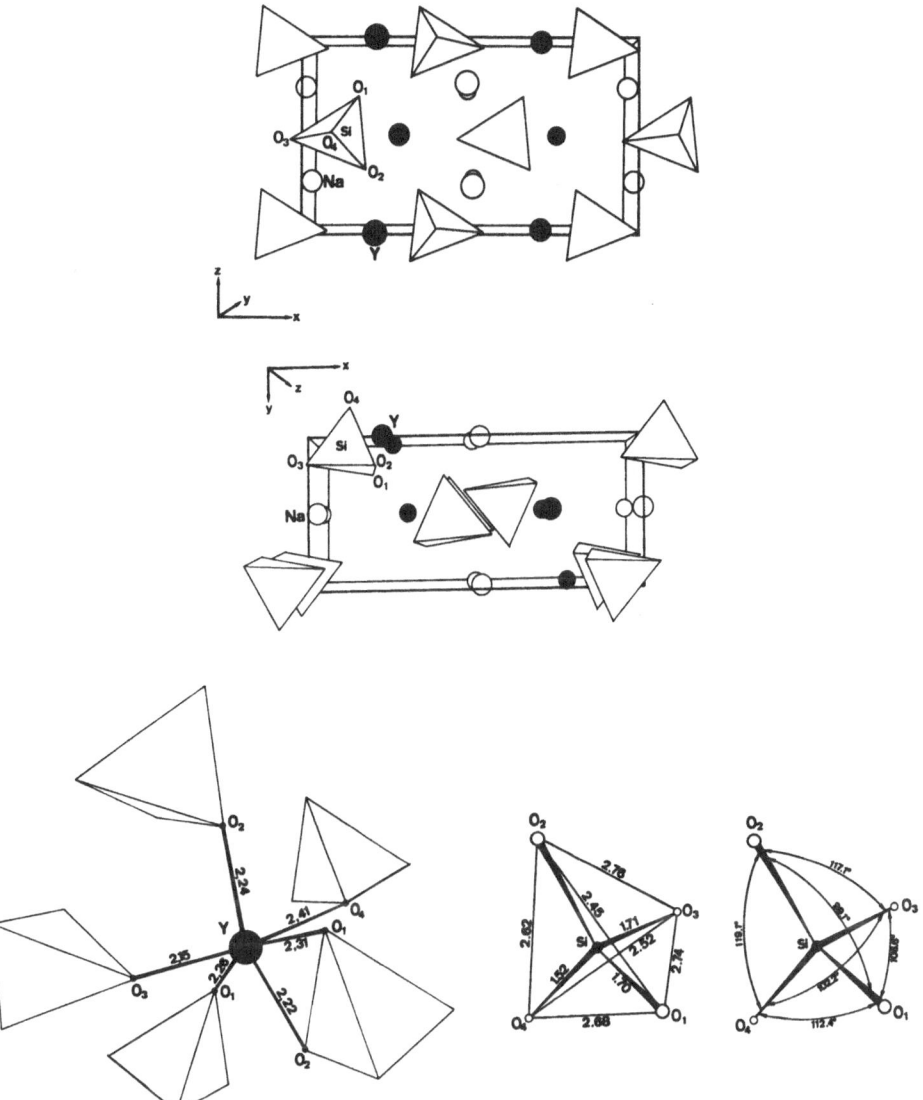

Fig. 17. Crystal structure of NaY(SiO$_4$)

The structure is illustrated in Fig. 17. This compound is isostructural with olivine-like γ-Ca$_2$(SiO$_4$). The two crystallographically independent Ca atoms are here replaced by the pair nearly equal in size, Na and Y. The strip pattern of Na/Y octahedra along [001] is comparable to the analogues in CaMg(SiO$_4$) (monticellite). Here, however, the open Na octahedra are located within the rods while the somewhat smaller Y ions lie in the projections. The reason for this apparently is that Na^{1+} differs considerably in charge from Y^{3+}, which means that the rare-earth cations must be as far apart as possible. Isostructural compounds of Li(Y,Ho ... Lu) (SiO$_4$) have been prepared by solid-state reaction of the corresponding oxide mixtures at 1050 °C (52). The cell dimensions given there for LiY(SiO$_4$) are $a = 4.94$ Å, $b = 10.68$ Å and $c = 6.29$ Å.

2.3.4. Stillwellite, Ce B(SiO$_4$)O

The crystal structure of stillwellite, a well-known rare-earth boron-silicate mineral, was determined by three-dimensional diffractometer single-crystal intensity data (58). The trigonal structure of space group P3$_1$ contains three formula units CeB(SiO$_4$)O in the unit cell of dimensions $a = 6.85$ Å, and $c = 6.70$ Å. The final R value in the refinement corresponding to the atomic parameters (given in Table 10) is 9.2% for (hko) and 12.8% for (okl) reflections.

Table 10. *Atomic parameters of CeB(SiO$_4$)O*

Atom	x	y	z
Ce	0.587	0.0	0.0
Si	0.585	0.0	0.500
B	0.113	0.0	0.973
O(1)	0.339	0.194	0.023
O(2)	0.195	0.339	0.310
O(3)	0.613	0.464	0.320
O(4)	0.464	0.614	0.014
O(5)	0.051	0.051	0.781

After Ref. (58).

The main structural elements are (SiO$_4$)tetrahedra, (BO$_4$) tetra-hedra and ninefold-coordinated Ce polyhedra, as illustrated in Fig. 18. The main 'architectural' detail of the structure is apparently determined by the infinite helical chains of (BO$_4$)tetrahedra parallel to the 3$_1$-axis. Each (BO$_4$)tetrahedron of the chain is connected by its two free vertices to two (SiO$_4$) tetrahedra. Furthermore, it shares two edges with the polyhedra of ninefold-coordinated Ce, which are also arranged in parallel [001] columns. Analogous compounds RE B(SiO$_4$)O of La, Ce, Pr, and Nd were prepared by solid-state reaction starting from the oxides at temperatures around 1100 °C (80).

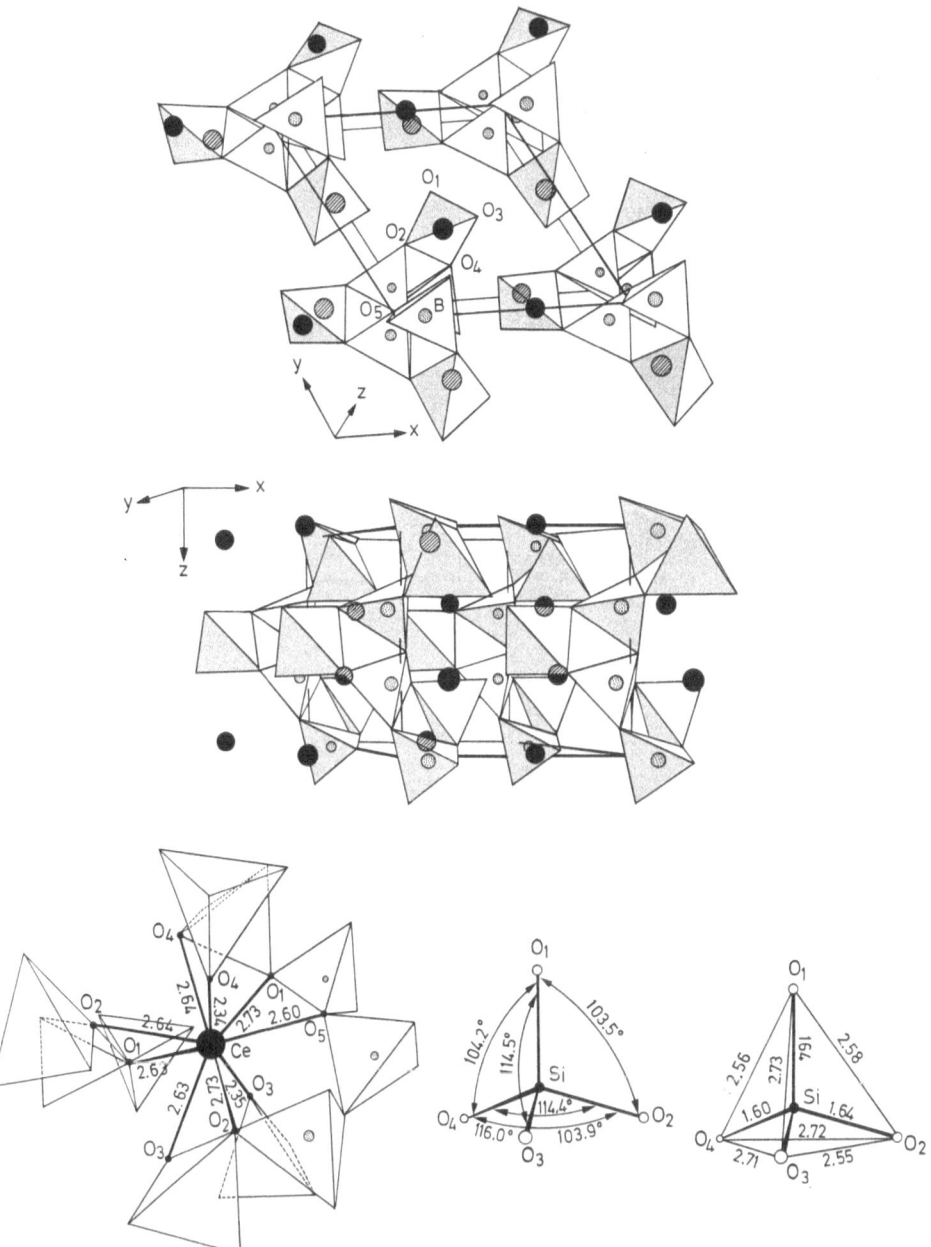

Fig. 18. Crystal Structure of CeB(SiO$_4$)O

135

J. Felsche

2.4. RE Mg Silicate Garnets

For the garnet structure type, various substitutions have been reported
into the three independent crystallographic sites c,a,d which show CN's
8,6,4, respectively. However, there is only one known group of rare-
earth silicate garnets. In the systems MgO—RE_2O_3—SiO_2 they were
synthesized hydrothermally at 720 °C under 2 kilobars H_2O pressure
(32). Unfortunately the chemical data on the 3—5μ diameter crystals
was too poor to yield a reliable definition of the chemical composition.
According to a private communication, the formula $RE_6Mg_5Si_5O_{24}$,
reported initially in (32). should probably be revised to $Mg_3RE_2(SiO_4)_3$.
This garnet structure obviously is only stable for the smaller rare earths
(Lu ... Ho,Y). The cell dimensions of the isomorphous compounds
$RE_2Mg_3(SiO_4)_3$ vary nearly linearly with the RE^{3+} radii, see Table 11.

Table 11. *Cell dimensions of $RE_2 Mg_3(SiO_4)_3$ compounds*

RE:	Y	Ho	Er	Yb	Lu
$a(\text{Å})$:	12.12	12.12	12.08	12.02	11.99

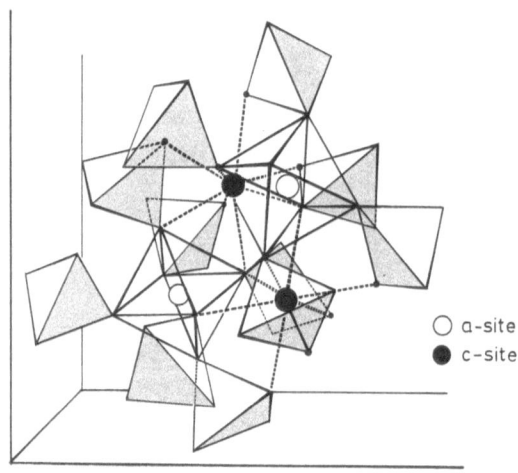

○ a-site
● c-site

Fig. 19. Cation-oxygen coordination in the garnet structure

No structural investigation, which might have confirmed the true
cell contents and cation distribution on the a, c, and d lattice sites, was
carried out because the crystals so far synthesized have been too small.

It is suggested that the rare-earth atoms are located on the eightfold coordinated c sites and the Mg atoms on the sixfold coordinated a sites. A section of the garnet cell, which illustrates these coordinations is shown in Fig. 19.

2.5. Compounds Containing Divalent Rare Earths

It is the special stability of the $4f^7$ and $4f^{14}$ electronic configuration, which is responsible for the occurence of the Sm^{2+}, Eu^{2+}, and Yb^{2+} ions. In the binary systems RE O–SiO_2, compounds $RE_2(SiO_4)$, $RE(SiO_3)$ of Sm, and additional $RE_3(Si_2O_7)$, $RE_3(SiO_4)O$ of Eu and Yb were described in terms of X-ray powder-, optical and infrared data (37). A strong similarity with alkaline earth silicates was found. A compound $Eu_3(Si_2O_7)$ is not confirmed, however, in the study of (36) (see Fig. 1). Single-crystal X-ray data exist so far only on ferromagnetic europium compounds, dimorphic $Eu_2(SiO_4)$ (34, 59, 60), and $Eu_3(SiO_4)O$ (62). $Eu_2(SiO_4)$ shows congruent melting at 1960 °C, $Eu_3(SiO_4)O$ ingruent melting at about 1800 °C. The crystals have paramagnetic Curie temperatures of 10 °K and 19 °K, respectively (36). Both compounds show a close structural relation to polymorphic Ca and Sr analogues. A silicate compound containing europium of mixed valence which crystallizes with the apatite structure $Eu_2^{2+} Eu_8^{3+} (SiO_4)_6O_2$ has already been mentioned in the chapter on oxyapatites.

2.5.1. Dimorphic $Eu_2(SiO_4)$

Single crystals up to several millimeters in size, of a high degree of purity, have recently been grown by chemical transport reaction at high temperatures (35). Unlike some earlier crystallographic data on the silicate of this composition (34) which had been prepared by solid state reaction of the oxides (33), these highly purified crystals showed monoclinic symmetry of space group $P2_1/c(P2_1/n)$ at room temperature (59). A displacive type of phase transition has been observed at 165 °C where the lattice symmetry changes to orthorhombic, corresponding to a distinct change in physical properties as shown in Fig. 20 b. The oxidation on $Eu_2(SiO_4)$ has recently been studied (90) by DTA/TGA and X-ray experiments. The formation of $Eu_2(SiO_4)O$ is topotactically controlled by intermediate apatite phases, which were observed at temperatures 600–1000 °C.

The monoclinic modification in the setting of $P2_1/c$ has cell dimensions $a = 5.661$ Å, $b = 7.101$ Å, $c = 11.518$ Å and $\beta = 122.01$ °C. There are four formula units $Eu_2(SiO_4)$ per unit cell based on experimental 6.67 g/cm^3 at 20 °C. The structural investigation carried out on these

J. Felsche

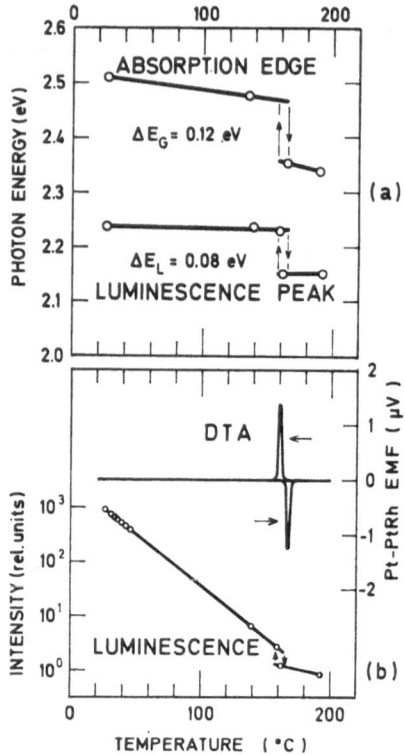

Fig. 20. Phase transition of dimorphic Eu₂(SiO₄)

Table 12. *Atomic parameters of monoclinic low Eu₂(SiO₄)*

Atom	x	y	z	B[Å²]
Eu(1)	0.9282(2)	0.0011(1)	0.6976(1)	0.42 (4)
Eu(2)	0.6604(2)	0.3424(1)	0.6976(1)	0.55 (4)
Si	0.676 (1)	0.767 (1)	0.419 (1)	0.39 (6)
O(1)	0.795 (4)	0.723 (2)	0.320 (2)	0.62(15)
O(2)	0.368 (4)	0.678 (2)	0.356 (2)	0.98(19)
O(3)	0.652 (4)	0.006 (2)	0.430 (2)	0.65(14)
O(4)	0.883 (4)	0.680 (2)	0.572 (2)	0.94(17)

After Ref. *(60)*.

crystals confirmed an isostructural relation with β-Ca₂(SiO₄) *(60)*. Because of the fact that no correction for absorption could be carried out, the R value of 8.9% indicates to a relatively high standard deviation (0.02 Å) for the mean cation-oxygen distances.

138

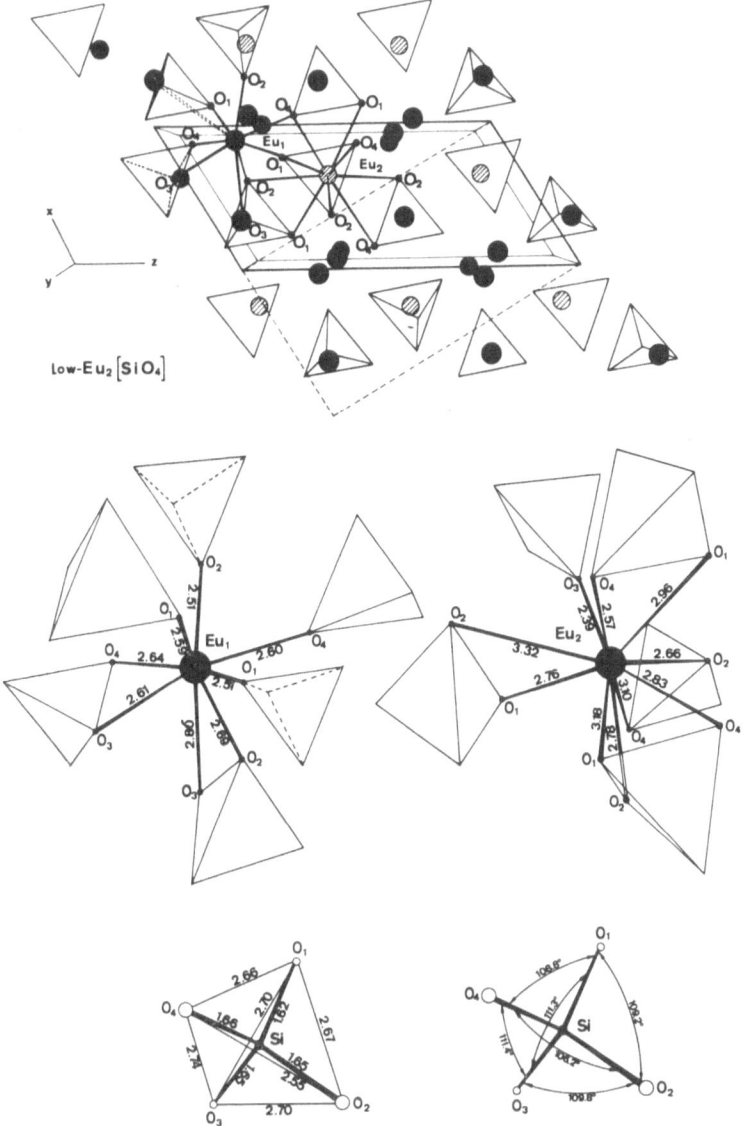

Fig. 21. Crystal structure of low Eu₂(SiO₄)

The structure consists of isolated (SiO₄)tetrahedra and two kinds of crystallographically independent Eu atoms. Viewing along [010], as in Fig. 21, reveals a pseudo-hexagonal arrangement of strings containing (SiO₄)tetrahedra alternating with Eu(2) atoms, which show CN 10.

139

Eu(1) atoms are accommodated in the channel between these strings. The oxygen-coordination polyhedron of Eu(1) has the shape of a strongly distorted cube with eight Eu—O distances ranging from 2.51–2.80 Å. The 'string Eu(2) atom' shows a rather peculiar oxygen coordination. Ten oxygens belong to its first coordination shell, which is unambigously defined by the distance of 0.32 Å to the second shell. Three Eu—O bonds in this polyhedron of 2.78, 2.83 and 3.19 Å are directed to the bottom face of the next (SiO$_4$)tetrahedron and one of 2.39 Å to the peak of the other tetrahedron within the same string. Three further pairs of bonds are clearly recognized, with the edges of the three nearly co-planar tetrahedra belonging to three different but parallel strings. The mean Eu(1)—O distance in the eightfold coordination is 2.62 Å and the mean Eu(2)—O distance of the tenfold-coordinated Eu is 2.86 Å. As shown in Fig. 22b each of the europium atoms is surrounded by eight nearest Eu metals in the shape of distorted cubes. Eu(1) has six Eu(2) neighbours at 3.63, 3.71, 3.75, 3.79. 3.89 Å and 4.08 Å and two Eu(1) at 3.70 Å. Eu(2) is surrounded by six Eu(1) at 3.63, 3.71, 3.75, 3.79, 3.89 and 4.08 Å and by two Eu(2) at 3.89 Å and 3.96 Å. Some appreciable differences from isotopic β-Ca$_2$(SiO$_4$) are demonstrated by the rather regular shape of the (SiO$_4$)tetrahedra with the mean Si—O distance of 1.64 Å (Fig. 21). Other isostructural compounds of this low-temperature form of Eu$_2$(SiO$_4$) are β-Sr$_2$(SiO$_4$), Ba$_2$(TiO$_4$) and β-Na$_2$(BeF$_4$).

The high-temperature modification of Eu$_2$(SiO$_4$) is orthorhombic with the possible space groups Pmnb/P2$_1$nb, according to single-crystal diffraction symmetry. Thus, it should correspond with the α_H—Ca$_2$(SiO$_4$) modification (61). The cell dimensions measured at 200 °C are $a = 5.665$ (2) Å, $b = 7.137(3)$ Å and $c = 9.767(3)$ Å. The crystal structure has not yet been determined. From the probable isostructural relation with α_H—Ca$_2$(SiO$_4$), only slight deviations in coordination are to be expected. There is probably a change in coordination from 8 to 9 of Eu(1) corresponding to a slight tilting of the (SiO$_4$)tetrahedra against each other. However, the coordination of Eu(2) is thought to remain unchanged. The high form of Eu$_2$(SiO$_4$) is also isotypic with the low temperature modification of K$_2$SO$_4$.

2.5.2. Eu$_3$(SiO$_4$)O

The crystal structure of Eu$_3$(SiO$_4$)O has tetragonal symmetry of space group I4c2. There are four formula units in the cell with $a = 6.004$ Å and $c = 10.760$ Å. The atomic parameters given in Table 13 were refined by means of three-dimensional intensity data (62), but no final R value nor ndividual temperature factors have been reported yet.

As observed in the oxyorthosilicate structure of the trivalent rare earths, there are also two types of anions present in this structure. One

Table 13. *Atomic parameters of Eu₃(SiO₄)O*

Atom	x	y	z
Eu(1)	0.0	0.5	0.0
Eu(2)	0.1851	0.1851	0.25
Si	0.0	0.0	0.0
O(1)	0.0	0.5	0.25
O(2)	0.111	0.153	0.092

After Ref. (*62*).

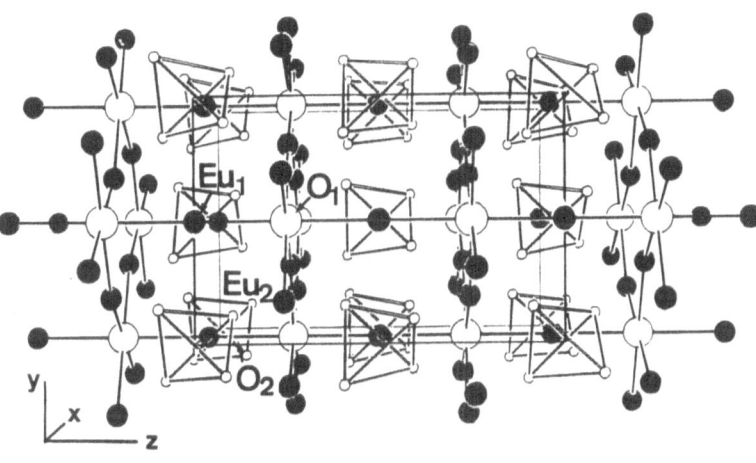

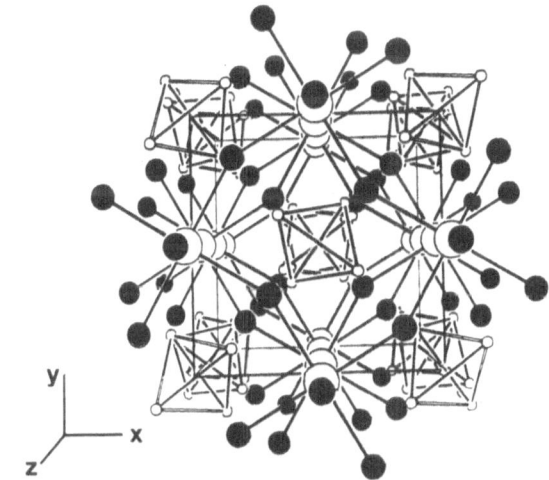

Fig. 22a. Crystal structure of Eu₃(SiO₄)O. After Ref. (*62*)

is the (SiO_4)tetrahedron and the other is the isolated, not silicon-bonded oxygen $O(1)$. There exist two crystallographically independent types of Eu cations which show not however eight- and tenfold oxygen coordination, as observed in the monoclinic structure of $Eu_2(SiO_4)$ (60) but octahedral coordination. $Eu(1)$ is surrounded by two $O(1)$ at 2.69 Å and by four $O(2)$ at 2.70 Å, $Eu(2)$ by two $O(1)$ at 2.52 Å and four $O(2)$ at 2.47 Å and 2.67 Å, respectively. As shown in Fig. 22a, a reasonable description of the structure is obtained in terms of (SiO_4) tetrahedra and $(O—Eu_6)$octahedra. In the latter, the 'extra' oxygen $O(1)$ is octahedrally surrounded by four $Eu(2)$ and two $Eu(1)$ cations.

Like $Sr_3(SiO_4)O$, which crystallizes in space group $P4/ncc$ (63), these $(O—Eu_6)$octahedra form a three-dimensional framework in which the (SiO_4)tetrahedra are located for charge balance. The corner-sharing

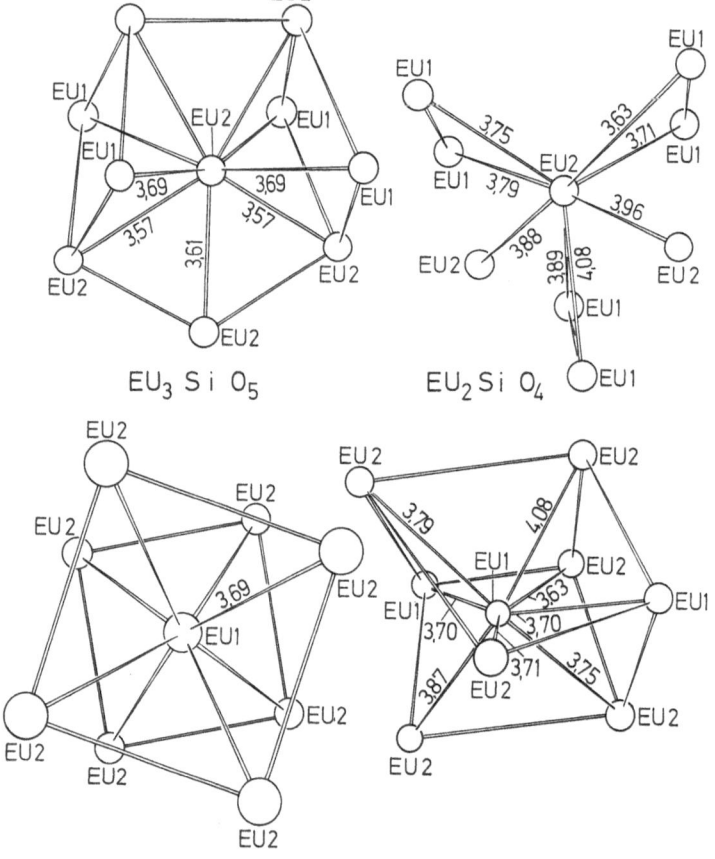

Fig. 22b. Metal to metal distances of closest neighbours in $Eu_3(SiO_4)O$ and $Eu_2(SiO_4)$

(O—Eu$_6$)octahedra run parallel to the fourfold axis. The silicon-bonded oxygens O(2) are in the general positions. The Si—O(2) distance is equal to 1.64 Å for all four bonds because Si is located in the special position on $(0,0,0)$. As shown in Fig. 22b the metal to metal distances are close to the values known from EuO (Eu—Eu = 3.67 Å), which is also ferromagnetic with a Curie temperature of 77 °K. Eu(1) has eight nearest Eu(2) neighbours at equal distances of 3.69 Å. The polyhedron is close to a antisquare prism. Eu(2) is surrounded by nine next metal neighbours in the shape of a distorted cube with one additional Eu(2) attached to one cube edge. Four Eu(2) are at 3.57 Å, four Eu(1) at 2.69 Å and one Eu(2) is at 3.61 Å.

3. Structures Containing Isolated Groups (Si$_2$O$_7$), or (Si$_3$O$_{10}$), + (SiO$_4$)

Compounds 1 RE$_2$O$_3 \cdot$ 2 SiO$_2$ are known from all binary rare-earth silicate systems. They show extensive polymorphism. The polymorphism is characterized by transition temperatures between 1300—1500 °C and by boundaries at europium and holmium along the series of trivalent rare earths. Seven polymorphic forms were observed; all are of type RE$_2$(Si$_2$O$_7$) with one exception, RE$_4$(Si$_3$O$_{10}$)(SiO$_4$), which occurs with the medium-sized rare earths. Ternary compounds of type Alk.$_3$RE(Si$_2$O$_7$) crystallize with two different structures, which also contain (Si$_2$O$_7$) double-tetrahedra groups.

It is the configuration of the (Si$_2$O$_7$)double-tetrahedra groups in all these disilicate structures which is of special interest from several points of view. One aspect concerns the systematics of all possible double-tetrahedra configurations of anions (X$_2$O$_7$)$^{-n}$ and the packing with a certain number of individual cations providing charge balance. This viewpoint has recently been followed up in (86). Another aspect involves the double π-bonding theory, as developed in (78) for (XO$_4$)$^{-n}$tetrahedra ions (X = Si,P,S,Cl). Many helpful quantitative data might be introduced into this discussion in terms of the bonding lengths and angles observed for the (Si$_2$O$_7$)groups in rare-earth disilicate structures. A third aspect of the disilicate polymorphs is that concerning lanthanide periodicity. The (Si$_2$O$_7$)configuration is apparently determined by the bonding and polarizing forces of the surrounding cations. Thus, any periodicity of the rare-earth electronic structures, such as variation in ionic size or discontinuity in chemical bonding properties, along the series of trivalent rare earths should be reflected in the changing configuration of the (Si$_2$O$_7$)

double tetrahedra. This might result in the formation of a new structural type. These are the main reasons for paying special attention to the polymorphism and structural data of this type of rare-earth compound.

3.1. *Polymorphic Disilicate Compounds 1 $RE_2O_3 \cdot 2SiO_2$*

The polymorphism of the binary disilicates, and crystal data for a collection of polymorphic forms representing the seven observed structure types, are shown schematically in Figs. 23 and 24, respectively. A complete set of crystal data on all 30 investigated polymorphs is given in Table 14. For X-ray powder patterns see also Appendix II. Following the nomenclature of the polymorphic forms of the rare-earth sesquioxides, the various structural types of the disilicates are designated by capital letters: A,B,C,D,E,F,G. The values of the ionic radii used throughout this chapter apply to octahedral coordination of the trivalent rare-earth ions in the tetragonal oxychlorides ($La^{3+} \rightarrow Nd^{3+}$) and in the cubic form of the sesquioxides ($Sm^{3+} \rightarrow Lu^{3+}$) (*48*).

The polymorphism of the disilicates at temperatures from 900° to 1800 °C (*27*) is shown in Fig. 23. The data relating to the smaller rare earths, Gd to Lu, are almost in agreement with those given in Ref. (*17*).

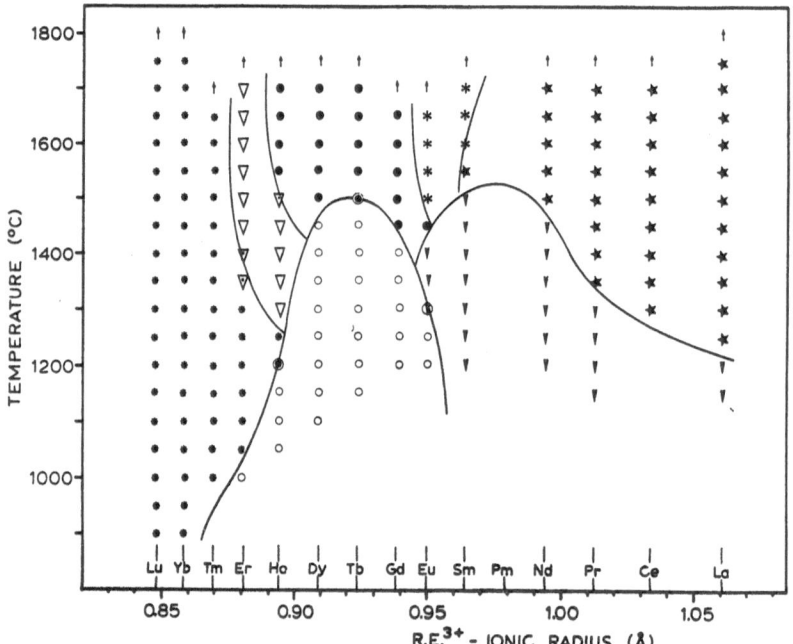

Fig. 23. Polymorphism of compounds $RE_2Si_2O_7$. After Ref. (*27*)

The limitation of the results in the lower temperature range is imposed by the method of preparation, that is, by the slow reaction rate of the starting materials. At higher temperatures, the values close to the melting point are somewhat in doubt relative to the results below 1600 °C because these experiments were carried out under high-vacuum conditions. The regions of stability of the polymorphs in Fig. 23 are based on X-ray powder data, obtained from samples of heated oxide mixtures 1 RE_2O_3 + 2 SiO_2. The transition temperatures of the polymorphs, observed during annealing experiments in which each phase was intended to be converted into the other, differ slightly from these data. From the results of the differential thermoanalysis carried out with all the polymorphic compounds it was concluded that the reconstructive type of transition, according to BUERGER's classification, exists for all transitions observed in this study. No thermal effect was found with the 10 μV full-scale sensitivity of the Pt 10% Rh—Pt thermocouple under various conditions. However, the energy barriers between the various modifications seem likely to be of rather different heights. This is suggested by the large variations in time, from 30 min. to ∞, which were observed for the individual transitions. '∞' means not converted within a reasonable

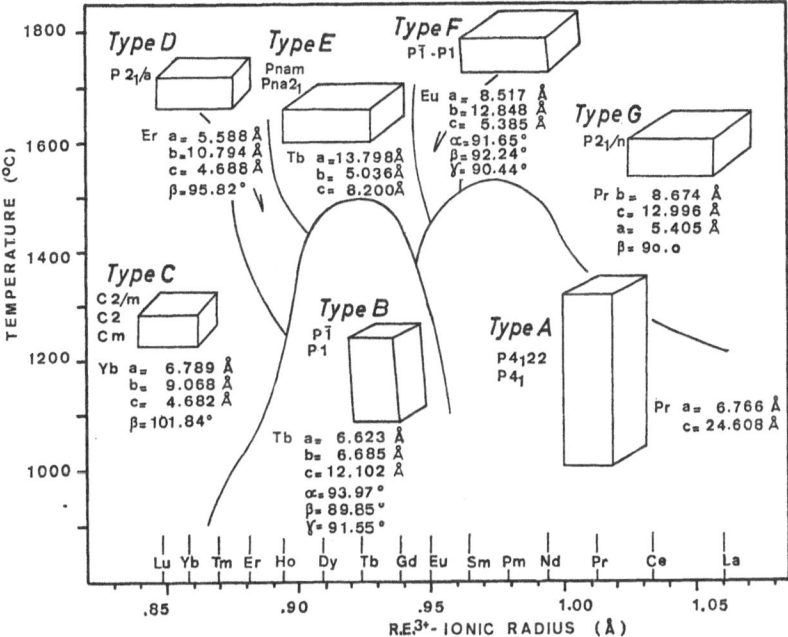

Fig. 24. Crystal data on the structure types A ... G of disilicate rare-earth compounds. After Ref. (21)

Table 14. Cell dimensions and densities of all observed polymorphic forms of the rare-earth disilicates. Standard deviations of the refined values are given in parentheses in units of the last decimal place. ϱ_o, pyknometer data at 20 °C. Z, formula units per unit cell volume, V.

Structure type A (tetragonal, P $4_1$22–P4_1)

	$La_2Si_2O_7$	$Pr_2Si_2O_7$	$Nd_2Si_2O_7$	$Sm_2Si_2O_7$	$Eu_2Si_2O_7$
a[Å]	6.7945(9)	6.7657(6)	6.7405(6)	6.6933(8)	6.6727(7)
c[Å]	24.871(8)	24.608(4)	24.524(4)	24.384(9)	24.338(3)
V[Å³]	1148.1(9)	1126.4(2)	1114.3(2)	1092.4(3)	1083.6(9)
Z	8	8	8	8	8
ϱ_o[g cm⁻³]	5.11(9)	5.26(6)	5.38(6)	5.67(7)	5.68(9)
ϱ_c[g cm⁻³]	5.15	5.30	5.44	5.70	5.79

Structure type B (triclinic, P$\bar{1}$–P1)

	$Eu_2Si_2O_7$	$Gd_2Si_2O_7$	$Tb_2Si_2O_7$	$Dy_2Si_2O_7$	$Ho_2Si_2O_7$	$Er_2Si_2O_7$
a[Å]	6.716(3)	6.624(5)	6.623(5)	6.639(2)	6.664(5)	6.583(5)
b[Å]	6.762(3)	6.679(5)	6.684(5)	6.691(2)	6.674(5)	6.609(5)
c[Å]	12.321(7)	12.132(9)	12.101(9)	12.152(3)	12.110(9)	12.000(9)
α[°]	94.36(4)	94.10(8)	93.97(7)	94.03(3)	94.07(8)	94.50(8)
β[°]	90.02(3)	89.79(9)	89.85(9)	89.81(2)	89.97(8)	90.57(8)
γ[°]	91.75(4)	91.60(7)	91.55(6)	91.69(3)	91.66(7)	91.79(9)
V[Å³]	557.7(1)	535.2(7)	534.4(6)	538.2(1)	537.1(6)	520.3(7)
Z	4	4	4	4	4	4
ϱ_o[g cm⁻³]	5.54(6)	5.82(7)	5.93(5)	6.06(7)	6.11(4)	6.28(9)
ϱ_c[g cm⁻³]	5.62	5.99	6.04	6.09	6.15	6.42

Structure type C (monoclinic, C2/m–C2–Cm)

	$Ho_2Si_2O_7$	$Er_2Si_2O_7$	$Tm_2Si_2O_7$	$Yb_2Si_2O_7$	$Lu_2Si_2O_7$
a[Å]	6.875(5)	6.841(5)	6.818(7)	6.789(6)	6.760(6)
b[Å]	9.184(9)	9.135(9)	9.104(9)	9.067(9)	9.051(9)
c[Å]	4.697(4)	4.694(6)	4.679(5)	4.681(4)	4.685(7)
β[°]	101.69(6)	101.70(7)	101.75(8)	101.84(7)	101.86(6)
V[Å³]	290.5(3)	287.3(2)	284.4(4)	282.1(3)	280.6(1)
Z	2	2	2	2	2
ϱ_o[g cm⁻³]	5.62(6)	5.78(5)	5.82(4)	6.01(7)	6.02(4)
ϱ_c[g cm⁻³]	5.68	5.81	5.91	6.06	6.10

Structure type E (orthorhombic, Pnam–Pna2₁)

	$Eu_2Si_2O_7$	$Gd_2Si_2O_7$	$Tb_2Si_2O_7$	$Dy_2Si_2O_7$	$Ho_2Si_2O_7$
a[Å]	13.9142(9)	13.8665(9)	13.797 (2)	13.7275(9)	13.7934(9)
b[Å]	5.0553(4)	5.0532(4)	5.036 (1)	5.0303(3)	5.0371(4)
c[Å]	8.3486(7)	8.3008(8)	8.200 (2)	8.2050(6)	8.2524(8)
V[Å³]	587.25 (3)	581.64 (4)	573.35 (9)	566.58 (4)	573.36 (5)
Z	4	4	4	4	4
ϱo[g cm⁻³]	5.22 (8)	5.38 (6)	5.56 (7)	5.75 (4)	5.69 (5)
ϱc[g cm⁻³]	5.33	5.51	5.63	5.78	5.82

Structure type D (monoclinic P2₁/a)

	$Ho_2Si_2O_7$	$Er_2Si_2O_7$
a[Å]	5.957 (2)	5.588 (2)
b[Å]	10.842 (3)	10.793 (3)
c[Å]	4.696 (2)	4.689 (2)
α[°]	90.0	90.0
β[°]	95.72 (3)	95.82 (4)
γ[°]	90.0	90.0
V[Å³]	283.6 (1)	281.4 (1)
Z	2	2
ϱo[g cm⁻³]	5.76 (4)	5.82 (7)
ϱc[g cm⁻³]	5.82	5.93

Structure type F (triclinic P1̄–P1),

	$Sm_2Si_2O_7$	$Eu_2Si_2O_7$
a[Å]	8.513 (3)	8.517 (1)
b[Å]	12.867 (4)	12.848 (2)
c[Å]	5.374 (2)	5.385 (1)
α[°]	91.34 (3)	91.65 (2)
β[°]	92.06 (4)	92.24 (2)
γ[°]	90.43 (3)	90.44 (2)
V[Å₃]	588.2 (2)	588.56 (9)
Z	4	4
ϱo[g cm⁻³]	5.10 (5)	5.30 (5)
ϱc[g cm⁻³]	5.26	5.33

Structure type G (pseudoorthorhombic, P2₁/n)

	$La_2Si_2O_7$	$Ce_2Si_2O_7$	$Pr_2Si_2O_7$	$Nd_2Si_2O_7$	$Sm_2Si_2O_7$
b[Å]	8.794 (2)	8.722 (1)	8.674 (1)	8.630 (2)	8.564 (7)
c[Å]	13.201 (2)	13.056 (2)	12.996 (2)	12.945 (2)	12.855 (9)
a[Å]	5.409 (1)	5.401 (1)	5.405 (1)	5.391 (1)	5.383 (5)
α,β,γ[°]	90.0	90.0	90.0	90.0	90.0
V[Å³]	627.95 (8)	615.09 (7)	609.40 (7)	602.37 (7)	592.61
Z	4	4	4	4	4
ϱo[g cm⁻³]	4.61 (6)	4.81 (7)	4.86 (6)	5.01 (6)	5.11 (7)
ϱc[g cm⁻³]	4.71	4.84	4.90	5.04	5.23

After Ref. (21)

J. Felsche

period of time, i.e. within 100 h of observation. A larger time constant was generally observed for the high→low form transitions. The low diffusion rate of the ions involved in the 'reconstructive transitions' sometimes resulted in the coexistence of two corresponding modifications. Hence, in some cases, mixed phases were observed over the remarkable temperature range of 50°—150 °C. This feature of the transition decreased with the ionic radii within the isostructural groups, e.g. for the dimorphic disilicates La to Sm and Gd to Dy. Mixed phases were further observed at transitions of the low-temperature forms of the compounds displaying 4 and 3 phases $Sm_2Si_2O_7$, $Eu_2Si_2O_7$, $Ho_2Si_2O_7$, and $Er_2Si_2O_7$, respectively. The following transitions were not observed: F→G of $Sm_2Si_2O_7$, E→A, F→A, E→B and F→B of $Eu_2Si_2O_7$, C→B and D→C of $Ho_2Si_2O_7$ and C→B of $Er_2Si_2O_7$. On transition G→A of $La_2Si_2O_7$, and all transitions E→B, mixed phases have been found below the transition temperature given in Fig. 23. The low-temperature form of $Ce_2Si_2O_7$ has not been obtained, even under reducing atmosphere, in long-time annealing experiments where the starting materials were either the oxides or the high-temperature form. In the latter case, on the thermobalance under high vacuum (10^{-5}torr), decomposition of G type $Ce_2Si_2O_7$, to CeO_2 and silica was observed below 1300 °C without any increase in weight, which may even imply the reduction of silicon to a lower oxidation state than Si^{4+}.

The crystal data given in Fig. 24 and Table 14 were obtained from single crystals and powder samples, respectively (27). When the regions of stability of each of the disilicate structure types A—G, as shown in Fig. 23, had been established by means of Guinier powder photographs, the cell geometry and diffraction symbols implying the possible space group were obtained from MoKα precession photographs. The crystals 0.02—0.4 mm in size, used in this procedure, were obtained by long-time annealing (20—100 hours) of powder samples at suitable temperatures in range of 1300—1600 °C. Flux was not employed in order to ensure maximum purity of the compounds. The chemical composition was confirmed by microprobe analysis. The cell dimensions of 30 $RE_2(Si_2O_7)$poly-morphs, shown in Table 14, represent refined values based on Guinier powder data, which had been provided with a Jagodzinski camera-type using FeKα$_1$ radiation. The data in Table 14 correspond closely to the cell dimensions given recently in (24) and (87), respectively. Small differences found in some cases might be caused by traces of the flux occasionally employed for the synthesis of these crystals.

Lacking any information about the disilicate structures, we may draw some preliminary conclusions on the extensive polymorphism from the data shown in Figs. 23 and 25. The arrangement of the stability regions in Fig. 23 is apparently determined by two parameters. One is

148

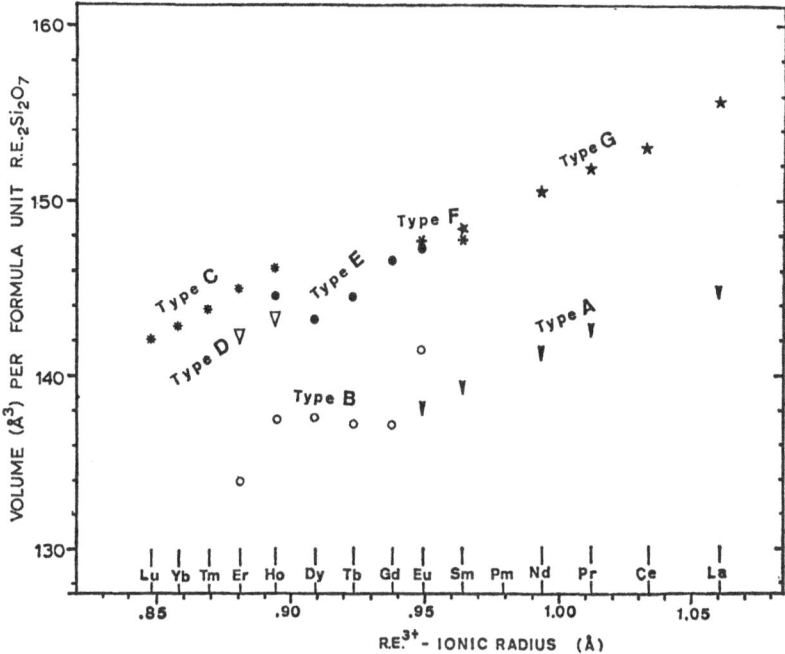

Fig. 25. Volume per formula unit $RE_2(Si_2O_7)$ of the polymorphic disilicate compounds

given by the properties of the rare earths, such as the RE^{3+} ionic size which might change gradually with the atomic number. The other parameter is the temperature. All phase transitions were found to occur within the temperature range 1200—1500 °C. On the other hand, the plot of volume ($RE_2Si_2O_7$) vs. ionic radius (RE^{3+}) in Fig. 25 reflects the lanthanide contraction. The two lines given by the volume of the high- and low-temperature forms, respectively, show about the same slope with increasing rare-earth atomic number. Special attention is drawn to the two intersections in Fig. 23, marked by polymorphic $Eu_2(Si_2O_7)$, $Sm_2(Si_2O_7)$, and $Ho_2(Si_2O_7)$, $Er_2(Si_2O_7)$, respectively. All the other compounds display dimorphism or show only one phase. This feature may reflect the preferred stability of the $4f^0$, $4f^7$, $4f^{14}$ electronic configurations — as described already in chapter 1.0 — or, as in this case, reflect the rather unstable $4f^5$, $4f^6$, and $4f^{11}$, $4f^{12}$ electronic configurations of the Sm^{3+}, Eu^{3+}, and Ho^{3+}, Er^{3+} cations, respectively.

All the structure types signified in Fig. 24 by crystal data have been investigated in the meantime by three-dimensional X-ray intensity data, with the exception of tpye F. Efforts to prepare crystals of F-type

Eu$_2$(Si$_2$O$_7$) and Sm$_2$(Si$_2$O$_7$) of sufficient size and quality have not been successful so far. Thus, this is the only structure in the field of disilicate polymorphs, which has not yet been worked out. However, its cell dimensions and powder diffraction intensities (27) suggest a strong similarity with structure type G, which has already been determined. A better approach to an understanding of the polymorphism of the rare-earth disilicates will be achieved by giving extensive information about the structural details of all the polymorphic forms. This is provided below following the alphabetical sequence of structure types from A to G.

3.1.1. Structure Type A, (La, ... Eu)$_2$(Si$_2$O$_7$), 'RE Di A'

Single crystals of the low-temperature form of Pr$_2$(Si$_2$O$_7$) were obtained by sintering the compound at temperatures at which the modification of type G is stable. These crystals were cooled to below the transition temperature of about 1350 °C. After annealing for a few hours, single

Table 15. *Atomic parameters of A-type Pr$_2$(Si$_2$O$_7$)*

Atom	x	y	z	$B[\text{Å}^2]$
Pr 1	0.76655(7)	0.29698(7)	0.99312(2)	1.09 (2)
Pr 2	0.52041(7)	0.16681(7)	0.14075(2)	1.01 (2)
Pr 3	0.33792(7)	0.91768(7)	0.99347(3)	1.02 (2)
Pr 4	0.12165(7)	0.76307(7)	0.13275(3)	1.06 (2)
Si 1	0.8522 (4)	0.7634 (4)	0.0085 (1)	0.78 (4)
Si 2	0.5985 (4)	0.6548 (4)	0.1067 (1)	0.71 (4)
Si 3	0.2623 (4)	0.3762 (4)	0.0147 (1)	0.79 (4)
Si 4	0.0091 (4)	0.2912 (4)	0.1141 (1)	0.67 (4)
O 1	0.8951(10)	0.6156(1)0	0.9578 (3)	0.89 (8)
O 2	0.7207(10)	0.9412(10)	0.9835 (3)	1.03 (9)
O 3	0.0458(10)	0.8439(10)	0.0404 (3)	0.87 (8)
O 4	0.7181(10)	0.6252(10)	0.0511 (3)	0.84 (8)
O 5	0.4799(13)	0.5167(13)	0.1333 (3)	1.48(11)
O 6	0.4326(10)	0.8590(10)	0.0877 (3)	1.10 (9)
O 7	0.7530(10)	0.8124(10)	0.1467 (3)	0.96 (9)
O 8	0.3262(10)	0.5715(10)	0.9834 (3)	1.10 (9)
O 9	0.4456(10)	0.2375(10)	0.0338 (3)	0.98 (9)
O 10	0.1249(10)	0.2404(10)	0.9746 (3)	1.05 (8)
O 11	0.1207(10)	0.4291(10)	0.0668 (3)	1.12 (9)
O 12	0.9685(10)	0.4559(10)	0.1604 (3)	1.03 (9)
O 13	0.1573(11)	0.1247(11)	0.1403 (3)	1.15 (9)
O 14	0.8124(12)	0.2054(12)	0.0867 (3)	1.34(10)

After Ref. (26), however more accurate values as obtained recently by a LSQS refinement on 1004 independent reflections measured with the same crystal and identical experimental conditions as described in Ref. (26)

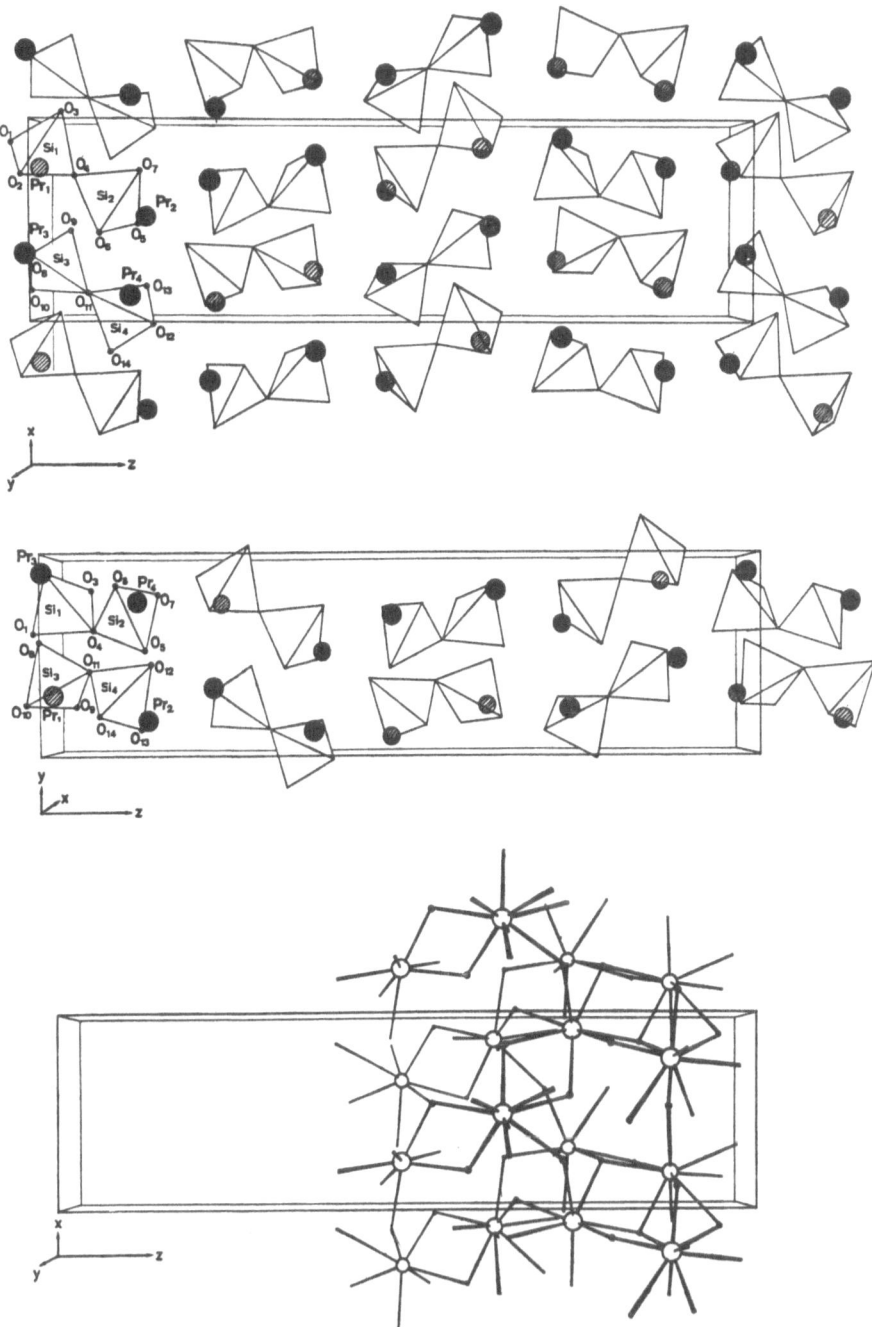

Fig. 26. Crystal structure of A-type $Pr_2(Si_2O_7)$

crystals of the low-temperature form were obtained of space group $P4_1$ or $P422$. The cell dimensions are $a = 6.769$ Å, $c = 24.607$ Å. There are eight formula units $Pr_2(Si_2O_7)$ in the cell, based on the experimental densities of 5.26 g/cm^3. Space group $P4_122$ was excluded because of the relationship $F_{(hkl)} \neq F_{(h\bar{k}l)}$ in the intensity data. From a set of 2318 three-dimensional intensity data which had been collected on a single-crystal diffractometer, the parameters of the 22 independent atoms per cell were determined unambiguously (26). The final R value of 3.0% corresponds to a mean standard deviation for cation-oxygen distances of 0.007 Å. The values of the refined atomic parameters are listed in Table 15.

The acentric structure consists of isolated (Si_2O_7) double tetrahedra and Pr ions, which are arranged in four sheets perpendicular to the c axis (Fig. 26). Within each of the four adjacent sheets, which are related through the 90° rotation of the 4_1 axis, the (Si_2O_7) units and the heavy atoms form parallel rows in the [110] direction. Within those rows there are always two of the double tetrahedra which are almost related by a centre of symmetry, with the corner oxygens forming some kind of closest packing. The heavy atoms are attached to each side of the sheets and provide a connection to the adjacent sheet.

The anion group $[Si_2O_7]^{-6}$ is free of symmetry restrictions since all atoms of the structure are in the general position. The tetrahedra pair is essentially of eclipsed configuration (Fig. 27), but the ideal point symmetry $C_{2v}/mm2$ is destroyed by a remarkable distortion of the tetrahedra in bond lengths and angles and a twist angle about the Si—Si direction of each of the double tetrahedra. The two independent $[Si_2O_7]^{-6}$ units are only slightly different if we compare the mean values of the terminal oxygen-to-silicon distances and the bridging oxygen-to-silicon bond lengths. However, in both cases the mean bridging O—Si distances are significantly longer: $(1.62 \gtrless 1.65 \lessgtr 1.63)$; $(1.62 \gtrless 1.65 \lessgtr 1.62)$ [Å].

If we consider the dimensions of the praseodymium-oxygen coordination polyhedra in Fig. 28, the general impression is that the lanthanide ion is seeking a compromise which will achieve spherical shielding for itself and minimize repulsion of the coordinating oxygen atoms. However, this compromise need not necessarily cause the bonding atoms to assume a geometry that can be described in Cartesian terms. Hence, the following assignment of Cartesian polyhedra should be viewed as an approximation.

Taking the next neighbours, i.e. the first coordination shell, into account there is no difficulty in defining the nine neighbours of the first shells of Pr(2) and Pr(4). There is a significant separation from the second coordination shell which starts at Pr—O distances of about 3.5 Å. The Pr(2)—O distances range from 2.39—2.87 Å and 2.39—2.94 Å for Pr(4) (e.s.d.'s for all values 0.007 Å) with an average of 2.58 Å in each of the polyhedra. The bridging oxygen atom O(11) of the second $[Si_2O_7]^{-6}$

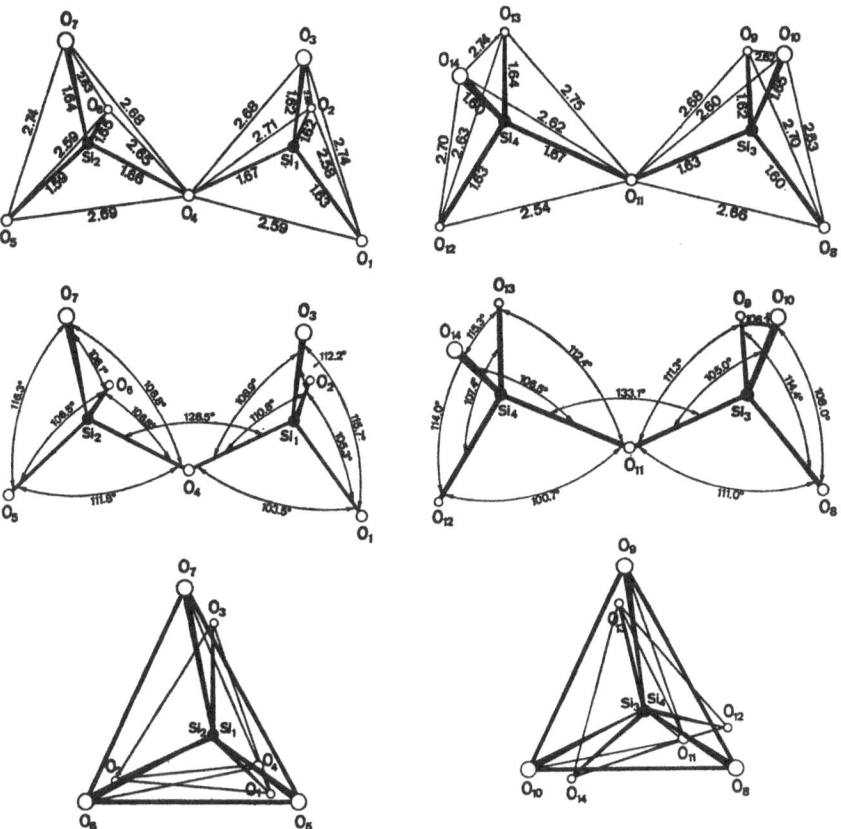

Fig. 27. Interatomic distances and angles in the double-tetrahedra (Si$_2$O$_7$) of A-type Pr$_2$(Si$_2$O$_7$)

double tetrahedron is included in the Pr(4) coordination shell at a quite regular distance of 2.78 Å. This seems worth mentioning in respect to the other members of the nine fold coordination, which are all terminal oxygen atoms. The polyhedra may be described as distorted square antiprisms with an additional oxygen atom jammed against one square face and one edge, respectively.

However, the low-coordinated heavy atoms Pr(1) and Pr(3) presented some difficulty in defining the first shell. In each of the coordination polyhedra there is one oxygen atom within the distance gap to the next shell.

In addition to the seven Pr(1)—O linkages, which are in the range 2.41—2.66 Å, there is one oxygen atom O(11) at 3.12 Å, about 0.3 Å away from the second shell which starts with Pr(1)—O(12') of 3.42 Å.

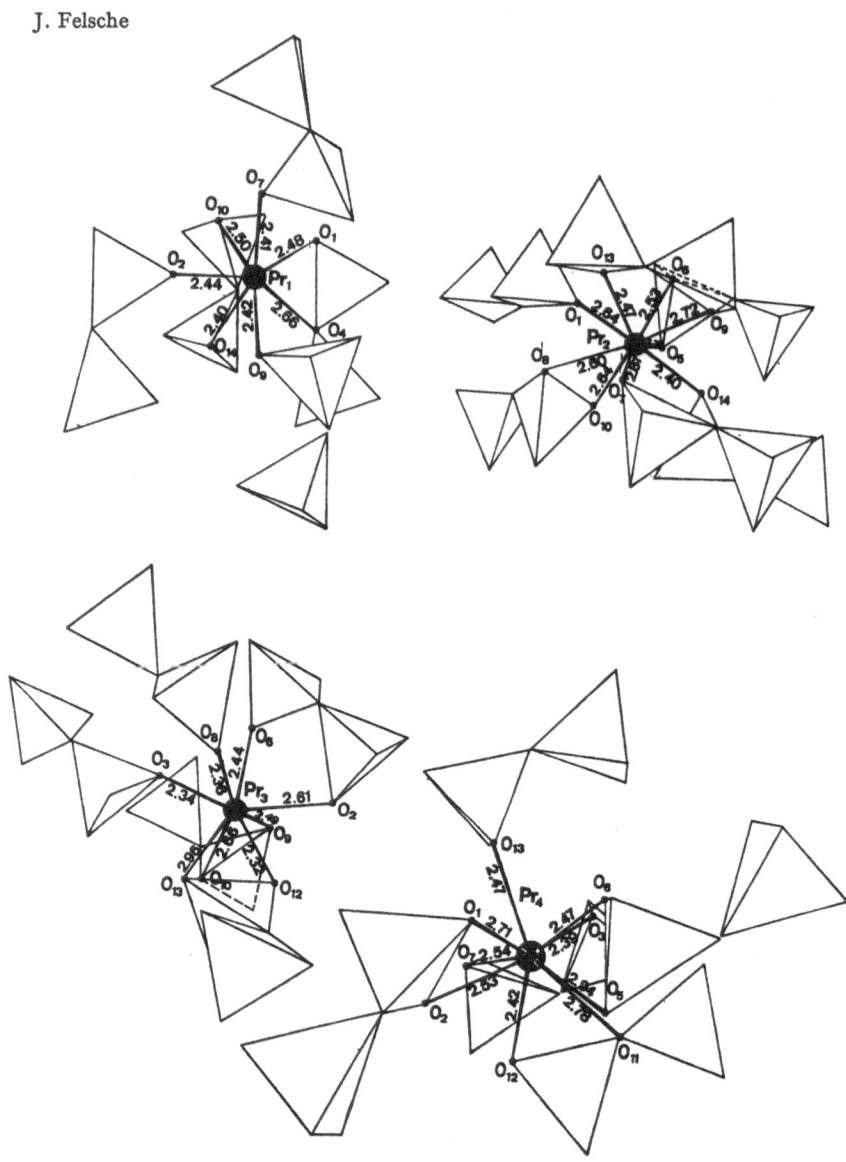

Fig. 28. Oxygen coordination around the rare earth cations in A-type $Pr_2(Si_2O_7)$

Therefore O(11) was considered to belong to the second shell; this is supported by its function as the bridging oxygen in a double tetrahedron with a reduced charge contribution as compared with the terminal oxygen atoms of the polyhedron. The arrangement of the closest seven oxygen atoms at a mean distance of 2.48 Å of the first shell of Pr(1) recalls the sevenfold-coordinated lanthanide ion in the B-type sesqui-oxide structure. It has the shape of a trigonal prism with the seventh oxygen atom coordinating through one face. An interesting feature of this polyhedron is the corner oxygen atom O(4) on the prism, which is the bridging oxygen of the first double-tetrahedra group. The distance Pr(1)—O(4) of 2.66 Å is about 0.16 Å larger than the next inner one O(10) with 2.52 Å.

In contrast to the result for Pr(1), the Pr(3)-oxygen coordination O(13) was considered to belong to the first shell because its distance of 2.96 Å is closer to the seven distances of the other oxygen atoms ranging from 2.32 to 2.66 Å than to the second shell which starts at 3.56 Å. Hence, Pr(3) has coordination number 8. The shape of the Pr(3) oxygen polyhedron is close to a dodecahedron, which is a common coordination type in lanthanide compounds.

This disilicate structure type A is also known from the corresponding pyrophosphate β—$Ca_2P_2O_7$ (64). A structure analysis was carried out also on the isomorphous compound $Sm_2(Si_2O_7)$ (25).

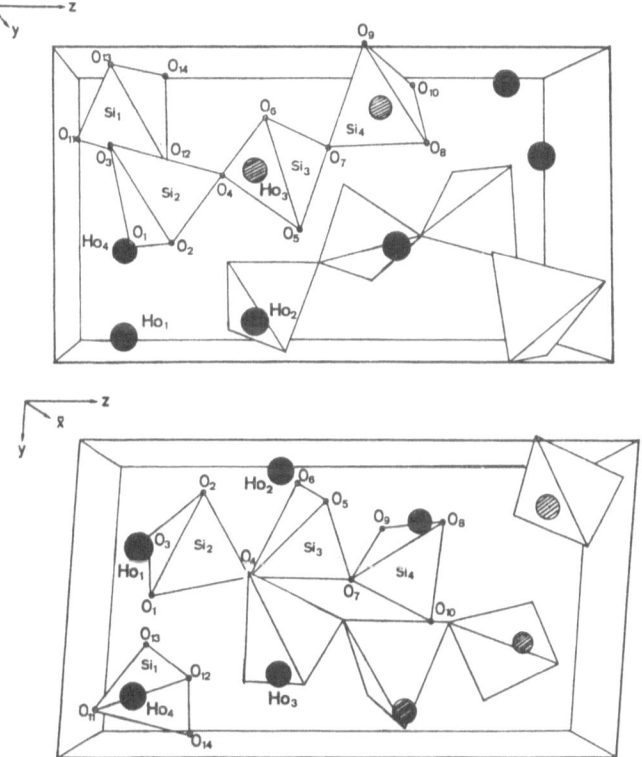

Fig. 29. Crystal structure of $Ho_4(Si_3O_{10})(SiO_4)$

3.1.2. Structure Type B, (Eu, ... Er)$_4$(Si$_3$O$_{10}$)(SiO$_4$), 'Re Di B'

Structure type B has been determined from three-dimensional single-crystal diffraction intensity data (27). The parameters of the 22 independent atoms in the triclinic unit cell have been determined from 4708 observations and refined to a residual $R = 5.8\%$, which corresponds to an average e.s.d. of 0.007 Å for the main cation-oxygen distances. Final atomic parameters are given in Table 16.

Table 16. *Atomic parameters of Ho$_4$(Si$_3$O$_{10}$)(SiO$_4$)*

Atom	x	y	z	$B[\text{Å}^2]$
Ho 1	0.9479 (1)	0.3310 (1)	0.11666(6)	0.87 (4)
Ho 2	0.8845 (1)	0.0908 (1)	0.35915(6)	0.75 (4)
Ho 3	0.3705 (1)	0.7756 (1)	0.36947(7)	1.02 (4)
Ho 4	0.6657 (1)	0.8280 (1)	0.10590(7)	1.22 (5)
Si 1	0.1539 (6)	0.8505 (6)	0.1168 (3)	0.30 (6)
Si 2	0.4862 (6)	0.3353 (6)	0.1761 (3)	0.37 (6)
Si 3	0.3781 (6)	0.2726 (6)	0.4051 (3)	0.28 (6)
Si 4	0.1457 (6)	0.3719 (6)	0.6179 (3)	0.39 (6)
O 1	0.6408(18)	0.4893(17)	0.1258 (9)	0.76(18)
O 2	0.6226(19)	0.1401(18)	0.2096(10)	0.77(19)
O 3	0.2968(19)	0.2956(18)	0.0948(10)	0.93(19)
O 4	0.4002(17)	0.4255(16)	0.3040 (9)	0.61(17)
O 5	0.5879(18)	0.1717(18)	0.4454(10)	0.74(18)
O 6	0.2240(18)	0.0950(18)	0.3785(10)	0.69(18)
O 7	0.2937(17)	0.4220(17)	0.5099 (9)	0.79(17)
O 8	0.2862(18)	0.2179(18)	0.6870(10)	0.81(18)
O 9	0.9626(18)	0.2281(18)	0.5714(10)	0.85(18)
O 10	0.0752(17)	0.5789(17)	0.6864 (9)	0.74(17)
O 11	0.2414(20)	0.9141(20)	0.0010(11)	1.22(21)
O 12	0.3397(20)	0.7841(19)	0.1903(10)	0.82(20)
O 13	0.9959(16)	0.6708(16)	0.0797 (8)	0.60(16)
O 14	0.0023(18)	0.0137(17)	0.1857(10)	0.79(18)

After Ref. (27).

The striking feature in this structure of triclinic space group symmetry P$\bar{1}$ is the isolated chain-like (Si$_3$O$_{10}$) groups plus additional (SiO$_4$) tetrahedra (Fig. 29), whereas all the other rare-earth disilicate structure types contain (Si$_2$O$_7$) double-tetrahedra groups either of the staggered or of the eclipsed configuration. These chains of (Si$_3$O$_{10}$) groups are parallel [$\bar{7}01$]. The Si atoms might be seen to be accomodated in holes left between the 'heavy cation'-oxygen polyhedra. These polyhedra

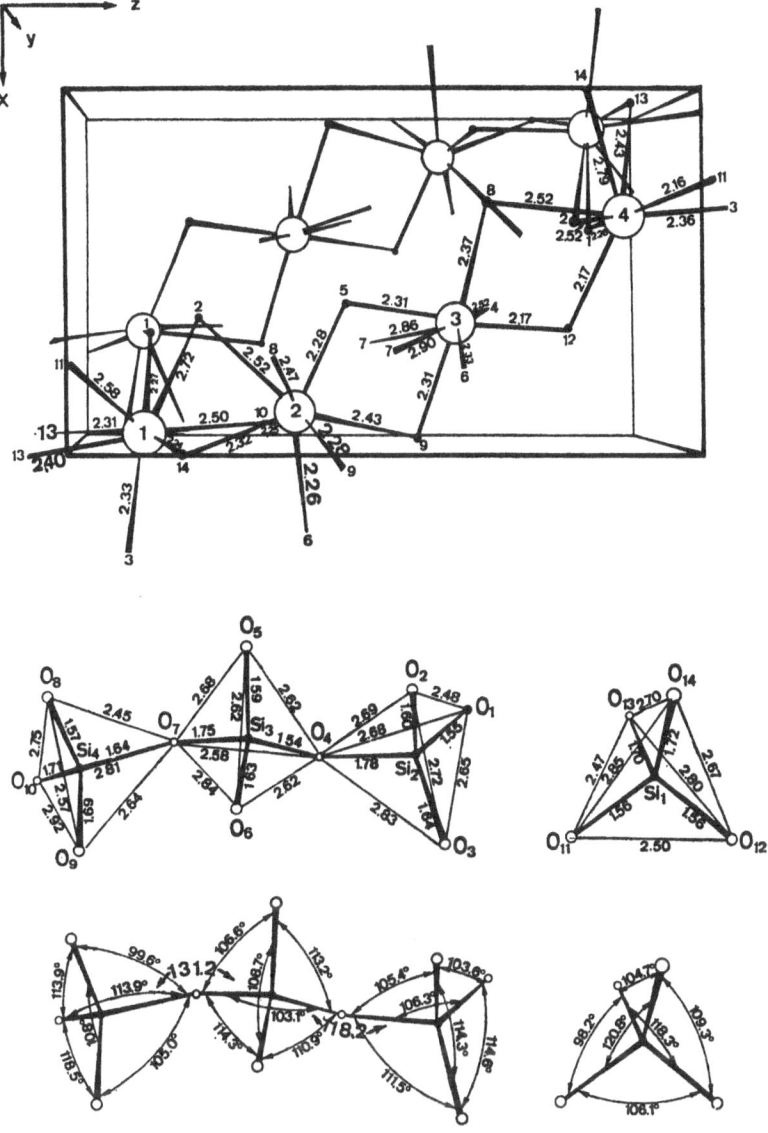

Fig. 30. Interatomic distances and angles in the crystal structure of $Ho_4(SiO_4)$-(Si_3O_{10})

all show eightfold coordination. They are also parallel [$\overline{7}01$], forming infinite chains by edge sharing. These (RE—O_8)chains are connected to each other by corner sharing at every second polyhedron of each chain.

The shape of the eightfold coordinated heavy atoms can hardly be described in terms of Cartesian geometry (Fig. 30). Their shape varies between a strongly distorted cube and a distorted type of dodecahedron. The mean values of the Ho—O distances are 2.42 Å, 2.42 Å, 2.47 Å, 2.46 Å, which gives an average value of 2.44 Å for all four crystallographically independent Ho cations.

The (SiO_4) tetrahedra show a high degree of distortion (Fig. 30) with Si—O distances varying between 1.78 and 1.54 Å. This is the largest difference ever observed in a (SiO_4) tetrahedron in silicate structures, in view of the fact that e.s.d. values are 0.007 Å in this investigation. These extreme Si—O values within the individual tetrahedra belonging to the (Si_3O_{10}) chain are fairly well balanced, however. The mean values of the four Si-O distances in the three tetrahedra are 1.64 Å, 1.63 Å, and 1.65 Å. Another extreme value seems to be given in this configuration by the 118.2° angle at the bridging Si(2)—O(4)—Si(3) between the first and the middle tetrahedron in the (Si_3O_{10}) chain-like group. The other bridging angle Si(3)—O(7)—Si(4) is 133.2°, thus quite regular as compared to the observed angles for other disilicate groups, e.g. 129° and 133° in the Pr Di A structure. The degree of distortion seems also to be extremely high in the isolated (SiO_4) tetrahedra in this structure with two long Si—O distances of 1.72 and 1.70 Å, which are compensated by two extremely short Si—O distances of 1.56 Å. This single tetrahedron also shows the largest deviations from the ideal tetrahedral angle with 120.8° for O(11)—Si(1)—O(14) and 98.2° for O(11)—Si(1)—O(13).

At this point, it seems worthwhile to comment on the transition characteristics observed for polymorphic compounds of types RE Di E and RE Di B. The reconstructive type of transition, which had been suggested because of the extremely low rate of transformation and the fact that the pure B-type phase had never been observed after transition from the E-type modifications, is best understood in the light of the given structural information. The E → B transition appears to be mainly determined by the breaking of Si—O—Si bonds of the double-tetrahedra groups in the E-type structure to achieve the (Si_3O_{10}) configuration present in the RE Di B-type structure, and vice versa, on the pattern $2 (Si_2O_7) \longleftrightarrow (Si_3O_{10}) + (SiO_4)$.

An isotypic compound is likely to exist with the digermanate structures of the large rare-earth cations, as has been described recently for the La analogue (65). These structural data on the La digermanate were extremely helpful to identify the oxygen positions in the Ho disilicate structure as given in Table 16. Earlier suggestions (27) that there might be a close structural relation to $Cd_2P_2O_7$ or $K_2Cr_2O_7$ have to be rejected from the present point of view since both of the latter structure types contain double-tetrahedra groups (X_2O_7).

3.1.3. Structure Type C, (Ho. ... Lu)$_2$(Si$_2$O$_7$), 'RE Di C'

Disilicate structure type C is the only one in the family of disilicate structures which is stable from room temperature up to the melting point of the compounds. Its range of stability along the rare-earth series is extended beyond the radius of the smallest rare-earth Lu^{3+}, to Sc^{3+} with $r = 0.68$ Å (17, 66). This structure type has also been named after the mineral thortveitite.

The latest crystal structure analysis, carried out on the Yb analogue (24), gave essentially the same result as that published for thortveitite, Sc$_2$(Si$_2$O$_7$) (67). These authors concluded, from arguments derived from crystal chemistry concerning the different bond lengths and temperature vibrations of individual oxygen atoms, that space group C2/m is the correct one rather than the other possible space groups C2 or Cm. The data on Sc$_2$(Si$_2$O$_7$) originally gave rise to a most stimulating discussion as to the possibility of 180° angles for the bridging Si—O—Si in silicate structures. This linear bridge has now been confirmed for Yb$_2$(Si$_2$O$_7$). Considerable thermal motion of the bridging oxygen is however indicated by the B-value of 1.02 Å^2 as compared to 0.50 Å^2 and 0.54 Å^2 of the terminal oxygens. The structural analysis was carried out with 1220 single-crystal diffractometer intensity data. The final R value of 5.4%, corresponds to e.s.d.'s for the cation-oxygen distances of about 0.005 Å. The refined values of the atomic parameters are given in Table 17.

Table 17. *Atomic parameters of Yb$_2$(Si$_2$O$_7$)*

Atom	x	y	z	$B[\text{Å}^2]$
Yb	0.5	0.80687(2)	0.0	0.25
Si	0.7189(3)	0.5	0.4125(6)	0.37
O(1)	0.5	0.5	0.5	1.02
O(2)	0.8831(5)	0.5	0.7151(15)	0.50
O(3)	0.7361(5)	0.6504(4)	0.2197(11)	0.54

After Ref. (24).

The wide range of stability of this structure type is likely to be explained by the nearly closest hexagonal packing of the oxygens (Fig. 31) containing rare-earth cations in the octahedra holes and silicons in the tetrahedra holes in alternating parallel layers (001). The (SiO$_4$) tetrahedra show a very low degree of distortion as compared to other disilicate configurations (Fig. 32). The mean value of the Si—O bond length is 1.63 Å and the variation is ±0.01 Å. Also the octahedral oxy-

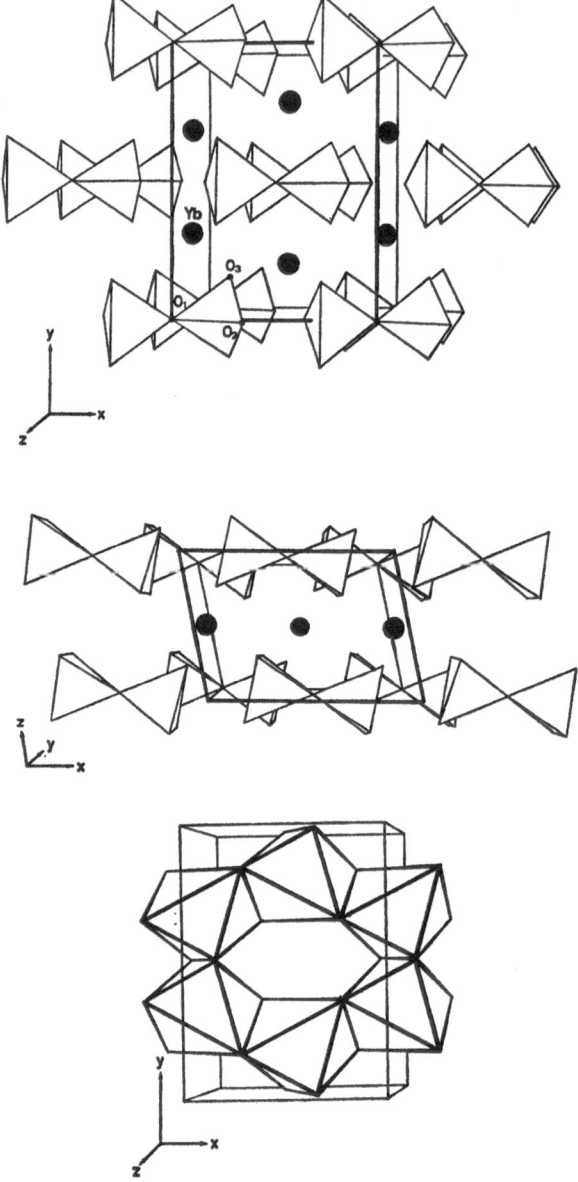

Fig. 31. Crystal structure of $Yb_2(Si_2O_7)$

gen surrounding the Yb atom is fairly regular. The mean Yb—O distance is 2.24 Å and the low degree of distortion is demonstrated by the small deviations of $+0.04$ Å and -0.03 Å.

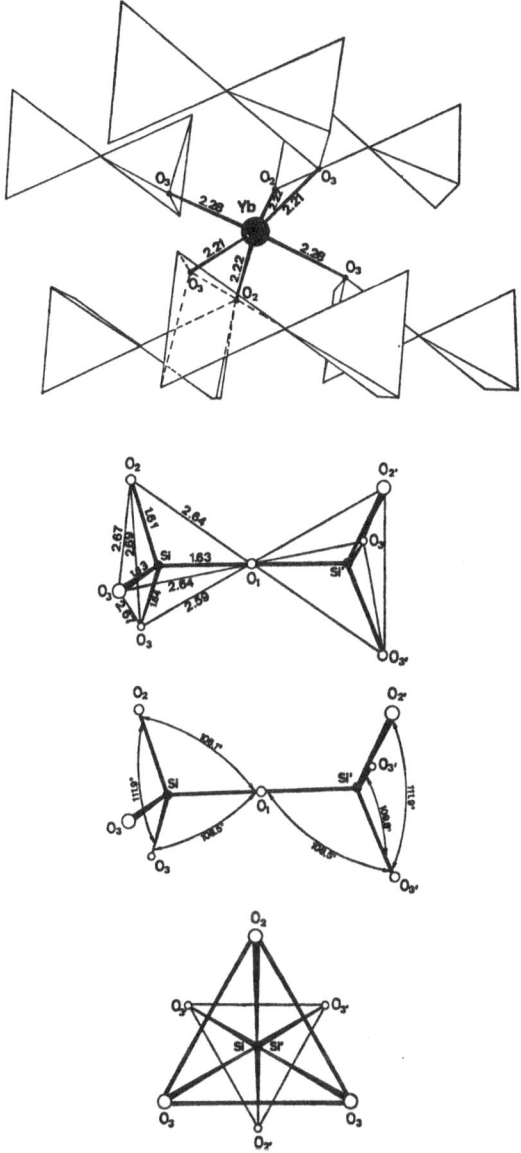

Fig. 32. Cation-oxygen coordination in the crystal structure of $Yb_2(Si_2O_7)$

3.1.4. Structure Type D, (Ho, ... Er)$_2$(Si$_2$O$_7$), 'Re Di D'

The crystal structure of type D disilicates, was first described for Y_2 (Si$_2$O$_7$) (22, 23), which is isostructural with (Ho,Er)$_2$(Si$_2$O$_7$). The struc-

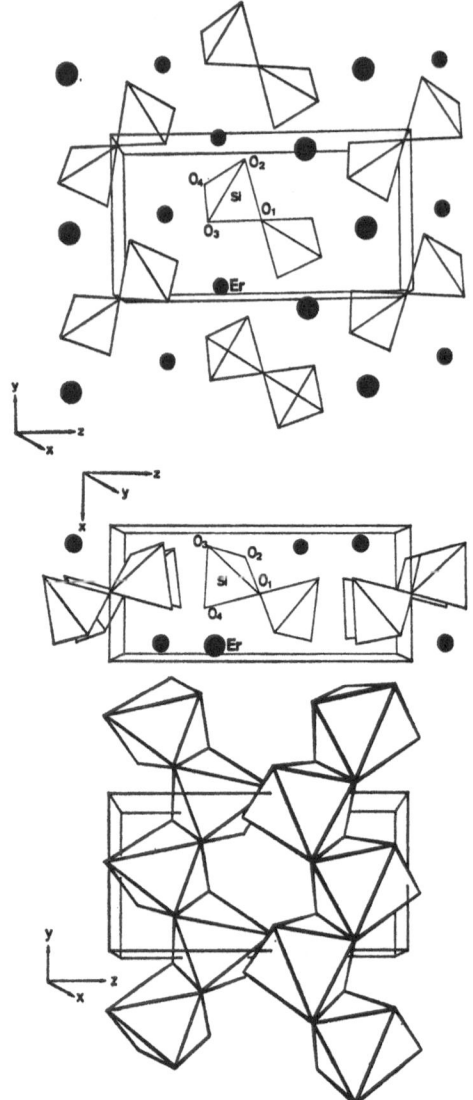

Fig. 33. Crystal structure of $Er_2(Si_2O_7)$ type D

ture has recently been refined from 1860 independent observations (*hkl*) for $Er_2(Si_2O_7)$ (*24*) which gave a final R value of 6.2% for the three-dimensional data. The atomic parameters and their standard deviations are listed in Table 18.

Table 18. *Atomic parameters of D-type $Er_2(Si_2O_7)$*

Atom	x	y	z	$B[\text{Å}^2]$
Er	0.88829(8)	0.09318(6)	0.34934(5)	0.29
Si	0.3601 (4)	0.6442 (3)	0.3871 (3)	0.33
O(1)	0.5	0.5	0.5	0.91
O(2)	0.2052 (8)	0.8653 (7)	0.4486 (6)	0.64
O(3)	0.1235 (9)	0.4583 (8)	0.3191 (6)	0.63
O(4)	0.6184 (9)	0.7522 (7)	0.2984 (6)	0.56

After Ref. (24).

The interesting unit in this structure (Fig. 33) is the (Si_2O_7) double-tetrahedra group. Its centrosymmetry and 180° Si—O—Si angle follow directly from the space group $P2_1/b$. Fourfold general positions and twofold special positions at the centers of symmetry are possible in this space group. Since the unit cell with $Z = 2$ contains only two pyrosilicate groups, they necessarily occupy these special positions. Evidence for the linearity of the Si—O—Si bond, therefore does not depend on the accuracy of the intensity data applied during the structure refinement, as was the case with C-type $Sc_2(Si_2O_7)$ and $Yb_2(Si_2O_7)$. Thus, the structure of $Er_2(Si_2O_7)$ appears to provide further evidence for the possible existence of (Si_2O_7)groups with a Si—O—Si angle equal to 180°.

Also in this case the thermal motion of the bridging oxygen is considerably stronger than of the terminal oxygens (see Table 18). The bond lengths and valence angles in the double-tetrahedra group are given in Fig. 34.

The mean Si—O terminal O length is 1.62 Å. These terminal oxygen atoms of the (Si_2O_7) group form strongly distorted octahedra around the Er atoms similar to those around the Yb atom in $Yb_2(Si_2O_7)$. The bridging oxygen atom O(1) is not bonded to the Er cation. The mean Er—O distance in the octahedron is 2.27 Å. With the subgroup relation of space group $P2_1/b$ to space group $C2/m$ of structure type C, some structural features are correlated, namely, the staggered orientation of the double tetrahedra with the 180° Si—O—Si bridging angle and the sixfold oxygen coordination around the heavy atoms. However, in structure type D the orientation of the (Si_2O_7) groups relative to each other, and the connection of the Er—O octahedra are different. The (Si_2O_7) groups in structure type D are mutually directed along $[0\overline{1}1]$ and $[0\overline{1}\overline{1}]$. Consequently, the coordination of the heavy atoms does not result in a network-forming feature, as in structure type C, but in a ribbon-like arrangement along the b axis.

163

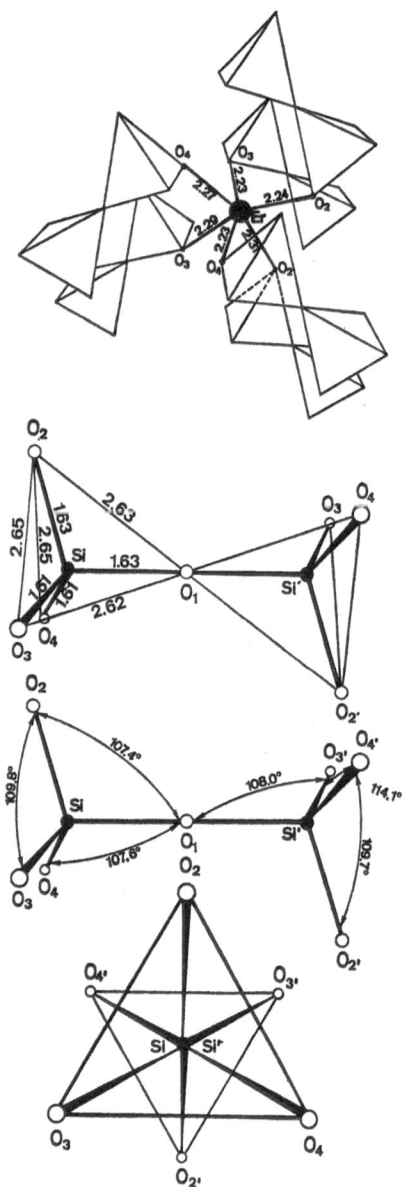

Fig. 34. Cation-oxygen coordination in D-type $Er_2(Si_2O_7)$

3.1.5. Structure Type E, (Eu, ... Ho)₂(Si₂O₇), 'RE Di E'

This rare-earth disilicate structure type was determined first for the Gd analogue (24) and later was also refined for the Eu disilicate compound (28). $Eu_2(Si_2O_7)$ is of special interest in the rare-earth disilicate series because of its large number of polymorphic forms. According to its position between the larger and the medium-sized rare earths, four of the seven rare-earth disilicate structure types might be studied in Eu_2 (Si_2O_7). The structure of E-type $Eu_2(Si_2O_7)$ has been refined from 2986 observations (hkl). The structure was confirmed to have space group symmetry $Pna2_1$ with four formula units per cell. The estimated standard deviations of the Eu-oxygen distances are 0.008 Å. The corresponding final atomic parameters refined from a set of three-dimensional data to a residual R value of 4.5% are listed in Table 19.

Table 19. *Atomic parameters of E-type Eu₂(Si₂O₇)*

Atom	x	y	z	$B[Å^2]$
Eu_1	0.12580(5)	0.33681(16)	0.99142(5)	0.44 (9)
Eu_2	0.12558(5)	0.33727(14)	0.50659(5)	0.39 (8)
Si_1	0.3208 (1)	0.3742 (3)	0.2499(10)	0.11 (1)
Si_2	0.5383 (1)	0.6232 (4)	0.2523(11)	0.14 (1)
O_1	0.2710 (6)	0.4747 (16)	0.0857(12)	0.28(11)
O_2	0.2673 (7)	0.4860 (18)	0.4059(13)	0.41(12)
O_3	0.3454 (4)	0.0653 (10)	0.2490(21)	0.41 (8)
O_4	0.4208 (4)	0.5550 (10)	0.2443(13)	0.28 (8)
O_5	0.5459 (7)	0.8003 (17)	0.0937(13)	0.53(13)
O_6	0.5469 (7)	0.7754 (16)	0.4221(13)	0.39(12)
O_7	0.5989 (7)	0.3508 (11)	0.2564(24)	0.58 (7)

After Ref. (28)

The orthorhombic structure consists of isolated (Si_2O_7) double-tetra-hedra groups which are in the eclipsed configuration with 158.3° of the Si(1)—O(4)—Si(2) angle, and two crystallographically independent heavy atoms which show sevenfold oxygen coordination. The main structural feature is given by a layering of the (Si_2O_7) groups and the rare-earth atoms alternating in parallel [001], as illustrated in (Fig. 35). This layering of both types of ions along the c axis with a period of c/2 and additional layering of the heavy atoms in parallel [100] with a period of about $a_0/4$, is responsible for a peculiar diffraction pattern on MoKα-precession photographs, suggesting a superstructure character. The pseudo-extinctions which are not defined by the common extinction

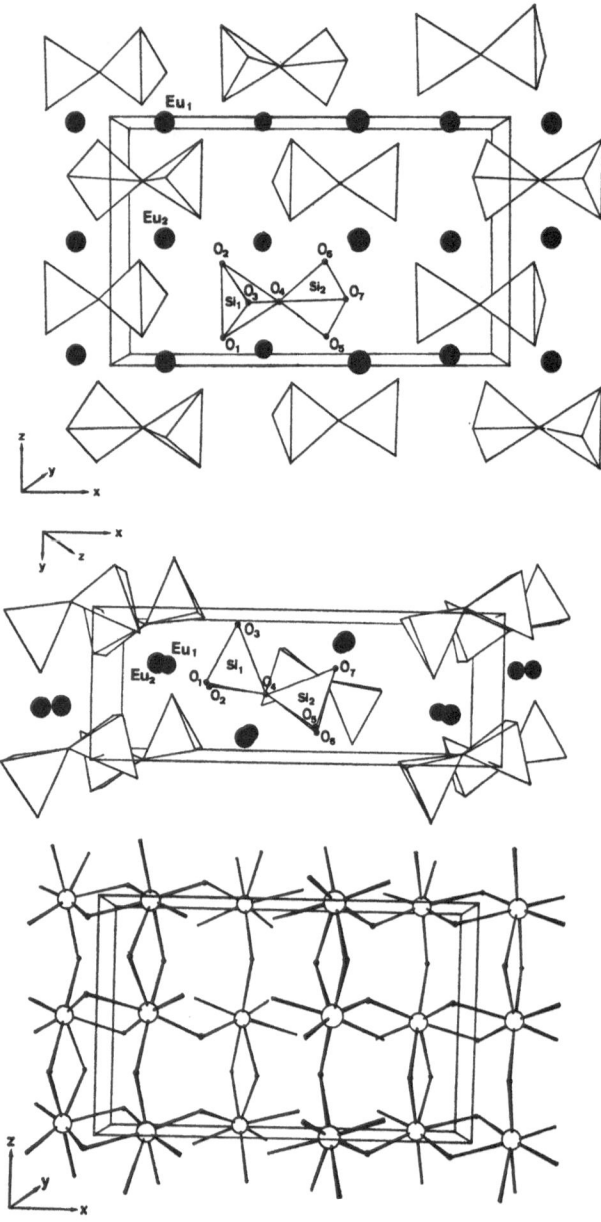

Fig. 35. Crystal structure of E-type $Eu_2(Si_2O_7)$

rules give the impression of a B-centred subcell which has just 1/4 of the volume of the supercell of corresponding space group symmetry Pna2₁-Pnam.

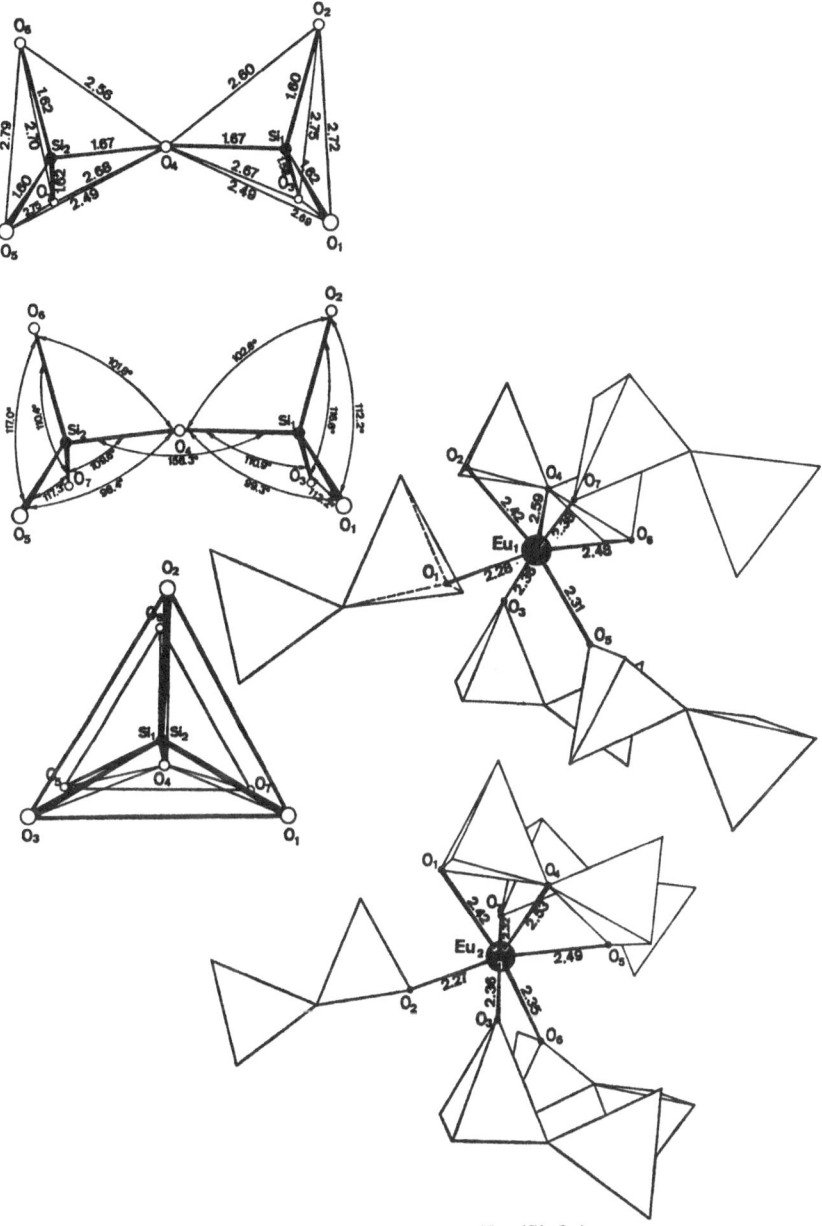

Fig. 36. Cation-oxygen coordination in E-type Eu₂(Si₂O₇)

An intersting feature of the structure which results from the layering is the participation of the bridging oxygen and two terminal oxygens of the same (Si_2O_7) group in the first coordination shell of both the heavy atoms. The double-tetrahedra group shows a distinct difference between the lengths of the terminal oxygens to the silicon atoms and that of the bridging oxygen to the silicon atoms with $1.61 \geqslant 1.67 \leqslant 1.62$ Å (Fig. 36). The sevenfold oxygen-coordinated heavy atoms showing a mean Eu—O distance of 2.41 and 2.40 Å, respectively, share two edges and three corners with each other. This arrangement of the (RE—O) polyhedra results in a wave-like packing of the polyhedra in parallel (010).

3.1.6. Structure Type F/G, (LA, ... Sm)$_2$(Si$_2$O$_7$), 'RE Di F/G'

The structure type G of the rare-earth disilicates has been determined once for $Nd_2(Si_2O_7)$ in the orthorhombic space group P2$_1$2$_1$2$_1$ (24) and again for $Pr_2(Si_2O_7)$ in the monoclinic space group P2$_1$/n with a monoclinic angle of 90.0° (26), thus indicating pseudo-orthorhombic symmetry. As the results of both structure determinations do not reveal any differences in crystal chemistry, we shall refer to the data for the $Pr_2(Si_2O_7)$ analogue in the following discussion. Single crystals of this compound were easily obtained by sintering the compound at temperatures of about 1550 °C. 1093 independent observations on (hkl) were used for the determination and refinement of the crystal structure. The pseudo-orthorhombic unit cell contains four formula units $Pr_2(Si_2O_7)$ It has been refined to a residual R value of about 3.0%, which corresponds to e. s. d.'s of 0.008 Å for the individual cation-oxygen bond lengths. The atomic parameters are given in Table 20:

Table 20. *Atomic parameters of G-type Pr$_2$(Si$_2$O$_7$)*

Atom	x	y	z	$B[\text{Å}^2]$
Pr(1)	0.52097(12)	0.80685(7)	0.76857(25)	0.604(25)
Pr(2)	0.84809(13)	0.60513(7)	0.59012(4)	0.647(25)
Si(1)	0.7968 (6)	0.2473 (3)	0.0252 (2)	0.68 (11)
Si(2)	0.9085 (6)	0.4964 (3)	0.1789 (2)	0.72 (11)
O(1)	0.8324 (15)	0.4220 (8)	0.0643 (5)	1.17 (36)
O(2)	0.0713 (16)	0.1434 (8)	0.0783 (50)	0.43 (32)
O(3)	0.5749 (16)	0.1548 (9)	0.0534 (5)	0.28 (32)
O(4)	0.7403 (16)	0.2391 (9)	0.9060 (5)	0.76 (35)
O(5)	0.7441 (15)	0.4167 (8)	0.2403 (5)	0.30 (32)
O(6)	0.2216 (16)	0.4744 (9)	0.2477 (5)	0.72 (35)
O(7-	0.8192 (17)	0.6747 (9)	0.1724 (5)	0.83 (34)

After Ref. (26).

The structure is built up of columns containing eclipsed double-tetrahedra groups (Si_2O_7), packed with their 'backbones' towards each other but at alternating heights of $a_0/2$ (Fig. 37). Hence, the terminal oxygen atoms are found at the surface of the columns. This arrangement leads to an eightfold oxygen coordination of both the heavy atoms. The Pr oxygen polyhedra in the shape of slightly distorted cubes share edges parallel to the [001], [011̄] and [100] directions. Thus, twice the mean of the cube edge lengths gives precisely the value of the a_0 dimension (5.40 Å) of the pseudoorthorhombic but monoclinic unit cell. The shape and the arrangement of the Pr-oxygen polyhedra reveal some structural relation to the rare-earth dioxide fluorite-type structure which is the basic unit of a series of suboxide structures in the system PrO_2—$PrO_{1.5}$. The dimension of the face-centred cubic dioxide unit cell is 5.39 Å and the C-type sesquioxide body-centered cubic unit cell of Pr_2O_3 shows a $= 11.15$ Å.

The [Si_2O_7] double tetrahedra (Fig. 38) show a significant difference in the length of the Si(1)—O(1) and Si(2)—O(1) bridging bonds which are 1.599 ± 0.009 Å and 1.638 ± 0.009 Å, respectively. This difference is compensated by silicon terminal oxygen bonds which show mean values of 1.590 Å for the first tetrahedron and 1.578 Å for the Si(2)—O distances. For detailed data on the geometry of the double tetrahedra see Fig. 38. The Si(1)—O(1)—Si(2) angle is $131.2 \pm 0.4°$ and the twist angle of the tetrahedra against each other about the Si(1)—Si(2) direction is $31.4 \pm 0.2°$. The cube-like praseodymium-oxygen polyhedra show remarkable variation of the Pr—O distances, as shown in Fig. 38. The bridging oxygen atom of the [Si_2O_7] group is not involved in the first coordination shells, however. Pr(1)—O distances range from 2.39 to 2.86 Å, resulting in a mean value of 2.57 Å. The variation is more restricted in the Pr(2) coordination with values 2.44—2.64 Å for seven oxygen distances. But the extremely short distance of 2.27 Å for the eighth oxygen atom of O(7) reduces the mean to 2.51 Å, which is considerably smaller than in the first cube. This experimental value of 2.27 Å is smaller than the sum of the ionic radii of 2.48 Å, according to data from (54). It accounts for the strong polarizing power of the lanthanide ion at this position.

The reversible transformation which was described for (La, ... Eu)$_2$ Si_2O_7 compounds at 1350 ± 50 °C (18) is best understood in terms of the structural data given for $Pr_2(Si_2O_7)$. The transformation is of the reconstructive type and demands a considerable rearrangement of the cations and anion groups. The sheet-like structure of low $Pr_2(Si_2O_7)$ has to assume the new arrangement of back-to-back packed double tetrahedra ordered in columns from the nearly closest packing in the sheets of the low form. The long diffusion lengths of the atoms account for the long time observed for the transformation.

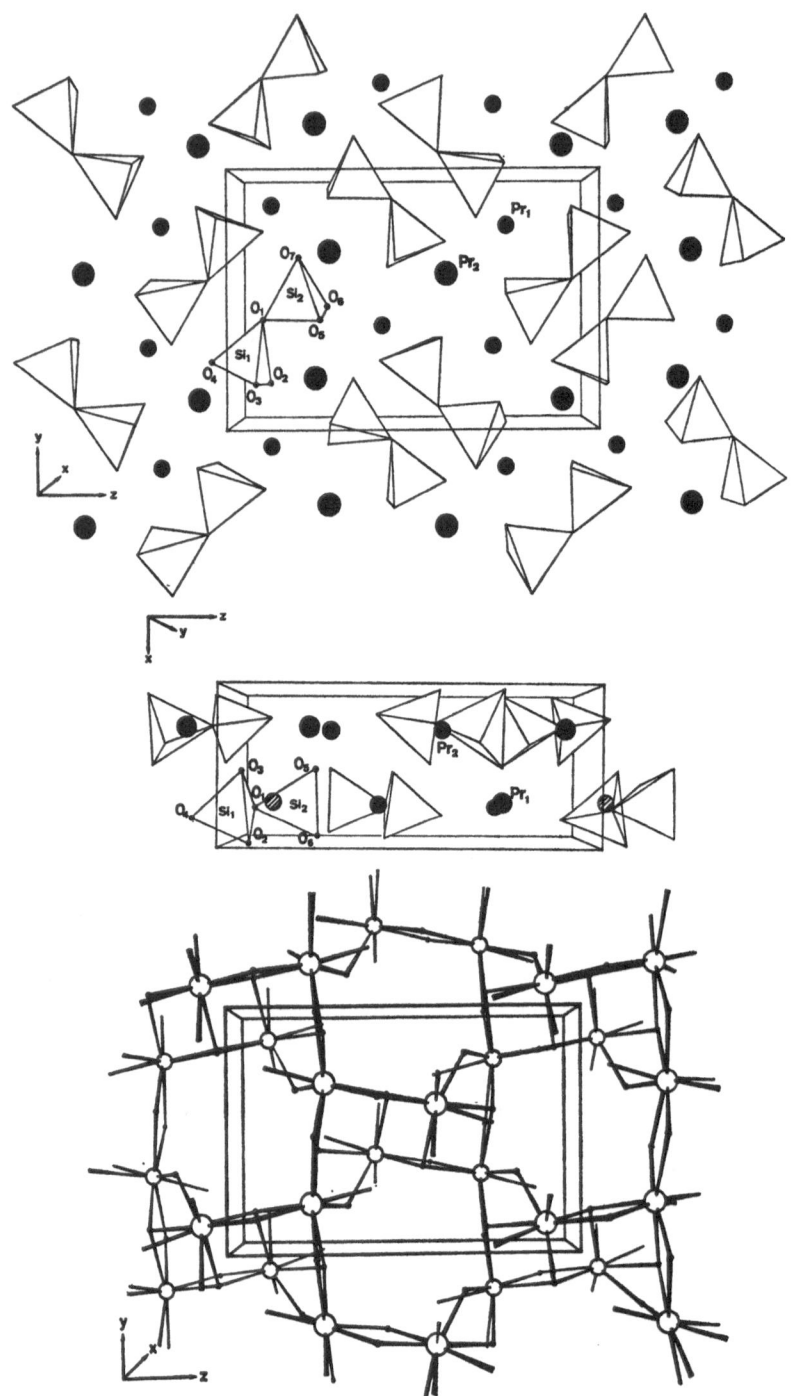

Fig. 37. Crystal structure of G-type Pr$_2$(Si$_2$O$_7$)

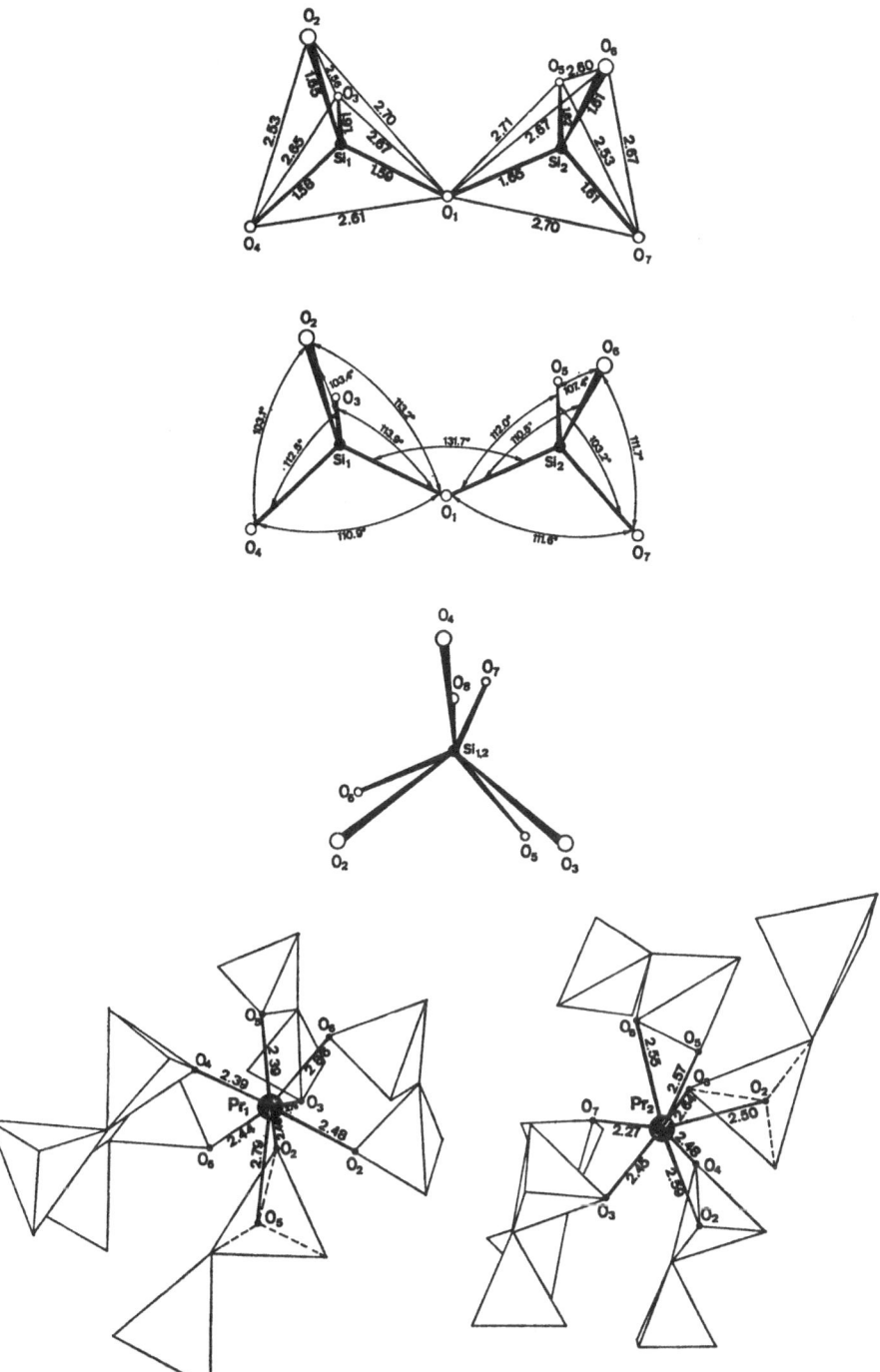

Fig. 38. Cation-oxygen coordination in G-type $Pr_2(Si_2O_7)$

The G-type disilicate structure is the preferred one during the phase formation on crystallization of Ce_2O_3–Li_2O–SiO_2 glasses (68). Tb_2 (Si_2O_7) of type G has also been observed during heating experiments in the corresponding system (69).

3.2. Ternary Compounds, $Alk_3RE(Si_2O_7)$

Compounds of this composition have been reported for Na and K analogues. The crystal structure of the compound $K_3Eu(Si_2O_7)$, which has been synthesized by solid-state reaction of the oxides using a flux of KF (70), has not yet been investigated. There are no isotypic rare-earth compounds known so far. However, the structure type of the Na—containing rare-earth disilicate is a common one, often observed during hydrothermal synthesis in systems Na_2O–RE_2O_3–SiO_2–H_2O (29–31). It obviously coexists with the C-type compounds $NaRE(SiO_4)$ of the smaller rare earths Er to Lu. The structure has been determined from a $Na_3Sc(Si_2O_7)$ crystal (77).

3.2.1. Structure Type A, $K_3Eu(Si_2O_7)$

Single crystals of $K_3Eu(Si_2O_7)$ have been synthesized by the flux method using KF in a long-time slow cooling experiment yielding a crystal up to several millimeters in size (70). Goniometric measurements showed that the crystals belong to the hexagonal system. The indices of refraction are $n_o = 1.713$, and $n_e = 1.709$. The observed density is 3.41 g/cm^3. IR absorption spectra have been reported in the range 400—1200 cm^{-1}. Weissenberg single-crystal diffraction patterns showed systematic absences of reflections $hhOl$ with $l = 2n + 1$ which suggested the space groups P6₃cm, P6c2 and P6₃/mcm. From a discussion of the possible lattice sites occupied by the different atoms, the space group P6₃/mcm was considered to be the most probable one. The unit cell dimensions are $a = 9.98 \pm 0.01$ Å and $c = 14.44 \pm 0.02$ Å. The authors of (70) suggest a close structural relation to compounds of type $X_2Pb_2(Z_2O_7)$ with X = K, Rb, Cs and Z = Si, Ge. A close relation with $K_2Pb_2(Si_2O_7)$ (72) is discussed by these authors. This structure is supposed to be characterized by layers of composition $Pb_2Si_2O_7$ containing centro-symmetric (Si_2O_7) groups which are linked by Pb atoms. The layers are likely to be separated by the larger potassium cations of CN 9.

3.2.2. Structure Type B, $Na_3(Y,Er \ldots Lu,Sc) (Si_2O_7)$

Well-formed transparent single crystals $Na_3Sc(Si_2O_7)$ of a distinct orthorhombic habitus were obtained during the study of the system Na_2O–Sc_2O_3–SiO_2–H_2O (31). The orthorhombic structure of space group Pbnm contains four formula units $Na_3Sc(Si_2O_7)$ in the unit cell with $a = 5.354(3)$, $b = 9.347(4)$, and $c = 13.089(4)$ Å. Three-dimensional X-ray intensity film data were used to determine the parameters of the eight independent ions per primitive unit cell (71). By means of 541 independent (hkl) observations, the atomic parameters (given in Table 21), have been refined in accordance with a residual $R = 11.3\%$. This might mean e.s.d. values of the cation-anion distances of about 0.03 Å.

Table 21. *Atomic parameters of $Na_3Sc(Si_2O_7)$*

Atom	x	y	z	$B[Å^2]$
Sc	0.0	0.0	0.0	0.77
Si	0.512	0.176	0.131	0.86
Na(1)	0.017	0.021	0.250	1.82
Na(2)	0.015	0.347	0.093	1.92
O(1)	0.442	0.122	0.250	1.15
O(2)	0.479	0.347	0.118	1.19
O(3)	0.320	0.090	0.062	1.32
O(4)	0.800	0.127	0.107	1.21

After Ref. (71).

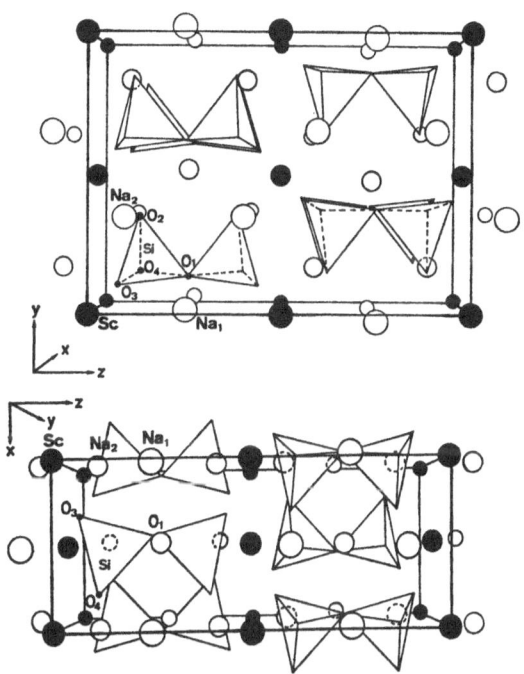

Fig. 39. Crystal structure of $Na_3Sc(Si_2O_7)$

The structure consists of isolated (Si_2O_7) groups, scandium-oxygen polyhedra in the shape of slightly distorted octahedra, and two independent sodium cations with Na(1) in a fivefold oxygen coordination and Na(2) in a fourfold coordination. The main structural motif is created by the isolated Sc—O octahedra with the Sc in the position 0,0,0 (Fig. 39). These octahedra are linked both by the (Si_2O_7) double-tetrahedra groups of eclipsed configuration and by Na—O polyhedra. The structure shows a high degree of similarity to the sheet-like silicates of vermiculite, nontronite and sanbornite, which also show two types of cores along the c axis. The degree of distortion in the Sc—O octahedron is fairly low (Fig. 40) with an average value for the Sc—O distances of 2.10 Å $\pm$ 0.03 Å. The (Si_2O_7) double-tetrahedra group is characterized by a plane of symmetry which relates the individual SiO_4 tetrahedra of the (Si_2O_7) group to each other. The mean Si—O bond length of the terminal oxygens is 1.62 Å, the distance to the bridging oxygen 1.68 Å. The angle Si—O—Si of the eclipsed (Si_2O_7) configuration is 136°.

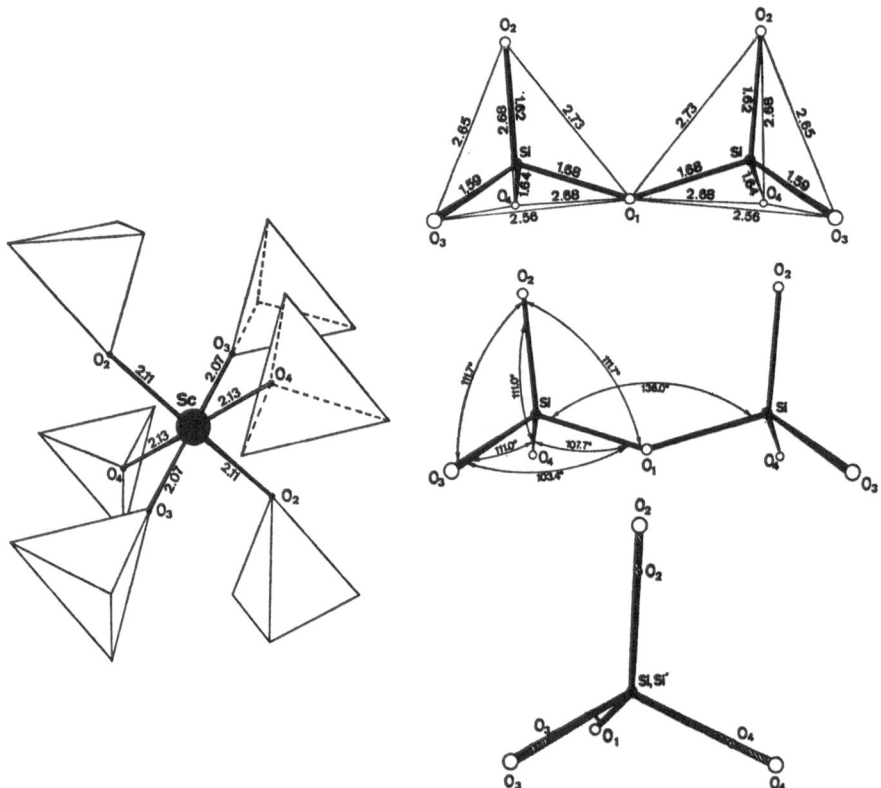

Fig. 40. Cation-oxygen coordination in $Na_3Sc(Si_2O_7)$

4. Polymorphism and Structural Data

Rare-earth silicates of composition $1\ RE_2O_3 \cdot 1\ SiO_2$, $7\ RE_2O_3 \cdot 9\ SiO_2$ and $1\ RE_2O_3 \cdot 2\ SiO_2$ were found in the binary systems RE_2O_3—SiO_2, with RE including the series of trivalent rare earths La^{3+} to Lu^{3+}. The disilicates $RE_2O_3 \cdot 2\ SiO_2$ show extensive polymorphism. Seven different structures, here named RE Di A, ... RE Di G, were observed of types $RE_2(Si_2O_7)$ and $RE_4(Si_3O_{10})(SiO_4)$. The structures contain (Si_2O_7)-douple tetrahedra of either the staggered or the eclipsed configuration and in one case, (RE Di B), almost linear (Si_3O_{10})groups plus isolated (SiO_4) tetrahedra. Oxyorthosilicates $1\ RE_2O_3 \cdot SiO_2$ show two different structures of type $RE_2(SiO_4)O$, named RE Oxy A and RE Oxy B. These structures contain isolated (SiO_4)tetrahedra plus isolated, not silicon-bonded oxygens. Compounds $7\ RE_2O_3 \cdot 9\ SiO_2$ were observed to crystallize in only one structure with the complete series of trivalent rare earths. This structure is of the apatite type $RE_{9.33}\square_{0.66}(SiO_4)_6O_2$, showing $2/3$ cation deficiency per cell.

The ranges of stability of the various structural types over the rare-earth series are shown below Fig. 41. Structures of the ternary rare-earth silicate compounds and the sesquioxides (85) are included for comparison.

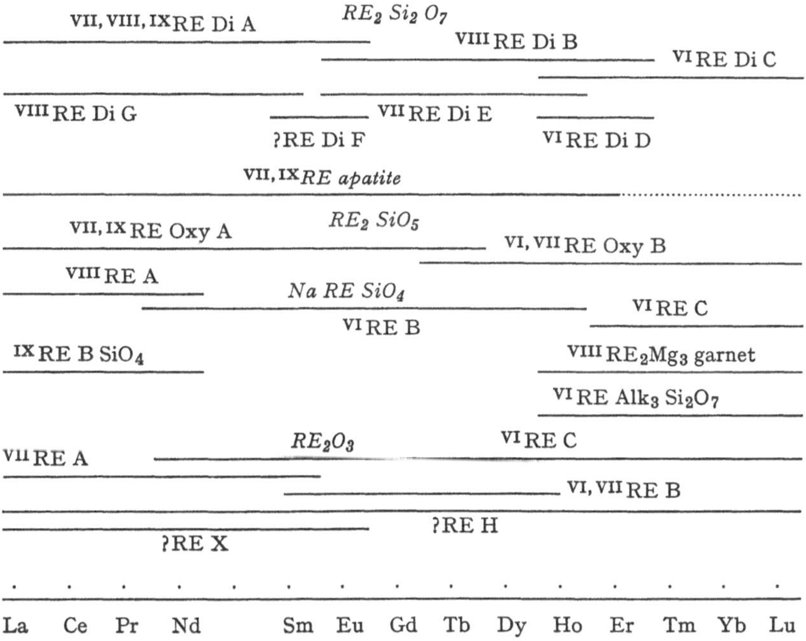

Fig. 41. Stability ranges of structural types along the rare-earth series

The Roman numbers at the upper left of RE are the oxygen coordination numbers.

A purely visual impression from this plot is the formation of three subdivisions along the rare-earth series, marked by the lengths of the horizontal lines. The boundaries occur at positions close to Sm/Eu and Ho/Er, respectively. With the additional information provided by the symbols given for each set of lines, the picture becomes more quantitative. In all structures the larger rare-earth cations show coordination numbers VII to IX, and the smaller ones coordination numbers VII to VI. Some structures, such as RE_2SiO_5 and apatite, deviate from this statistical model. The main reason for this might lie in the mixed coordination numbers (VII, IX) or in the different qualities of the anionic units of the structures. Outsiders, such as the apatite structure, might also occur because of its special situation of cation deficiency. It is interesting to see, however, that the Ho/Er intersection corresponds quite well to the upper limit of the radius ratio for octahedral coordination (0.73), e.g. $r(Ho^{3+})/r(O^{2-}) = 0.93\ Å/1.32\ Å = 0.71$. The Sm/Eu intersection also corresponds to the upper limit (0.84) for the 7-fold coordination (7 balls arranged at the surface of a sphere touching each other), e.g. $Sm^{3+}/O^{2-} = 1.05\ Å/1.32\ Å = 0.81$. With respect to the appreciably ionic nature of the rare earth — oxygen bonds ($\sim75\%$) this would suggest that the sequence of structures along the rare earth series is largely controlled by the radius ratio between cations and anions following Pauling's First Rule (73). However, besides this geometrical principle, which implies the tendency to minimize the over-all potential energy, another aspect concerning the changing degree of covalency in the rare earth series might also be of influence. The covalency of the rare earth is strongly correlated to the energy of the $4f$–$5d$ orbitals and the spin-orbit interaction, both of which show also extreme values around Sm/Eu and Ho/Er (88). Correlations of this type have previously been discussed qualitatively for 154 series of metallic and non-metallic compounds (88). Another aspect in understanding the crystal structure sequence might also be the approach recently published for RE metals (89). There the structural sequence, melting points and heat of sublimation of the lanthanides are explained by a 'dual $4f$ electron model'. This involves the varying density of the 'atomic $4f$ electrons' and 'band or bond $4f$ electrons'. With respect to the ionic rare earth compounds, however, especially to the more complex compounds such as the silicates, some more theoretical and experimental efforts are necessary to estimate the influence of electronic structure and covalency on the structural sequence and polymorphism.

Regarding the appreciable ionic nature of rare earth oxycompounds the following conclusions might be drawn on the silicate structures.

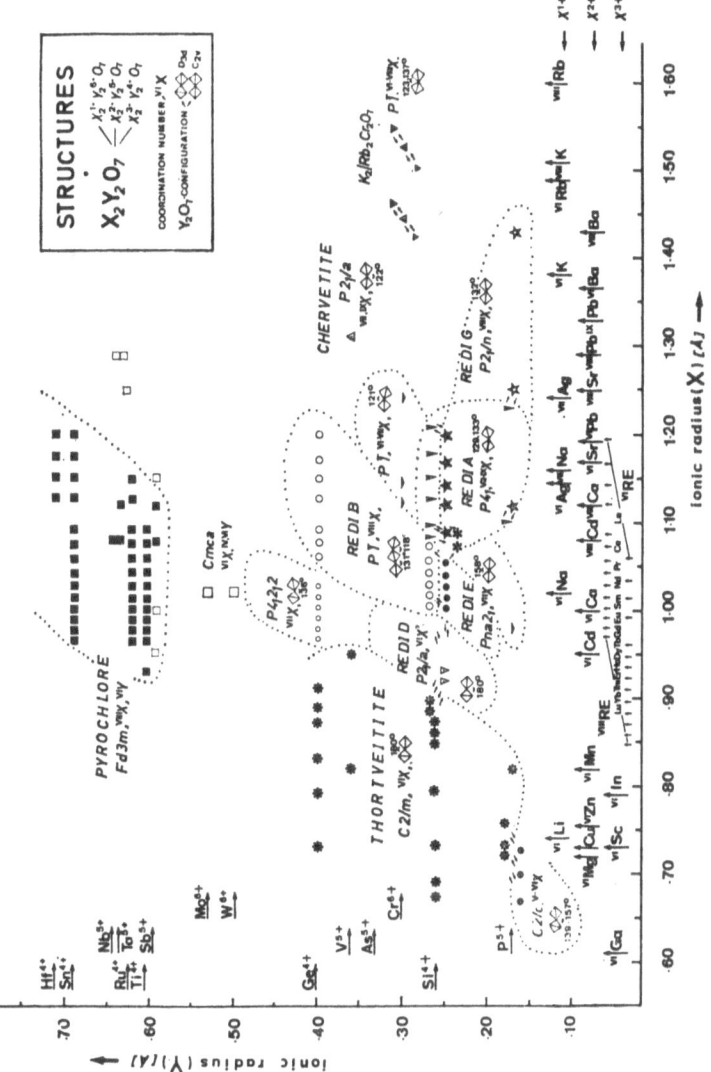

Fig. 42. Stability fields and some structural data on compounds $X_2Y_2O_7$. Dotted lines describe areas of isotypic compounds

The stability of a given structure along the rare-earth series apparently is dependent on the shrinkage in RE^{3+} ion size as the $4f$ subshell is filled. Speaking in terms of a simple electrostatic model, the repulsion forces

J. Felsche

Table 22. *Data on cation-oxygen coordination in rare-earth disilicate crystal structures cor taining (Si_2O_7) or $2(Si_2O_7) \equiv (Si_3O_{10}) + (SiO_4)$ configurations. For symbols used se definitions as given in e. g. chapter 4.1*

Compound				RE—O coordina·	
Mol. composition	Structure type	Coordination formula	mean e.s.d. (M—O)	^{CN}RE	$<\bar{d}>$ [Å] (RE—O)
$1 Pr_2O_3 \cdot 2 SiO_2$	RE Di A	$^{VII}Pr\ ^{VIII}Pr\ ^{IX}Pr_2(Si_2O_7)_2$	0.007 Å	^{VII}Pr	2.48 Å
				^{IX}Pr	2.58
				^{VIII}Pr	2.52
				^{IX}Pr	2.59
$1 Ho_2O_3 \cdot 2 SiO_2$	RE Di B	$^{VIII}Ho_4(Si_3O_{10})(SiO_4)$	0.009 Å	^{VIII}Ho	2.42 Å
				^{VIII}Ho	2.42
				^{VIII}Ho	2.47
				^{VIII}Ho	2.46
$1 Yb_2O_3 \cdot 2 SiO_2$	RE Di C	$^{VI}Yb_2(Si_2O_7)$	0.005 Å	^{VI}Yb	2.24 Å
$1 Er_2O_3 \cdot 2 SiO_2$	RE Di D	$^{VI}Er_2(Si_2O_7)$	0.005 Å	^{VI}Er	2.26 Å
$1 Eu_2O_3 \cdot 2 SiO_2$	RE Di E	$^{VII}Eu_2(Si_2O_7)$	0.008 Å	^{VII}Eu	2.41 Å
				^{VII}Eu	2.40
$1 Pr_2O_3 \cdot 2 SiO_2$	RE Di G	$^{VIII}Pr_2(Si_2O_7)$	0.009 Å	^{VIII}Pr	2.57 Å
				^{VIII}Pr	2.51

between the oxygens in the coordination spheres increase as the size of the rare-earth cations decreases. These forces eventually become large enough to make the structure energetically unstable. At this point the coordination number of the cation is reduced in order to minimize the potential energy from its short-range configuration and a new structure is initiated. This electrostatic principle of a compromise between a maximum of spherical shielding for the cation and a minimum of repulsion amongst the coordination anions, obviously results in a different number of stable crystal structures for compounds of different composition. As

tion					Si—O coordination				
$\Delta d_{max.}$ [Å] $(d-\langle\bar{d}\rangle)$	^{cn}O	$\langle\bar{d}\rangle$ [Å] (Si—O)	Δd_{max} $(d-\langle\bar{d}\rangle)$	^{cn}O	$\bar{d}_{ter}$ [Å]	$\bar{d}_{bri}$ [Å] / Si—O—Si [°]	$\bar{d}_{ter}$ [Å]	^{cn}O	^{cn}O
+0.18 Å / −0.07	3.7	1.63 Å	+0.04 Å / −0.02	3.5		1.65 Å			
					1.62	128.5°	1.63	3.7	3.0
+0.14 / −0.19	3.7	1.63	+0.03 / −0.04	3.5					
+0.44 / −0.19	3.6	1.63	+0.01 / −0.04	3.5		1.65 Å			
					1.62	133.1°	1.62	3.5	3.0
+0.35 / −0.20	3.2	1.63	+0.04 / −0.03	3.3					
+0.30 Å / −0.18	3.5	1.64 Å	+0.08 Å / −0.08	3.5					
+0.42 / −0.17	3.6	1.64	+0.14 / −0.08	3.3	1.60	1.66 Å / 118.2°		3.5	3.0
+0.43 / −0.30	3.5	1.63	+0.12 / −0.09	3.3		1.69 Å			4.0
+0.33 / −0.30	3.3	1.65	+0.06 / −0.08	3.8		132.2°	1.65	3.5	
+0.04 Å / −0.03	3.0	1.63 Å	+0.01 Å / −0.01	2.8	1.63	1.63 Å / 180.0°	1.63	3.0	2.0
+0.05 Å / −0.03	2.3	1.62 Å	+0.01 Å / −0.01	2.3	1.62	1.63 Å / 180.0°	1.62	2.3	2.0
+0.18 Å / −0.13	3.1	1.62 Å	+0.05 Å / −0.02	3.3		1.67 Å			
					1.61	158.3°	1.62	3.0	4.0
+0.12 / −0.13	3.1	1.63	+0.06 / −0.01	3.3					
+0.29 Å / −0.17	3.8	1.61 Å	+0.04 Å / −0.03	3.3		1.62 Å			
					1.62	131.2 Å	1.61	3.7	2.0
+0.13 / −0.24	3.8	1.62	+0.03 / −0.01	3.3					

was shown in detail, silicates of composition 1 $RE_2O_3 \cdot 2\,SiO_2$ show the highest number of polymorphic forms amongst all the rare-earth silicates. The very sensitive configuration of the (Si_2O_7) groups is likely to be responsible for the seven different structure types. The double tetrahedra generally have a less spherical character than isolated (SiO_4) tetrahedra or isolated oxygens, even when these include all possibilities of distortion between staggered and eclipsed orientation. This shortcoming prohibits to provide extensive spherical shielding around the larger rare-earth cations and coordination with oxygens distributed almost equi-

Table 23. *Coordination data on binary rare-earth silicate structures containing isolated* $(SiO_4)^{-4}$ *groups. For symbols used, see definitions as given in e. g. chapter 4.1*

Compound			mean e.s.d. M—O	RE—O coordina-	
Mol. composition	Structure type	Coordination formula		RE	$<\bar{d}>$ (RE—O)
1 Gd_2O_3 1 SiO_2	RE Oxy A	$^{IX}Gd\ ^{VII}Gd(SiO_4)O$	0.009 Å	^{IX}Gd	2.49 Å
				^{VII}Gd	2.39
1 Yb_2O_3 1 SiO_2	RE Oxy B	$^{VII}Yb\ ^{VI}Yb(SiO_4)O$	0.008	^{VII}Yb	2.33
				^{VI}Yb	2.23
7 $Gd_2O_3 \cdot$ 9 SiO_2	Apatite	$^{IX}(Gd_{3.33}+\square_{0.67})$ $^{VII}Gd_6(SiO_4)_6O_2$	0.006	^{IX}Gd	2.53
				^{VII}Gd	2.40
2 $EuO \cdot$ 1 SiO_2	β-Ca_2SiO_4	$^{VIII}Eu\ ^{X}Eu(SiO_4)O$	0.02	^{VIII}Eu	2.62
				^{X}Eu	2.86
3 $EuO \cdot$ 1 SiO_2	Sr_3SiO_5	$^{X}Eu\ ^{VIII}Eu(SiO_4)O$	∼0.02	^{X}Eu	2.83
				^{VIII}Eu	2.64

distantly on the surface of a sphere. Thus, any arrangement of (Si_2O_7) or even (Si_3O_{10}) groups with the large RE^{3+} cations, showing oxygen coordination > 6 should result in less stable structures which are more sensitive to a variation in cation size and to thermal vibration of the individual ions. This seems to be essentially confirmed by the data in Fig. 42. Minimizing their free (*Gibbs*) energy, these disilicate compounds also display polymorphism with transition temperatures in the range of 1250° to 1500 °C. All transitions have been found to be of the reconstructive type because of the extensive structural rearrangements necessitated by the special (Si_2O_7) configurations.

The smaller rare earths, however, which prefer sixfold oxygen coordination, form the only disilicate structure which is stable up to the melting point. Structure type RE Di C is found also for cations, such as

tion		Si—O coordination			O_{ex}—RE coordination		
$\Delta d_{max.}$ $(d-<\bar{d}>)$	cnO	$<\bar{d}>$ Si—O	$\Delta d_{max.}$ $(d-<\bar{d}>)$	cnO	cnO	$<\bar{d}>$ (O—RE)	$\Delta d_{max.}$ $(d-<\bar{d}>)$
+0.27 Å / −0.22	4.0	1.63 Å	+0.04 Å / −0.03	4.0	$^{IV}O_5$	2.32 Å	+0.03 Å / −0.03
+0.14 / −0.09	4.0						
+0.30 Å / −0.05	3.6	1.63	+0.01 Å / −0.02	3.3	$^{IV}O_5$	2.24	+0.10 Å / −0.07
+0.03 / −0.05	3.5						
+0.25 Å / −0.17		1.64	+0.04 Å / −0.02	3.5	$^{III}O_4$	2.23	+0.00 Å
+0.28 / −0.17							
+0.23 Å / −0.12	5.3	1.65	+0.00 Å / −0.01	5.5			
+0.46 / −0.37	5.9						
+0.22 Å / −0.14	6.0	1.64	±0.00 Å	6.0	$^{VI}O_1$	2.58	+0.11 Å / −0.06
+0.29 / −0.17	6.0						

Sc^{3+} and In^{3+}, which are even smaller than Lu^{3+}. The reason for the extensive stability area of this thortveitite structure (see also Fig. 42) might be seen in the almost closest packing of the oxygens with the silicons and rare earths on the tetrahedral and octahedral sites respectively. On the other hand, the smaller stability range of the larger rare-earth disilicate structures is best demonstrated by the large variance in coordination numbers and in RE—O distances found in the individual coordination shells. This is also shown by the data of, e.g. Pr Di A, as compared to Yb Di C in Table 22. Tables 22 and 23 present the coordination data on (RE—O_n)polyhedra and (SiO$_4$)tetrahedra of the individual structures. Structural data of compounds other than the binary compounds have not been taken into consideration because of their relatively poor accuracy.

4.1. Configuration of the (SiO₄)Tetrahedra

The size of most of the trivalent rare-earth ions is far beyond what is required for the stability of octahedral oxygen coordination. For this reason, closest packing of the oxygens with Si and RE cations accommodated in tetrahedral and octahedral holes was not found in any of the larger rare-earth silicate structures. The anionic part of the structures represented by single oxygens, (SiO_4)tetrahedra or by (Si_2O_7)-, (Si_3O_{10})-groups seems to be adapted by the rare-earth cations in such a way as to achieve as much spherical shielding as is allowed by the degree of distortion possible in the individual (SiO_4)tetrahedra. The electrostatic bonding character (75%) and the strong polarizing forces of the rare-earth cations against oxygen result in extreme distortion of the individual (SiO_4)tetrahedra from the ideal tetrahedral configuration. The ideal symmetry of point group T_d, which would correspond to a tetrahedron showing equal Si—O distances and 109° 28' of the O—Si—O angles, was always found to be reduced to C_1.

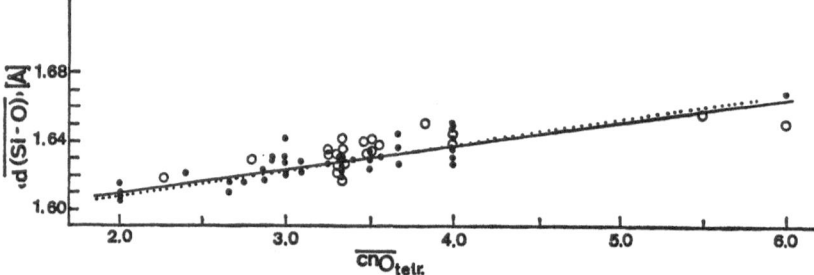

Fig. 43. Variation of mean bond lengths $<\overline{d(\text{Si—O})}>$ with mean coordination number of the oxygens $\overline{^{cn}O}_{\text{tetr.}}$ of individual (SiO_4) tetrahedra in the asymmetric unit of RE-silicate structures, which have been refined to a level corresponding to e.s.d.'s (Si—O) ≤ 0.01 Å (white balls). Black balls and solid line represent corresponding data from (82), dotted line data from (54)

In order to achieve a better understanding of the appreciable variation in Si—O bond lengths, the following correlations were carried out. The average coordination numbers of the silicon-bonded oxygens of individual (SiO_4)tetrahedra were correlated with the mean Si—O distances in the tetrahedra. As Fig. 43 shows, there is rather poor correlation between these values in most cases. Better correlation was found between the variation in the individual Si—O bond lengths Δd, and the variation in charge balance Δp_0 on the oxygens. This correlation is based on the following assumptions. According to Pauling's electrostatic valence principle (73), the individual bond strength of an oxygen surrounded by 'cn' cations of valency z_i to a single cation i with coordination number CN at the center of a polyhedron, is given by $s_i = \dfrac{z_i}{CN_i}$. The degree of charge balance p_0 on any individual oxygen is then given by the 'cn' contributions s_i reaching the oxygen from 'cn' cations of the coordination shells to which it belongs. Thus, the sum over the single bond strengths $p_0 = \overset{cn}{\underset{i=1}{\Sigma}} s_i$ should show a value for oxygens in ionic crystals close to 2.0.

The electrostatic valence principle recently was applied to a large number of silicate structures with some success (83, 84). There, the variations Δd (Si—O) and Δp_0 of the individual Si—O bonds were correlated. However, in the computations it is assumed that all contributions to the charge balance of the oxygen are controlled solely by the quotient $\dfrac{z_i}{CN_i}$. The author neglects the fact that the spherical density distribution of electrostatic forces, arising from the $z \cdot$ e-fold charged cation, decreases with $\dfrac{z \cdot e}{d^2}$, where d is the individual oxygen-cation distance in a given coordination polyhedron. This means that the local charge balance on any oxygen should be strongly dependent on the individual cation-oxygen distances. This point must not be neglected in the case of the rare-earth silicate structures. The RE—O distances have been shown in Tables 22 and 23 to vary considerably within a given coordination polyhedron.

Following Gauss' law, a qualified correction term for Pauling's electrostatic valence rule seems to be given by the quotient $\dfrac{<\bar{d}>^2}{d_i^2}$. It presents a simple dimensionless number. In this term d_i is the individual oxygen-cation distance and $<\bar{d}>$ is the mean cation-oxygen distance in a given coordination polyhedron. Thus, the charge balance on an individual oxygen is given by the value.

$$p_0 = \overset{cn}{\underset{i=1}{\Sigma}} s_i \quad [\text{v.u.}]$$

in valence units [v.u.] with

$$s_i = \frac{z_i}{CN_i} \left(\frac{<\bar{d}>}{d_i} \right)^2 \quad [\text{v.u.}]$$

A strong and nearly linear correlation is given in Fig. 44 for all binary rare-earth silicate structures on the basis of this term for the variation

183

of Δp_0 with Δd values. In the procedure to evaluate the p_0 values, the contribution reaching the individual oxygen from the tetrahedrally coordinated Si^{4+}, have been normalized to 1.0.

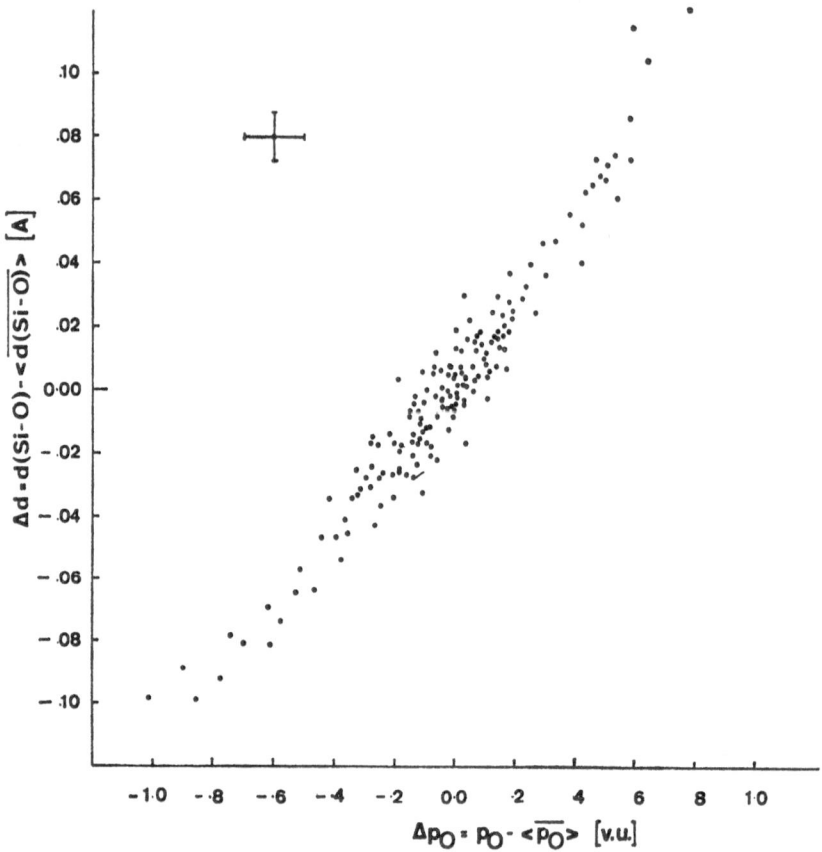

Fig. 44. Bond length variation Δd[Å] versus relative charge balance Δp_0 [v. u.] on the individual oxygens of (SiO_4) tetrahedra in RE silicate structures, which have been refined to an accuracy of ≤ 0.01 Å for the e. s. d.'s of corresponding Si—O distances

The correlation in Fig. 44 implies that the extreme bond-length variation, which was observed principally in the (Si_2O_7) and (Si_3O_{10}) tetrahedra groups, is largely controlled by (electrostatic) overbonding or underbonding on the individual oxygens. For this the strong polarizing forces of the RE^{3+} cations and their highly electropositive bonding

184

character seem likely to be responsible. The data shown in Fig. 44 were evaluated from different sources, as mentioned in the previous sections on structures, which were determined to an accuracy corresponding to e.s.d.'s of the Si—O distances of ≤ 0.01 Å.

4.2. RE—O Coordination Polyhedra

As mentioned above, the oxygen coordination around a rare-earth cation is mainly controlled by the principle of achieving as much spherical shielding as is allowed by the repulsion forces of the coordinating oxygens and by the degree of distortion possible in the $(SiO_4)^{-4}$ and $(Si_2O_7)^{-6}$ anionic units. This means that the coordination numbers of the rare-earths are determined by the size of the rare-earth cations and the geometrical quality of the silicate anions.

Coordination numbers observed in the rare-earth silicate structures vary from 6 to 9 for the trivalent rare-earth cations, and from 7 to 10 for the divalent rare earths, in so far as reliable data on these compounds are available. The variation in RE—O distances is rather large because the coordinating oxygens in the first coordination shell of the rare-earth cations do not represent independent spheres, but usually participate in the complex anions $(SiO_4)^{-4}$ or $(Si_2O_7)^{-6}$. These represent relatively rigid units. Hence, the assignment of Cartesian polyhedra to the distorted spherical coordinations may always be expected to involve a greater or lesser degree of approximation.

Empirical values of effective ionic radii of the trivalent rare earths have been obtained from the quasi-linear relation, unit-cell volume vs. $(RE^{3+}$ ionic radius$)^3$, for some isostructural series. The data used to construct Fig. 45 are the values V/Z (Å)^3, as listed in Tables 1, 4 and 14, for unit cell volume per formula unit, together with experimentally confirmed interatomic distances, i.e. the mean values $<\overline{d\,(\text{RE}-\text{O})}>$ for a given rare-earth coordination. Since the oxygen radius is known to be dependent upon its coordination number, too (Fig. 43: $^{II}O = 1.32$ Å, $^{III}O = 1.34$ Å, $^{IV}O = 1.36$ Å), this had to be taken into account in evaluating the experimental values $<\overline{d\,(\text{RE}-\text{O})}>$ which represent the sum of the ionic radii. Mean coordination numbers of the oxygens surrounding the i rare-earth cations are listed in Tables 22/23. Thus,

$$<\overline{d\,(\text{RE}_i-\text{O})}> - \overline{r\,(^{cn}\text{O})} = r\,(^{\text{CN}}\text{RE}_i)\ [\text{Å}],$$

gave empirically the effective radii r (^{CN}RE) of the rare-earth cations. This represents one fix point in the relation V/Z vs. r^3 for a given isostructural series of compounds. However, to fix the slope of the corre-

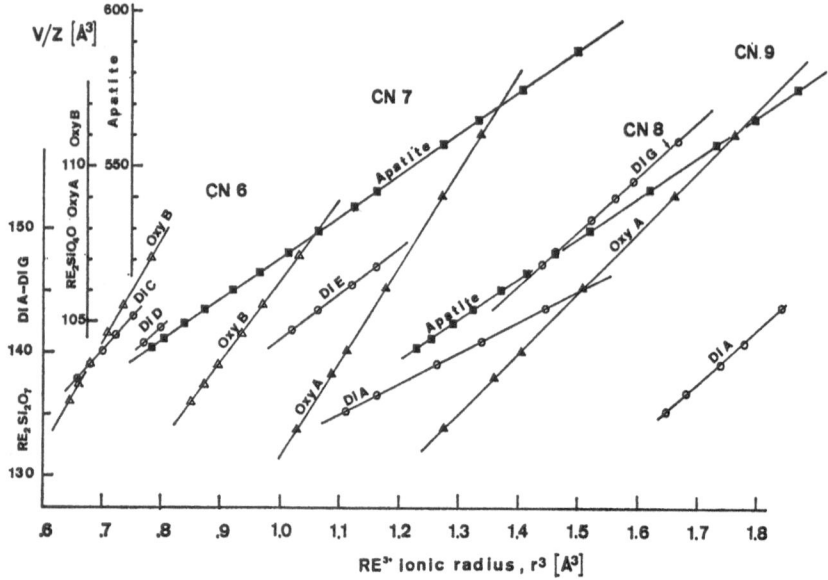

Fig. 45. Plot mol. volume, $V/Z[\text{Å}^3]$ vs RE^{3+} ionic radii, $r^3[\text{Å}^3]$ of isostructural series of RE-silicate compounds

sponding line in Fig. 45, a second known value r ($^{CN}RE_{i+n}$) had to be introduced. In isostructural series where experimental structural data are available on one compound only, the value V/Z $[\text{Å}]^3$ for the remaining members of the series has been made consistent on the assumption that the relative size of the RE^{3+} ions within a given series of identical coordination is the same as in other isostructural compounds of different coordination number. This seems to be proved by the two series for CN 6 and CN 8 of cubic C-type RE_2O_3, 'RE'OCl (48) and RE-pyrochlores (54), respectively. Thus, the unknown ionic radii r (RE_{i+n}) were supplied in terms of the quotients $r^3(RE_i)/r^3(RE_{i+n})$ from other isostructural series with different coordination numbers, which are fully supported by experimental data on structure, i.e. for at least two compounds in a given series. In this way, the slope of the linear relations in Fig. 45 has been determined for the isostructural series of compounds $RE_2(SiO_4)$ O (Oxy A, Oxy B) and oxyapatites $RE_{9.33}\square_{0.67}(SiO_4)_6O_2$. For polymorphic disilicates $RE_2Si_2O_7$ the following structural data have been taken into account: $DiA = (Pr_2(Si_2O_7)$ (26), $Sm_2(Si_2O_7)$ (25), $DiC = Yb_2(Si_2O_7)$ (24), $Sc_2(Si_2O_7)$ (67); $DiD = Er_2Si_2O_7$ (24), $Y_2(Si_2O_7)$ (22, 23); $DiE = Gd_2(Si_2O_7)$ (24), $Eu_2(Si_2O_7)$ (27); $DiG = Nd_2(Si_2O_7)$ (24), $Pr_2(Si_2O_7)$ (20)).

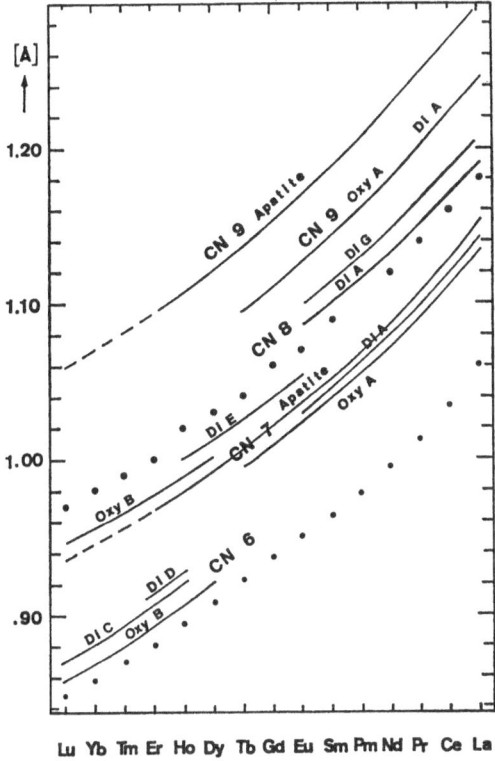

Fig. 46. Effective ionic radii [Å] of trivalent rare earths in different oxygen coordination, CN 6, 7, 8, 9, as derived from structural data, employed in the relation mol. volume vs. r^3 (RE^{3+}) (Fig. 45) for isostructural series of compounds RE$_2$(SiO$_4$)O (OxyA, OxyB), RE$_2$(Si$_2$O$_7$) (DiA, ... DiG) and apatite-like (RE$_{9.33}$ $\square_{0.67}$) (SiO$_4$)$_6$O$_2$. The symbols • and ● represent the r(RE^{3+}) values for CN 6 and CN 8 from Refs. (48) and (54), respectively

The empirical set of RE^{3+} ionic radii with different coordination numbers resulting from this procedure is presented in Fig. 46. The lines representing individual coordination in a given silicate structure type correspond to an experimental error of ± 0.01 Å. Thus, the agreement between r (RE^{3+}) values of the same coordination number in different structure types is fairly good. The data on the ninefold coordinated RE^{3+} cations of apatite (RE$_{3.33}$ + $\square_{0.67}$)RE$_6$(SiO$_4$)$_6$O$_2$ are of special character because of the cation deficiency ^{IX}(3$^1/_3$ RE + $^2/_3$ $\square$) in the (4f) position of this structure. The rare-earth silicate lines are on a generally higher level than the corresponding data from the simpler oxides (48) (54) of CN 8 and CN 6.

5. Appendix I

Silicates and other Minerals Containing Rare Earths

Mineral	Formula	Space Group	Z	Cell dimensions					
				a[Å]	b[Å]	c[Å]	α[°]	β[°]	[°]
Allanite	$(Ca,RE)_2 Fe^{2+} (Al,Fe^{3+}) O(OH) [SiO_4] [Si_2O_7]$	$P2_1/m$	2	8.98	5.75	10.23		115.0	
Britholite Beckelithe Ressingite	$(RE,Na,Ca)_5 [SiO_4,PO_4]_3 (O,OH,F)$	$P6_3/m$	2	9.63		7.03			
Cerite	$(RE)_8 (Ca,Mg)_2 [SiO_4]_7 \cdot 3\ H_2O$	$R\bar{3}c, R3c$	6	10.78		38.03			
Gadolinite	$(RE)_2 Fe^{2+} Be_2 [SiO_4]_2 O_2$	$P2_1/a$	2	9.89	7.55	4.66		90.5	
Götzenite	$(Ce,Ti)_{0-1} (Na,Ca)_3 [Si_2O_7] F_2$	triclinic	2	9.65	5.74	7.32	90.0	101.1	101.3
Mosandrite	$(RE,Ca,Na)_3 (Ti,Zr,RE) [Si_2O_7] (O,OH,F)_2$	triclinic	4	18.41	5.64	7.43		93.1	
Rinkite	$(Ca,RE)_4 Ca(Ti,Nb) Na_2 [Si_2O_7]_2 (O,F)_2$	$P2_1/a$	4	18.83	5.66	7.44		101.4	

Stillwellite	(RE) B [SiO$_4$] O	P3$_1$	3	6.85		6.70			
Thalenite	(RE,Y)$_2$ [Si$_2$O$_7$]	P2$_1$/n	6	10.38	11.16	7.32		97.2	
Thortveitite	(RE,Sc)$_2$ [Si$_2$O$_7$]	C2/m	2	6.57	8.60	4.75		103.1	
Titanite	(Ca,RE) Ti [SiO$_4$]O	C2/c	4	6.56	8.72	7.44		119.7	
Tundrite	(RE)$_2$Na$_2$Ti (SiO$_4$)O$_4$ · 4 H$_2$O	tricline		7.54	13.98	5.02	101.5	70.5	101.3
Bastnaesite	(RE) [Co$_3$]F	P$\bar{6}$2c	6	7.16	9.79				
Euxenite	(RE,U,Pb,Ca) (Fe^{3+},Nb,Ta,Ti)$_2$(O,OH)$_6$	Pbcn	4	14.57	5.52	5.17			
Fergusonite	(RE,Y) (Nb,Ta) O$_4$	I4$_1$/a	4	5.16		10.89			
(Cer-, Yttro-) Fluorite	(Ca,RE) F$_{2-2.17}$	Fm3m	4	5.50					
Monazite	(RE,Ca,Th) [PO$_4$]	P2$_1$/n	4	6.79	7.04	6.47		104.4	
Rhabdophane	(RE,Ca,Th) [PO$_4$]. 0—0.5 H$_2$O	P6$_2$22	3	7.06		6.44			
Xenotim	(RE,Y) [PO$_4$]	I4$_1$/amd	4	6.89		6.04			

After Ref. (87).

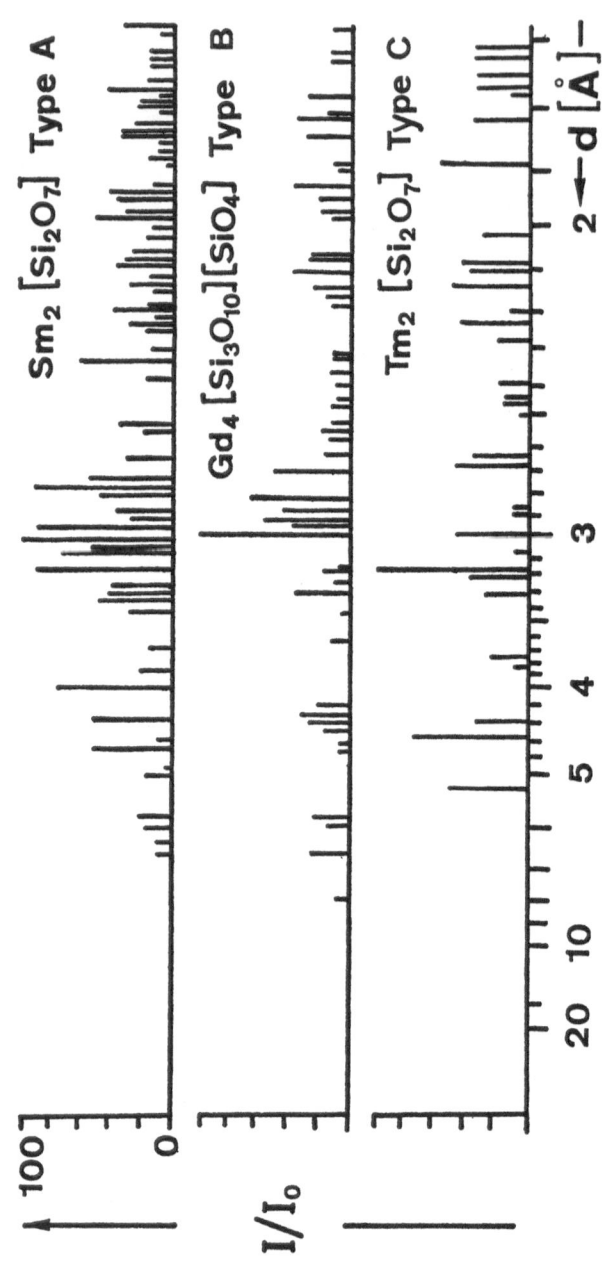

Fig. A 1

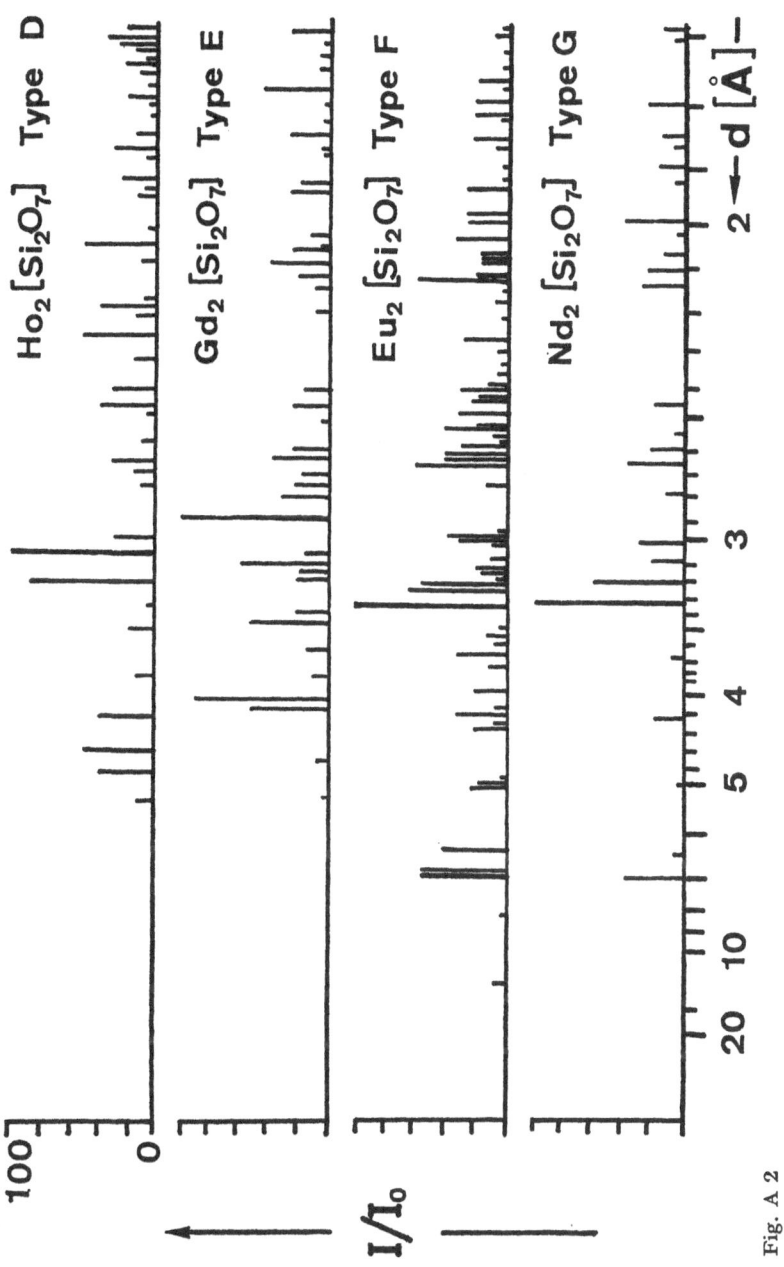

Fig. A 2

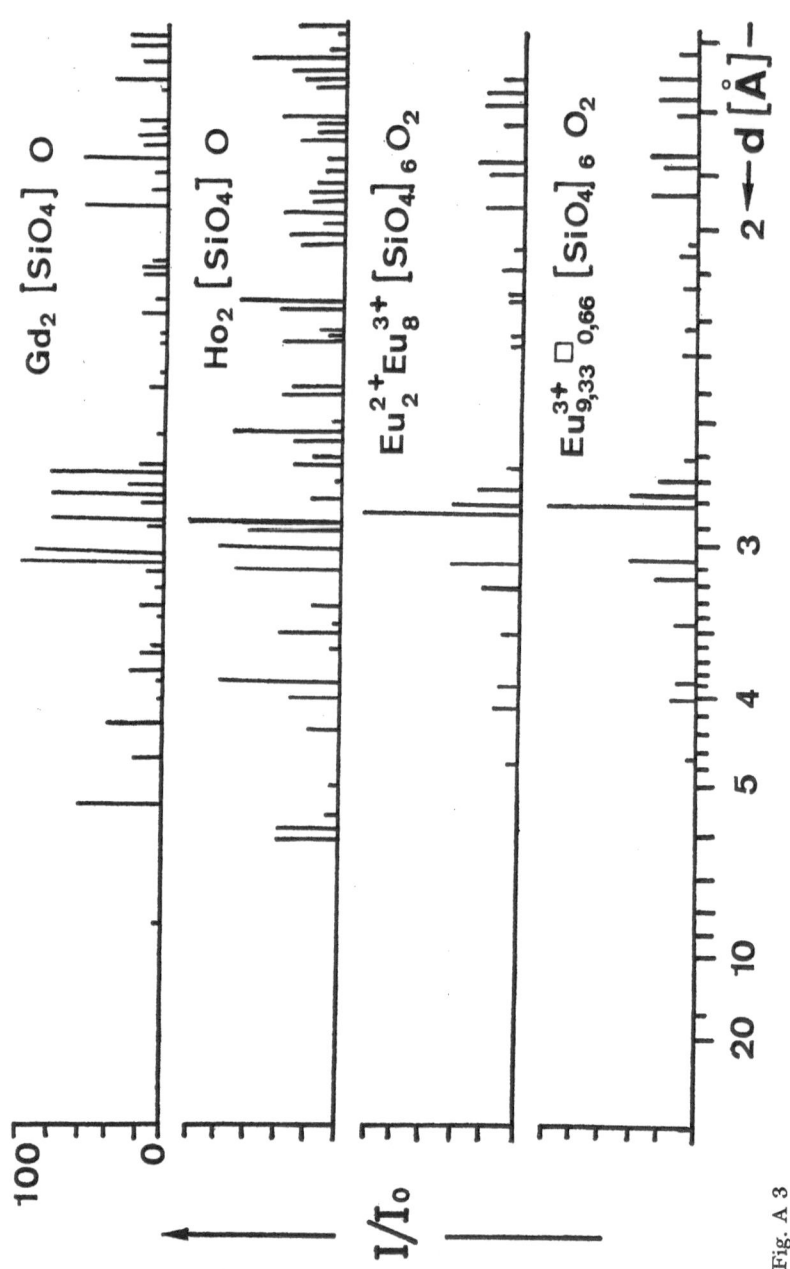

Fig. A 3

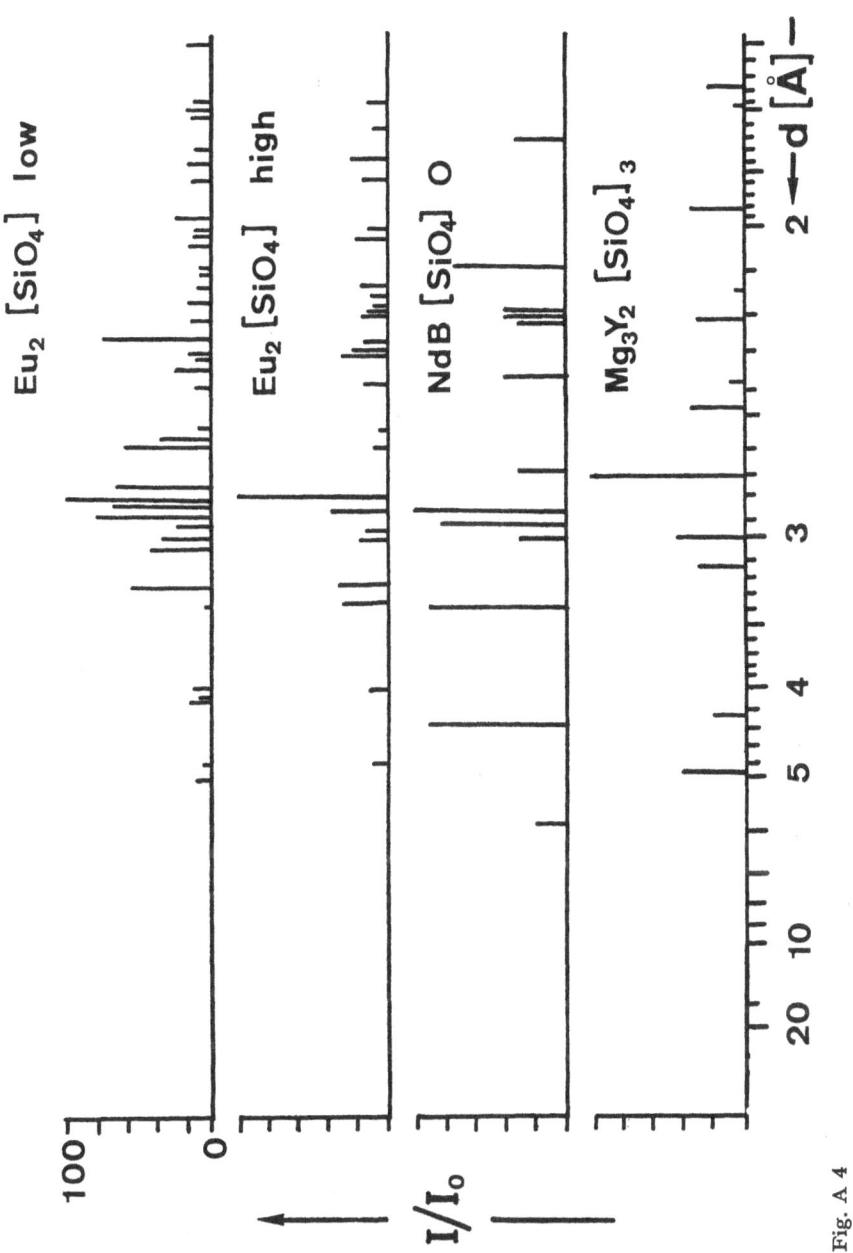

Eu$_2$ [SiO$_4$] low

Eu$_2$ [SiO$_4$] high

NdB [SiO$_4$] O

Mg$_3$Y$_2$ [SiO$_4$]$_3$

2 ←— d [Å] —

100

0

I/I$_o$

20 10 5 4 3 2

Fig. A 4

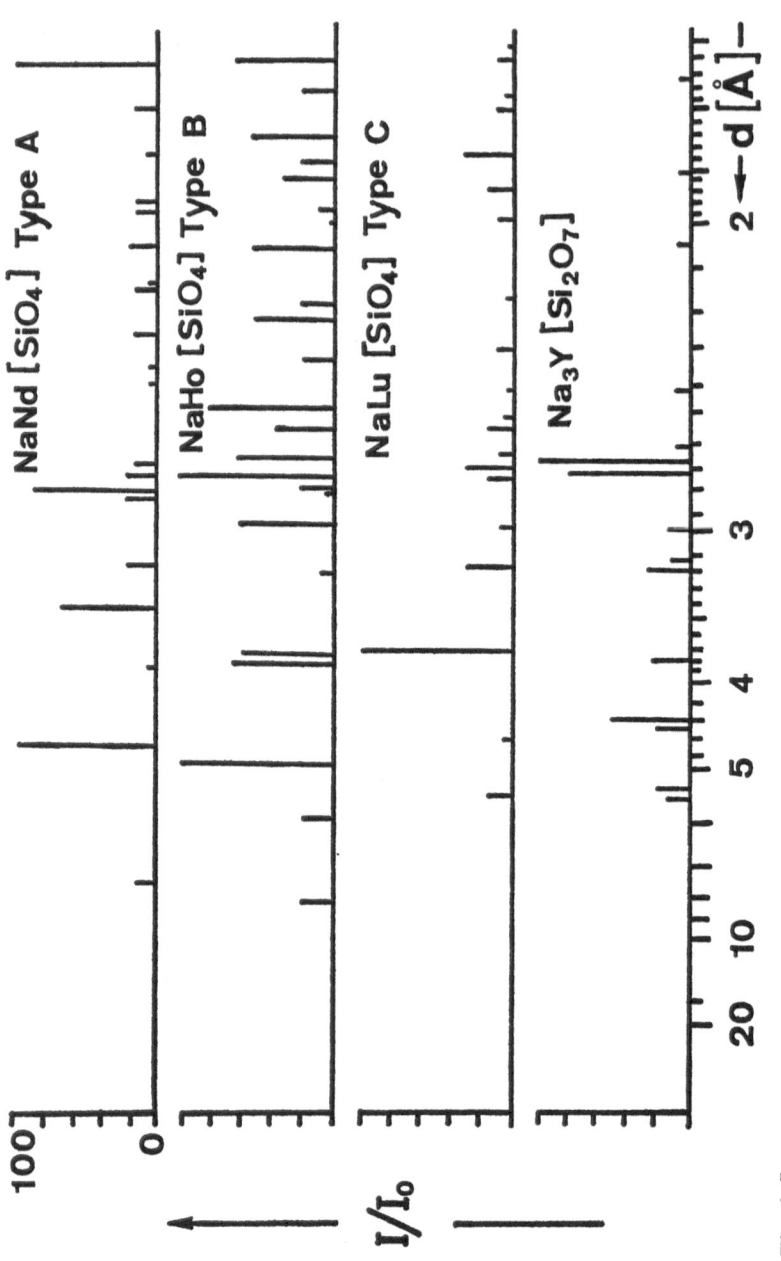

Fig. A 5

6. Appendix II

X-ray powder diffraction patterns of rare earth silicates. Data from FeKα guinier photographs taken with a camera of radius 57.3 mm, I/I_0 densitometer data. Data of compounds $NaRE(SiO_4)$ are from (30), of the garnet $Mg_3Y_2(SiO_4)_3$ from (32) and of the compound $Na_3Y(Si_2O_7)$ are from (31).

Acknowledgements. The author thanks Professor *F. Laves* for his stimulating interest in this study, his colleagues at the 'Institut für Kristallographie und Petrographie' as well as those at the 'Laboratorium für Festkörperphysik' for helpful discussions. The assistance given by *W. Hirsiger, H. Hertig* and *E. Fejer* for the 'air-brush' art work, by Dr. *W. Petter* and Dr. *D. Schwarzenbach* für the single crystal X-ray data collection and instructions for the X-ray 67 crystallographic computer programs, and furthermore by Prof. *P. Ribbe*, and Dr. *T. Woodman* for their critical reviews of parts of the manuscript is gratefully acknowledged. This work was financially supported by Owens Illinois, Inc., Toledo (Ohio/USA).

7. References

1. *Warshaw, J., Roy, R.:* Am. Ceram. Soc. Bull. *38*, 169 (1959).
2. *Wanmaker, W. L., de Graaf, W. P., Spier, H. L.:* Physica *25*, 1125 (1959).
3. *Toropov, N. A., Galakhov, F. Ya., Konovalova, S. F.:* Izv. Akad. Nauk SSSR, Otd, Knim. Nauk *5*, 739 (1961).
4. *Leonov, A. I., Rudenko, V. S., Keler, E. K.:* Izv. Akad. Nauk SSSR, Otd. Khim. Nauk *11*, 1925 (1961).
5. *Keler, E. K., Godina, N. A., Savchenko, E. R.:* Izv. Akad. Nauk SSSR, Otd. Khim. Nauk *10*, 1735 (1961).
6. — — — Izv. Akad. Nauk SSSR, Otd. Khim. Nauk *10*, 1728 (1961).
7. *Toropov, N. A., Galakhov, F. Ya., Konovalova, S. F.:* Izv. Akad. Nauk SSSR, Otd. Khim. Nauk *4*, 539 (1961).
8. — — — Izv. Akad. Nauk SSSR, Otd. Khim. Nauk *8*, 1365 (1961).
9. *Miller, R. O., Rase, D. E.:* J. Am. Ceram. Soc. *47*, 653 (1964).
10. *Warshaw, J., Roy, R.:* In: Progress in Science and Technology of the Rare Earths, Vol. I., p. 215. New York: Pergamon 1964.
11. *Kuzmin, E. A., Belov, N. V.:* Dokl. Akad. Nauk SSSR *165*, 88 (1965).
12. *Michel, C., Buisson, G., Bertrant, E. F.:* Compt. Rend. *264B*, 397 (1967).
13. *Smolin, Yu. I., Tkachev, S. P.:* Kristallografiya *14*, 22 (1969).
14. *Lazarev, A. N., Tenisheva, T. F.:* Izv. Akad. Nauk SSSR, Otd. Khim. Nauk *6*, 964 (1961).
15. *Batalieva, N. G., Kriwokoneva, G. K., Pjatenko, Jo. A.:* Dokl. Akad. Nauk SSSR *176*, 1146 (1967).
16. *Batalieva, N. G., Bondar, I. A., Sidorenko, G. A., Toropov, N. A.:* Dokl. Akad. Nauk SSSR *173*, 339 (1967).
17. *Ito, J., Johnson, H.:* Am. Mineralogist *53*, 1940 (1968).

18. *Felsche, J., Hirsiger, W.:* J. Less-Common Metals *18*, 131 (1969).
19. — J. Appl. Cryst. *2*, 1380 (1969).
20. — Naturwissenschaften *57*, 127 (1970).
21. — J. Less-Common-Metals *21*, 1 (1970).
22. *Batalieva, N. G., Pyatenko, Yu. A.:* Zh. Strukt. Khim. *8*, 548 (1967).
23. — — Zh. Strukt. Khim. *11*, 921 (1968).
24. *Smolin, Yu. I., Shepelev, Yu. F.:* Acta Cryst. *B26*, 484 (1970).
25. — — Kristallografiya *15*, 256 (1970).
26. *Felsche, J.:* Z. Krist. *133*, 304 (1971).
27. — Naturwissenschaften *59*, 35 (1972).
28. — Acta Cryst. 1973 (in the press).
29. *Chichagov, A. V., Litvin, B. N., Belov, N. V.:* Kristallografiya *14*, 119 (1969).
30. — — — Geokhimiya *9*, 1044 (1968).
31. *Maksimov, B. A., Litvin, B. N., Ilyukhin, V. V., Belov, N. V.:* Kristallografiay *14*, 498 (1969).
32. *Ito, J.:* Mat. Res. Bull. *2*, 1093 (1967).
33. *Shafer, W. M., McGuite, T. R., Suits, J. C.:* Phys. Rev. Letters *11*, 251 (1963).
34. *Rau, R. C.:* Acta Cryst. *17*, 1483 (1964).
35. *Kaldis, E., Verreault, R.:* J. Less-Common Metals *20*, 177 (1970).
36. *Shafer, W. M.:* J. Appl. Phys. *36*, 1145 (1965).
37. *Bondar, I. A., Toropov, N. A., Koroleva, L. N.:* Izv. Akad. Nauk SSSR, Neorg. Mat. *1*, 561 (1965); *3*, 2034 (1967).
38. *Maksimov, B. A., Kharitonov, Yu. A., Ilyukhin, V. V., Belov, N. V.:* Dokl. Akad. Nauk SSSR *183*, 1072 (1968).
39. *Buisson, G., Michel, C.:* Mat. Res. Bull. *3*, 193 (1968).
40. *Felsche, J.:* Naturwissenschaften *58*, 565 (1971).
41. *Smolin, Yu. I.:* Kristallografiya *14*, 985 (1969).
42. *Harris, L. A., Finch, C. B.:* Am. Mineralogist. *50*, 1493 (1965).
43. *Felsche, J.:* J. Sol. State Chem. *5*, 266 (1972).
44. *Ito, J.:* Am. Mineralogist *53*, 890 (1968).
45. *Cockbain, A. G., Smith, G. V.:* Mineral. Mag. *36*, 411 (1968).
46. — Mineral. Mag. *36*, 654 (1968).
47. *Smolin, Yu. I., Shepelev, Yu. F.:* Izv. Akad. Nauk SSSR, Inorg. Mater. *5*, 1823 (1969).
48. *Templeton, D. H., Dauben, C. H.:* J. Am. Chem. Soc. *76*, 5237 (1954).
49. *McCarthy, G. J., White, W. B., Roy, R.:* J. Inorg. Nucl. Chem. *29*, 253 (1967).
50. *Grisafe, D. A., Hummel, F. A.:* Am. Mineralogist *55*, 1131 (1970).
51. *Schwarz, H.:* Inorg. Nucl. Chem. Letters *3*, 231 (1967).
52. *Blasse, G., de Vries, J.:* J. Inorg. Nucl. Chem. *29*, 1541 (1967).
53. *Posner, A. S., Perloff, A., Diorio, A. F.:* Acta Cryst. *11*, 308 (1958).
54. *Shannon, R. D., Prewitt, C. T.:* Acta Cryst. *B25*, 925 (1969).
55. *Chichagov, A. V., Belov, N. V.:* Geokhim. SSSR *12*, 1456 (1968).
56. *Chichagov, A. V., Ilyukhin, V. V., Belov, N. V.:* Dokl. Akad. Nauk SSSR *177*, 574 (1967).
57. *Maksimov, B. A., Ilyukhin, V. V., Belov, N. V.:* Kristallografiya *12*, 214 (1967).
58. *Voronkov, A. A., Pyatenko, Yu. A.:* Kristallografiya *12*, 214 (1967).
59. *Busch, G., Kaldis, E., Verreault, R., Felsche, J.:* Mat. Res. Bull. *5*, 9 (1970).
60. *Felsche, J.:* Naturwissenschaften *58*, 218 (1971).
61. *Eysel, W., Hahn, Th.:* Z. Krist. *131*, 322 (1970).
62. *Weidenborner, J. E., Stemple, N. R., Okaya, Y.:* ACA-Abstract, p. 69 (1965).
63. *Glasser, L. S. D., Glasser, F. P.:* Acta Cryst. *18*, 453 (1965).
64. *Webb, N.:* Acta Cryst. *21*, 942 (1966).

65. *Smolin, Ju. I., Shepelev, Yu. F., Upatova, T. V.:* Soviet Phys. "Doklady" (English Transl.) *14*, 630 (1970).
66. *Shannon, R. D., Prewitt, C. T.:* J. Sol. State Chem. *2*, 199 (1970).
67. *Cruickshank, Lynton, H., Barclay, G. A.:* Acta Cryst. *15*, 491 (1962).
68. *Bayer, G., Felsche, J., Hirsiger, W.:* Glastechn. Ber. *42*, 317 (1969).
69. *Felsche, J., Hirsiger, W.:* Glastech. Ber. *45*, 173 (1972).
70. *Bondar, I. A., Tenisheva, T. F., Toropov, N. A., Shepelev, Yu. F.:* Dokl. Akad. Nauk SSSR *160*, 1069 (1965).
71. *Skshat, S. M., Simonov, V. I., Belov, N. V.:* Dokl. Akad. Nauk SSSR *184*, 337 (1969).
72. *Naray-Szabo, T.:* Phys. Chem. Glass *4*, 38A (1963).
73. *Pauling, L.:* J. Am. Chem. Soc. *51*, 1010 (1929).
74. *Zachariasen, W. H.:* Acta Cryst. *16*, 385 (1968).
75. *Pant, A. K., Cruickshank, D. W. J.:* Z. Krist. *125*, 1 (1967).
76. *Clark, J. R., Appleman, D. E., Papike, J. J.:* Contr. Mineral. Petrology *20*, 81 (1968).
77. *Voronkov, M. G.:* Dokl. Akad. Nauk SSSR *138*, 106 (1961).
78. *Cruickshank, D. W. J.:* J. Chem. Soc. *1077*, 5486 (1961).
79. *Brown, G. E., Gibb, G. V., Ribbe, P.:* Am. Mineralogist *54*, 1044 (1969).
80. *Neumann, H., Bergstol, S., Nilssen, B.:* Norsk Geol. Tidsskr. *46*, 327 (1966).
81. *Toropov, N. A., Bondar, I. A., Lazarev, A. N., Smolin, Yu. I.:* Rare Earth Silicates and Their Analogues, Akad. Nauk SSSR ('NAUKA'), Leningrad, 1971 (in russ.).
82. *Brown, G. E., Gibbs, G. V.:* Am. Mineralogist *54*, 1528 (1969).
83. *Baur, W. H.:* Am. Mineralogist *56*, 1573 (1971).
84. — Trans. Am. Cryst. Ass. *6*, 125 (1970).
85. *Foex, M., Traverse, I. P.:* Bull. Soc. Mineral. Crist. *89*, 184 (1966).
86. *Brown, I. D., Calvo, C.:* J. Sol. State Chem. *1*, 173 (1970).
87. *Strunz, H.:* Mineralogische Tabellen, Leipzig: Akad. Verlagsgesellschaft 1970.
88. *Bandurkin, G. A., Dzhurinskii, B. F.:* Dokl. Akad. Nauk SSSR *168*, 1315 (1966).
89. *Gschneider, K. A.:* J. Less-Common Metals *25*, 405 (1971).
90. *Felsche, J., Kaldis, E.:* J. Sol. State Chem. *5*, 49 (1972).

Received July 11, 1972.

The Inner Mechanism of Rare Earths Elucidated by Photo-Electron Spectra

C. K. Jørgensen

Département de Chimie physique, Université de Genève,
1211 Geneva 4, Switzerland

Table of Contents

I. Historical Introduction

A. From Meta-Elements to Isotopes

An earth was defined around 1800 as an oxide (usually trivalent) which cannot be reduced by heating with hydrogen or carbon. The rare earths turned out to be a particularly complicated family. When isolated from one type of minerals (such as the phosphate xenotime, the mixed niobate and tantalate fergusonite, or the beryllio-silicate gadolinite) the main constituent is yttrium ($Z = 39$) accompanied by the heavier lanthanides having Z above 64, whereas another more common type of minerals (such as the phosphate monazite) has cerium ($Z = 58$) as the major constituent with large quantities of lanthanum, praseodymium, neodymium and samarium having $Z = 57, 59, 60$ and 62. Very frequently, the discovery of new elements was announced (1) following long years of fractional crystallization or fractional precipitation, and the history is extraordinarily confused. However, we must admit that when the last non-radioactive lanthanide, europium, was discovered in 1896 by *Demarçay* (we are here excepting promethium ($Z = 61$) of which only short-lived radioactive isotopes are known) all fifteen elements from lanthanum to $Z = 71$ (called cassiopeium by *Auer von Welsbach* and lutetium by *Urbain*) were indeed characterized though this has not been done without a considerable amount of background noise. It is obvious that, under these circumstances, almost every element has been discovered several times in the sense of being separated from a new neighbour each time, and we no longer hear anything about denebium, carolinium and many other mixtures of our elements, nor about florentinium and illinium (supposed to be stable $Z = 61$) nor celtium (before the isolation of hafnium, $Z = 72$ was thought to be trivalent and possibly identified in lutetium fractions). It is worthwhile to memorize C. *James* who spent almost all his life in recrystallizing $Tm(H_2O)_9(BrO_3)_3$ in order to prove that thulium is indeed a single element. In a way, it is surprising enough that the constant trivalency became so well established in textbooks almost by default. Thus, europium was separated from its neighbours samarium and gadolinium by the rather ingenious trick of recrystallizing Mg_3M_2 $(NO_3)_{12}, 24 H_2O$ (by the way, these crystals contain $M(O_2NO)_6^{-3}$ with the central atom surrounded by an almost regular icosahedron of twelve oxygen atoms coordinated from the six bidentate nitrate ligands) containing $M = Bi$ which is easy to remove as the sulphide, though the solubility of the double salt is between those of $M = Sm$ and Eu. Only 13 years later, *Urbain* established that $EuCl_3$ is readily reduced to $EuCl_2$ and various devices using precipitation of $EuSO_4$ or amalgamation by electrolysis made europium(II) useful for separation techniques. The only exception to the constant trivalency recognized early was the existence of both cerium(III) and cerium(IV). Whereas the latter oxidation state is highly oxidizing in acidic aqueous solution (and used like permanganate for titrations) the yellow $Ce(OH)_4$ is readily precipitated in alkaline solution, as well as certain red-orange peroxides, and in mixed oxides formed by calcining mixed hydroxides, Ce(IV) is by far the most stable and difficult to reduce. It must be recognized that even the trivalency was not accepted without some dissent. La_2O_3 is as strong a base as MgO, and it was argued by some chemists that the whole series was constantly bivalent. *Wyruboff* went so far as to suggest that the higher oxidation number of cerium is $2\frac{2}{3}$ in analogy to Mn_3O_4 (now known to contain Mn(II): 2 Mn(III) in the tetragonal hausmannite and 2 Mn(II): Mn(IV) in the high-temperature cubic spinel). Fe_3O_4, Co_3O_4, Pb_3O_4 (containing 2 Pb(II): Pb(IV)) and U_3O_8.

It is difficult for us to recognize how the chemically autonomous methods failed in the special case of lanthanides (which IUPAC would like to call lanthanoids). For us, the atomic number Z is derived from nuclear physics and X-ray spectra and the atomic weight determined by mass spectrometers. Obviously, the latter technique is more accurate for a single isotope in a high-resolution instrument, whereas mixtures of several isotopes have to have their abundances determined as well. The gravimetric techniques of determining atomic weights were not of great use in lanthanides with a slowly increasing value from one element to the next, in particular because of the common yttrium impurities with about half the atomic weight. Actually, the chemical properties determined predominantly by the monotonically decreasing ionic radius going from La(III) to Lu(III) put yttrium somewhere between $Z = 66$ and 67 for all practical purposes. This is how dysprosium got its name ("difficult to find" in Greek) and holmium was also quite a problem before *Spedding* around 1942 developed the separation on ion-exchange resins (originally sulphonic acids of polymerized styrene) on a kilogram to ton level. Nowadays, one can buy the rare earths in 99% or better purity in drug-stores at a price comparable to gold for the three rarest (Eu, Tm and Lu) down to well below the price of silver for most of the even atomic numbers. For comparison, it may be noted that the writer at school in 1950 still recrystallized double nitrates from technical mixtures and separated rare earths from Norwegian gadolinite.

One of the reasons why physical properties of rare-earth mixtures sometimes looked confusing was that some measured quantities are not at all proportional to the concentration of the individual constituents. Sir *William Crookes* was a most extraordinary scientist, who also discovered thallium by flame emission spectra (as *Bunsen* and *Kirchhoff* did in the case of rubidium and caesium), but for our purposes his main achievement was the invention of the cathode-ray tube. It can hardly be realized what impact this invention would have had on chemistry if it had already been made by *Benjamin Franklin* establishing isolated phlogiston (later to be found in the blue solutions of alkali metals in liquid ammonia). Though *Lavoisier* was perfectly correct in considering the metallic elements as simpler than their oxides, the phlogistonists were projecting another part of the truth when saying that redox reactions are the transfer of phlogiston, our electrons. Anyhow, *Crookes* got interested in the emission spectra of the colourless solid materials bombarded with electrons in vacuo. In most cases, only broad unstructured emission bands were observed but in calcium oxide or calcium sulphate, several lanthanides produced narrow line spectra of the same kind as later found for *Geissler* tubes containing neon or other monatomic gases. Interestingly enough, the *cathodo-luminescence* of lanthanides diluted in colourless solids frequently have many more lines, also in the green and the blue, than the photo-luminescence of the compounds of the same elements when they are fluorescent in ultra-violet radiation. Whereas the well-known green luminescence of many terbium(III) compounds are due to transitions from the long-lived (about 1 millisecond) level $^5D_4$ to the seven levels of ^7F belonging to the same electron configuration $4f^8$, and the red fluorescence of europium(III) due to transitions from $^5D_0$ (and to a much smaller extent from the shorter lived, slightly higher $^5D_1$) to the seven ^7F levels of $4f^6$, the cathodo-luminescence of Eu(III) shows emission lines from higher levels populated by energy transfer from the rapid electrons to the solid. Recently, electrochemical processes (2) involving terbium(III) in the hydrogen-free solvent $POCl_3$ have been shown to have a low yield of similar luminescence. The screen of television sets shows cathodo-luminescence, and it is comparatively easy to obtain broad-band emission in the blue (such as zinc sulphide). When the first apparatus for colour television was developed, the red colour (due to broad-band emission of cadmium selenide) was dull and very unsatisfactory, and the only acceptable solution turned out to be the narrow bands of europium(III) cathodo-luminescence described by *Georges Urbain* (3) in 1909. The intensity distribution on the three transitions from $^5D_0$ to $^7F_0$ in the yellow (highly forbidden), to $^7F_1$ in the orange and to $^7F_2$ in the red [strong in the same crystals which would show strong hypersensitive pseudoquadrupolar absorption bands (4)] depends rather strongly on the surrounding atoms (5). As regards colour television, the colourless host crystal first used was yttrium orthovanadate YVO_4 (6) isotypic with tetragonal $ZrSiO_4$ and xenotime, but later, Y_2O_2S turned out to give a deeper red colour because of its pronounced nephelauxetic effect (7) [usually almost absent in europium(III) mixed oxides (8)] and strong hypersensitive transitions.

Urbain (*3*) also showed that very small traces of europium mixed with somewhat larger amounts of other rare earths in a cathodo-luminescent material exhibits intense line emission in the red (demonstrating effective energy transfer to the long-lived excited level $^5D_0$) which may or may not be accompanied by the weaker lines in the green and the blue. Hence, the spectrum cannot always be used to indicate two or more elements, and another non-existent rare earth, victorium, suggested by *Crookes*, went down the drain in this way. The energy transfer between different lanthanides in the same host lattice has been thoroughly, studied (*9, 10*) and explains why there is so little correlation between the concentration and the emission intensity of a given element.

Crookes, seeing more and more apparent elements in cathodo-luminescent experiments and realizing how difficult it is to achieve the perfect separation by fractional crystallization, suggested that the rare earths are a mixture of about a hundred *meta-elements* having *almost* the same chemical properties. This suggestion was made at a time when it was not perfectly certain that the *Döbereiner* differences between atomic weights of three consecutive analogous elements (such as 49 between calcium and strontium and between strontium and barium) were not of the same kind as the difference 14 between two consecutive homologous organic compounds obtained by adding a CH_2 link. Not until *Rutherford* was it realized that nuclear effects are about a million times stronger than chemical effects, both as regards exothermic reactions and reciprocal radii. One of the many aspects offering an analogy to the decline of the Western Roman Empire are the many paperbacks on alchemy recently published in France. One of the horrifying aspects of this literature is the extent to which it encourages a radical scepticism towards history. Many events are far less well documented than the twenty descriptions of how a demonstrator transmutes (in the presence of the notabilities of the town and a large number of other people) mercury or molten lead with a small amount of a brillant orange semi-conductor-type catalyst to give a yield of 100 to 105 percent of gold (it is perfectly untrue that the additivity of masses of metals in alloys was not recognized before *Lavoisier*; before *Marco Polo* related how the Chinese emperors printed banknotes, alloying was the only known technique of government-induced inflation, and docimastic analysis goes far back in time). However much the writer is willing to accept the argument that nuclear reactors operated at low temperature are transmutation catalysts, the absence of stronger exothermic effects in alchemistic experiments would need something like *Fermi*'s neutrinos running away without detection, or a most careful balance of exothermic and endothermic nuclear reactions. Though this would not be absolutely incompatible with residual technology from friendly visitors from other

Solar systems, or direct diabolic intervention, the writer feels that the exothermic argument is so strong that he, hesitantly, must look for other explanations such as collective hypnosis. Anyhow, for *Crookes* living 100 years ago, there was still enough of a connection with pre-scientific ideas that he suspected a "maturation" of the meta-elements in minerals. By the way, this is one of the rare cases of older history being optimistic and newer ideas pessimistic; with exception of nuclear physics directed by human beings, the only transmutation of elements on Earth is the irreversible and uninfluenced natural radioactivity of which the chains uranium 238 to lead 206, thorium 232 to lead 208, and the rare isotope potassium 40 to argon 40 (1 percent of the atmosphere) and calcium 40 have a perceptible influence on the heat balance of the crust.

The meta-elements of *Crookes* anticipated the idea of isotopes. The most readily available sources of separated stable isotopes are lead 206 from uranium minerals and lead 208 from thorium minerals, and it is interesting enough that before 1910 several successful chemical separations of radioactive isotopes of the same element were reported, involving both thorium 230 (ionium) and lead 210 (radium D). In our opinion, this can only be due to kinetically metastable chemical non-equivalency in the mixture, for instance, due to colloidal or oligomeric complexes. The valuable conclusion of this story is that the chemical similarity of trivalent rare earths is so striking that doubts have been expressed whether they deserved more than one place in the Periodic Table, a situation isotopes later had to accept. Such a doubt has never been expressed for any other elements, not even for a pair of elements like vanadium and chromium, which were confused at the time of their discovery (*1*). Nevertheless, studies based on the possibility of meta-elements continued rather late; for instance, *Debierne* attempted to separate neo-radium from conventional radium 226 and to perform nuclear reactions on charcoal cooled with liquid helium (*11*).

B. The Number 14 from *Rydberg* to *Stone*

Dimitri Mendeleev had considerable troubles with the rare earths. They might even have discouraged somebody not having his strong belief in a more fundamental reason than increasing atomic weights behind the periodicity of chemical properties. Originally, the Periodic Table was presented in the short form of octaves, which has had a most undeserved success in school textbooks. The starting point of the difficulties here were the "triad elements", such as iron, cobalt and nickel, admittedly having a closely similar chemistry, in particular when they remain bivalent. The two other triads (Ru, Rh, Pd and Os, Ir, Pt) owe their appeal to their promiscuity in geological platinum. As regards chemistry, the two series Mo, Tc, Ru and W, Re, Os are far more closely related and differ from the pairs Rh, Ir and Pd, Pt. Anyhow, the octave analogy to the triad elements became the noble gases in 1898. As a further illustration that even *Homer* sometimes nods, *Mendeleev* suggested that argon obtained from electric discharges fixing atmospheric nitrogen was N_3, like ozone O_3 obtained under similar circumstances, but he soon, accepted the noble gases as a magnificient confirmation of chemical periodicity. *Mendeleev* went out of his way to demonstrate that the region between barium and tantalum contains three or four octaves with all the alkaline metals and quinquevalent and higher columns as yet unknown. For this purpose, it was necessary to emphasize any evidence in favour of oxidation states differing from III. It had been known since the time of *Mosander* that certain mixed rare earths have yellow to orange colours which disappear on heating in hydrogen and reappear on heating in air. Besides the trivial case of iron(III) impurities, it became established that praseodymium(IV) and terbium(IV) produce such strong colours. Actually, thorium oxide containing 100 to 1000 parts per million of these two elements is purple, and the broad absorption band in the green can be ascribed (*12*) to electron transfer from a linear combination of $2p$ orbitals of the eight oxide neighbours (arranged in a cube) to the $4f$ shell containing one and seven electrons, respectively, in the ground state. However, when the multiple octave hypothesis was given up, it was argued that the lanthanides are a massive exception of the same kind as the triads. This opened up a certain credibility gap between the Periodic Table and the chemists, and many other attempts were made to isolate unexpected elements.

The first author to state clearly how many lanthanides exist, was *Rydberg* in 1913 (*13, 14*). His previous studies of atomic spectra made him search for a fundamental symmetry in the Periodic Table, and he then suggested that the long periods each ending with a noble gas contain $2, 2, 8, 8, 18, 18, 32, 32, 50, \ldots, 2n^2, \ldots$ elements. When *Moseley* in 1913 determined the atomic number Z from X-ray spectra, this conjecture turned out to be perfectly correct, except that two elements, nebulium and coronium, had been ascribed $Z = 3$ and 4 with the result that all the subsequent elements had Z two units higher according to *Rydberg* than according to *Moseley*. In analogy to helium ($Z = 2$), identified in the outer layers of the solar atmosphere during a solar eclipse in 1868 by *Janssen* and *Lockyer* and not found in the laboratory until *Ramsay* in 1895 extracted it from the uranium mineral cleveite, the spectral lines of nebulium and coronium were known only from astrophysical observations and could not be reproduced in the laboratory. However, they turned out to be transitions in highly ionized iron, such as Fe^{+13}, and in O^{+2}. This shows the danger of a certain platonism in the natural scientist, but at the same time, it shows how remarkably successful mathematical intuition can be. Ironically enough, *Rydberg* would have been correct all the way up to $Z = 120$ (but according to relativistic calculations not for $Z = 170$) if he had been perfectly consistent and let each long period end with the alkaline earth *(15)*. This is the first time where a distinction develops between the *spectroscopic* and *chemical* versions of the Periodic Table. The two versions are *almost* identical, but differ on certain points, such as the spectroscopic alkaline earths helium and ytterbium, which caused the whole 5f controversy *(16)*. *Rydberg* also emphasized the striking differences between atomic spectra of an even number of electrons (singlets, triplets, . . .) and odd number of electrons (doublets, . . .). This is not without reminiscences of the precursor Periodic Table of *Hinrichs (17)* which had the distinction between odd and even (valences) as the primordial classification of the elements. However, the great difference in atomic spectroscopy was that the first spark spectrum corresponding to the singly ionized atom has an odd number of electrons if the neutral atom contains an even number, and *vice versa*.

A corollary of the statement that the long period starting with caesium and ending with radon contains 32 elements is that lanthanum is followed by at most fourteen trivalent elements and that a quadrivalent element occurs before tantalum. This was not foreign to *Mendeleev*'s thoughts, and, as a matter of fact, zirconium was one of the elements known to vary its atomic weight outside the probable experimental limit of error. Such irregularities were eagerly watched for, in particular because of the inversions argon/potassium, cobalt/nickel and tellurium/ iodine in the Periodic Table, but frequently shown to be spurious by subsequent careful work. It is usually argued that *Coster* and *Georg von Hevesy* (working in Copenhagen in 1923), discovering from 1 to 5 percent of hafnium $(Z = 72)$ in zirconium from various minerals, confirmed the spectroscopic version of the Periodic Table based on the *Aufbau principle*.

This statement needs some clarification. The total of the wave functions of a many-electron system is not exactly an anti-symmetrized *Slater* determinant corresponding to a well-defined electron configuration, but one can classify the ground state *and* the excited levels by a *preponderant electron configuration (18)* predicting the correct symmetry types (parity, S, L, J for monatomic entities possessing spherical symmetry). The Aufbau principle of the Periodic Table applies to the preponderant electron configurations of consecutive elements. It is interesting to note that, in the July 1923 issue of Naturwissenschaften commemorating the ten years of N. *Bohr*'s hydrogen atom model, both *Niels Bohr (19)* and *Coster* (*20*) still accepted the hypothesis that the n groups of electrons (having what we would call differing l and j) of a given n value (the principal quantum number) did not immediately achieve their maximum number ($2n$) of electrons even when other shells are in the process of filling. Though this state of theory agreed with *Rydberg*'s long periods containing $2n^2$ electrons and also accepted that the outermost electron in the caesium atom has $n=6$ (this is not a trivial point because the ionization energy 0.28 rydberg seemed more fit for a system with $n=2$ and a small Rydberg defect 0.13), it seemed somewhat accidental that $2n^2$ repeated itself during the process of intermediate filling, and it was difficult to understand what exactly was going on from $Z=57$ to 71. *Stone* was the first, in 1925, to suggest our form of the Aufbau principle. Each n value is compatible with a lower non-negative integer $l=0$, 1, 2, 3, 4, 5, 6, ... having the spectroscopic trivial name s, p, d, f, g, h, ... and *each shell nl contains at most* $(4l+2)$ *electrons*. Hence, the fact that ten elements occur between the spectroscopic alkaline earths calcium and zinc, or between strontium and cadmium, is an expression of the Aufbau principle that the 3d or 4d shell having $l=2$ is able to contain ten electrons. The octave mystique originates in the combined s- and p-shell having $2+6=8$ electrons, which also contributes to *Abegg*'s rule that no element changes its oxidation number by more than eight units. Exceptions to this rule are furnished by ruthenium and osmium (*18*) varying from M(−II) in $M(PF_3)_4^{-2}$ to M(VIII) in MO_4. There is no reason to be worried about the triad elements representing a false problem. According to the new Aufbau principle, the 4f shell contains from zero to fourteen electrons, corresponding to the variation from Ba(II) to Yb(II) or from La(III) to Lu(III). It is not overwhelmingly imperative, but rather the most convenient choice to say that the 5d group starts with lutetium and goes on without interruption, rather than to say that lanthanum is the first 5d element followed by the fourteen lanthanides. This is particularly appealing to the chemist who can even ask the legitimate question whether scandium, yttrium, zirconium, lutetium and hafnium are indeed transition elements in the chemical sense (*21*).

C. The Arguments from Atomic Spectra

Hund (*22*) described the relations between atomic spectroscopy and the Periodic Table. This was a most important pioneer work but made one uncertain extrapolation: the almost invariant trivalency made it highly probable that the groundstate of the neutral atoms from lanthanum to lutetium belong to the electron configuration [Xe] $4f^q$ 5d $6s^2$ where [Xe] represents the 54 core electrons. We write [68] for the closed shells [Xe] $4f^{14}$ which are the ground states of Yb^{+2}, Lu^{+3}, Hf^{+4}, ...

and also of the corresponding Yb(II), Lu(III), Hf(IV), ... compounds, whereas no neutral atom has this configuration; the ground state of gaseous Tm^+ belongs to [Xe] $4f^{13}$ 6s and the neutral erbium atom to [Xe] $4f^{12}$ $6s^2$. Other systems containing [Xe] $4f^{12}$ include Er^{+2} and erbium(II) in fluorite crystals, whereas [Xe] $4f^{13}$ is represented by Tm^{+2}, Yb^{+3} and thulium(II) and ytterbium(III) compounds.

We run into the important problem for transition group chemistry that the exact form of the Aufbau principle depends on the *ionic charge*. Actually, gaseous M^{+3} and M^{+4} do not seem to present exceptions to the preponderant configuration of the ground state, being obtained by consecutive filling of the shells

$$1s \lessgtr 2s < 2p \lessgtr 3s < 3p \lessgtr 3d < 4s < 4p \lessgtr 4d < 5s < 5p \lessgtr$$
$$4f < 5d < 6s < 6p \lessgtr 5f < 6d < 7s < 7p \lessgtr \dots \tag{1}$$

where the double inequality signs indicate the systems isoelectronic with the noble gases containing 2, 10, 18, 36, 54, 86, 118, ... electrons. For the ionic charge M^{+2}, four exceptions to the series (1) are known (La^{+2} [Xe] 5d; Gd^{+2} [Xe] $4f^7$ 5d; Ac^{+2} [Rn] 6d and Th^{+2} [Rn] $6d^2$ followed very closely by [Rn] 5f 6d). In M^+ and neutral atoms, the s orbitals become competitive, and it is well known from the nd groups how the preponderant configuration of the ground state frequently contains one or two $(n + 1)$s electrons. It is important for chemists to realize that compounds are almost always well described by (1) if the transition element has a well-defined oxidation state (*18*). The metallic 4f group compounds are discussed below in section III B and the only other well-characterized exceptions are certain M(II) in fluorite crystals, and gaseous monoxides and monohalides MX where the ground state sometimes contains one or two electrons in a lone pair concentrated on the rear side of M and being a mixture of s and pσ character, in addition to the partly filled d or f shell (*23, 24*).

Unfortunately, Volume 4 of *Charlotte Moore*'s: Atomic Energy Levels has not yet been published, but a complete bibliography of 4f and 5f group atomic spectra has appeared (*25*). We are not here giving references to individual entities, but merely summarizing:

The neutral atoms La, Ce, Gd and Lu have ground states belonging to the preponderant configuration [Xe] $4f^q$ 5d $6s^2$ (q = 0, 1, 7 and 14), whereas Pr, Nd, Pm, Sm, Eu, Tb, Dy, Ho, Er, Tm and Yb all are known to belong to [Xe] $4f^{q+1}$ $6s^2$. It is particularly interesting to note that terbium has the lowest level of [Xe] $4f^9$ $6s^2$, only 286 cm^{-1} below the lowest level of [Xe] $4f^8$ 5d $6s^2$ according to unpublished measurements by *Klinkenberg* (*26*). This point shows that the exclusive emphasis on the ground state may be somewhat exaggerated; as we discuss below, the 4f → 5d substitution follows quite clear-cut rules across the lanthanides.

We discussed M^{+2} above. Though the only higher ionic charges which have been thoroughly studied are Pr^{+3} (27), Pr^{+4} (28) and Lu^{+3} (29), there is not the slightest theoretical doubt that all the ground states belong to [Xe] $4f^q$.

II. Spectra of Solutions and Crystals

A. Internal Transitions in the Partly Filled 4f Shell

Three years before *Bunsen* and *Kirchhoff* wrote their paper in 1860 about spectral analysis, *Gladstone* noted that an aqueous solution of didymium salts has several narrow absorption bands in the visible, reminiscent of the absorption lines in the solar spectrum first detected by *Wollaston* in 1802 and carefully studied by *Fraunhofer* since 1814. Didymium had been separated from lanthanum by *Mosander* in 1842, its name (meaning "twin" in Greek) turned out slightly ironical, since *Auer von Welsbach* in 1885 finally separated it in praseodymium and neodymium, using the relative intensity of the absorption bands as a semi-quantitative control of the extent of separation. *Becquerel* (30) studied various crystals and aqueous solutions containing rare earths and concluded that each (trivalent) element from praseodymium to thulium (with the exception of gadolinium, shown by *Urbain* (31) have narrow bands in the ultra-violet, and ytterbium, later shown to have narrow bands in the near infra-red) has characteristic line groups, some four to ten in the visible region, where the fine structure of each group depends to some extent on the anions and water of hydration. It is true that the octahedral chromophore Cr(III) O_6 produces absorption and emission lines in the red when it occurs in the ruby $Al_{2-x} Cr_x O_3$ and that certain other d group compounds show such lines (an extensive list (32) may include several experimental errors) as well as the collision complex of two O_2 molecules in liquid air, but, on the whole, it is quite clear that the coloured rare earths form a quite specific class of materials having narrow absorption bands in the condensed states of matter. After *Stone's* interpretation of shells containing $(4l + 2)$ electrons was accepted, the excited levels as well as the ground state were supposed to belong to an electron configuration where the partly filled shell $4f^q$ of gaseous M^{+3} is in some way conserved in the compounds. *Ephraim* and *Mezener* (33) even argued that the relatively narrow absorption bands of uranium(IV) compounds indicate the presence of two 5f electrons in agreement with *Goldschmidt's* hypothesis that a new rare earth series (but now predominantly quadri-valent) starts with thorium (34).

A long time before the detailed nature of the electrons producing the narrow absorption bands of lanthanides was recognized, the systematic shift as a function of the neighbour atoms was studied. Thus, in 1910 (three years before N. Bohr's hydrogen atom) Hofmann and Kirmreuther (35) argued that the orbits have larger radii in $Er_2 O_3$, showing about 1.5 percent lower wave numbers than hydrated erbium(III) salts. These studies were continued on a large scale by Ephraim and Bloch (36) with a large number of Pr(III), Nd(III) and Sm(III) compounds and later with other rare earths (8, 37—40). Today, it is recognized that the parameters of interelectronic repulsion are smaller in the oxides, sulphides, iodides and complexes of ligands with low electronegativity, than in the fluorides, aqua ions and double nitrates. We are not going to discuss in detail this nephelauxetic (cloud-expanding) effect; we only mention this since the interelectronic repulsion is decreased partly by a modification of the central field corresponding to a fractional atomic charge lower than the oxidation number (41) and partly by a weak delocalization on the ligands corresponding to the formation of anti-bonding M.O. The main result is that the nephelauxetic effect is comparatively weak (an order of magnitude smaller than typical d group complexes) thus showing the weak influence of covalent bonding on the partly filled 4f shell containing from 2 to 12 electrons.

It has been believed since 1926 (36) that gaseous M^{+3} would have wave numbers even larger than those in the fluorides. This statement was not verified until Sugar (27) demonstrated that Pr^{+3} indeed has parameters of interelectronic repulsion (producing the major part of the J-level energy differences for q between 2 and 12) 4 to 5 percent larger than the fluoride and ennea-aqua ion and less than 10 percent larger than the most covalent praseodymium(III) complexes. Further on, these values are unusually large; the typical variation between two compounds of a given M(III) is 1 percent.

Though there is little doubt that the phenomenological parameters of interelectronic repulsion are decreased because the average radius of the partly filled shell expands (18), one has to recognize that the Hartree-Fock wave functions produce integrals of interelectronic repulsion [derived from the theory of Slater, Condon and Shortley (42) and Racah (43); cf. also the books by Wybourne (44) and Judd (45)] which are too large by approximately the factor $(z + 3)/(z + 2)$ where z is the ionic charge (41, 46). This Watson effect (15) is a part of the correlation effects induced by the two-electron operator but shows regularities of such a kind as to suggest, colloquially speaking, an internal dielectric constant in the partly filled shell.

The first suggestion that the three absorption bands in the blue of praseodymium(III) compounds are due to transitions from $^3H_4$ to $^3P_0$, $^3P_1$ and $^3P_2$ of [Xe] $4f^2$ was made by *Ellis* (47). This area was opened up by the studies in the infra-red of *Gobrecht* (48) showing also the extent of spin-orbit coupling by identifying the only band of Yb(III) at 10 300 cm^{-1} as the transition from $^2F_{7/2}$ to the only other level $^2F_{5/2}$ of [Xe] $4f^{13}$. At one time, there had been attempts (49) to apply a naïve version of *Hund's* rules on $4f^3$ of Nd(III) with the quartets ($S = ^3/_2$) below the doublets ($S = ^1/_2$) and each series in order of highest L values at lowest energy. It was soon recognized (50) that this is mainly a question of $4f^q$ in spherical symmetry treated by the theory applied to monatomic ions (42, 43). Though the quartet levels of neodymium(III) had been correctly identified by *Satten* (51) and the triplet levels of thulium (III) by *Bethe* and *Spedding* (52), a controversy broke out regarding the uncritical assumption of small F^6 from La$^+$ (42) taken over to M(III). If plausible values of F^6 are assumed, the doublets of Nd(III) (50) and singlets of thulium(III) can be identified (53). Related results were obtained for europium(III) (54) and neodymium(III) and erbium (III) (55). Lists of the ten to thirty identified excited levels of the various M(III) below a lower limit situated somewhere between 30 000 and 60 000 cm^{-1} have been published (56, 57), the most recent including arguments about relative intensities and parametrization of hypersensitivity (58). Whereas the original approach (50, 51, 53) was to obtain agreement with predicted positions, safer methods have now been elaborated using, as *Hellwege* first did, spectra in polarized light of crystals of known structure and symmetry and the *Zeeman* effect at low temperatures. It is surprising how much better the phenomenological theory of interelectronic repulsion and spin-orbit coupling works for $4f^q$ [also for cases such as gaseous Pr^{+2} (59)] than for d^q and (the even worse case of) p^q (18). Correlation effects on term distances of f^q have been introduced *via* the three additional *Racah-Trees* parameters, which are a phenomenological representation of configuration interaction with f^{q-2} d^2, f^{q-2} p^2 and f^{q-2} s^2 (44, 58). These effects are numerically small compared with the *Watson* effect.

The individual J-levels are split to the extent of 100 to 300 cm^{-1} with the exception of a single state for $J = 0$ (or a *Kramers* doublet for an odd number of electrons). This splitting is connected with small energy differences between the seven 4f orbitals, typically amounting to 500 cm^{-1}. The sub-levels of a given J-level normally correspond to strongly mixed configurations of the seven orbitals, and hence the observed energy differences are smaller and a highly intricate function of the one electron energy differences. From 1929 to 1962 the non-spherical part of the *Madelung* potential was assumed to produce these differences, and it is possible to consider this electrostatic model of the "ligand field" as an exclusively phenomenological description including the group-theoretically necessary results. However, there is little doubt that the origin of the one-electron energy differences is the formation of slightly anti-bonding M. O. (60) and the *angular overlap model* (61, 62) is applicable as in the d groups. This model has parameters of anti-bonding strongly increasing with shorter internuclear distances, but, when these distances are kept constant the parameters are transferable from one chromophore MX_N to another with differing symmetry or differing coordination number N. Since the energetic effects are so small, we do not further discuss the "ligand field" treatment (15) of 4f group compounds here, but draw attention to specific applications of the angular overlap model to such problems (63, 64, 65). It is obvious that the hybridization description is of absolutely no use (66) in connection with 4f group complexes, because the incompatibility of the radial functions of the other orbitals being far more extended.

B. 4f → 5d Transitions

In molecular spectroscopy (67) one speaks about *Rydberg* transitions in analogy to the external electron of alkali metal atoms studied by *Rydberg*. The excited state has a large average radius compared with the occupied M. O. in the ground state. Transition group complexes may show transitions from the partly filled shell to an empty shell with another nl. There is one well-established case of $3d^6 \to 3d^5\,4s$ transition of $Fe(H_2O)_6^{+2}$ at 40 500 cm^{-1} but, on the whole, the $4f^q \to 4f^{q-1}\,5d$ and corresponding $5f \to 6d$ transitions in transthorium compounds have been a much more fertile class. *Freed* (68) found five broad absorption bands of cerium(III) salts such as $Ce(H_2O)_9(C_2H_5\,SO_4)_3$ between 33 700 and 47 400 cm^{-1} not present in the other lanthanides. What is somewhat worrying is that the aqueous solution of cerium(III) perchlorate has a sixth band (69) at 50 000 cm^{-1} most probably ascribed to the transition 4f → 6s, since arguments based on the angular overlap model (15) suggest that even the weak band at lowest wave-number is due to the same $Ce(H_2O)_9^{+3}$ as the four strong bands. An alternative which was preferred previously (38) was a minor constituent with another coordination number. Thus, europium(III) ethylenediaminetetraacetate occurs in two forms (70) differing in their content of constitutional water. Anyhow, *Freed* made the correct suggestion that the cerium(III) bands have the five *Kramers* doublets of 5d (formed by the combined "ligand field" and spin-orbit effects) as excited states. *Lang* (71) found the excited levels of Ce^{+3} belonging to [Xe] 5d at 49 737 and 52 226 and of [Xe] 6s at 86 602 cm^{-1}. In aqueous solution, praseodymium(III) and terbium(III) have 4f → 5d transitions at higher wave-numbers than cerium(III) though lower than

Pr^{+3}, having the levels of [Xe] 4f 5d distributed between 61 171 and 78 777 cm^{-1} (27). CaF_2 crystals are transparent to about 90 000 cm^{-1}, much higher than water, and *Loh* (72) succeeded in measuring the 4f→5d transitions of all the M(III) up to Lu(III) in this lattice, thus confirming the theory of spin-pairing energy discussed below.

The general impression that compounds have 4f→5d transitions some 10 000 cm^{-1} below the corresponding gaseous ion [though dependent on the ligands (38)] is confirmed by a comparison between M(II) and M^{+2}. Though much of the work on M(II) in the fluorites CaF_2, SrF_2 and BaF_2 and other ionic crystals has been done by *Feofilov* and *Kaplyanski*, we refer to here the compilation (18) and the general comparison of all M(II) by *McClure* and *Kiss* (73). It is not surprising that Gd(II) has the ground state belonging to [Xe]$4f^7$ 5d like the gaseous Gd^{+2}, but Ce(II) in CaF_2 has the ground state belonging to [Xe] 4f 5d with low-lying levels (74) of [Xe] $4f^2$, whereas it is the other way round for Ce^{+2}.

C. Electron Transfer Spectra

For many years, there has been no doubt that the yellow or orange colour of most cerium(IV) complexes is due to electron transfer bands of the kind reviewed elsewhere (75) where an electron is transferred from M. O. mainly localized on reducing ligands to the empty or partly filled nl-shell of the oxidizing central atom. Thus, the octahedral complexes orange $CeCl_6^{-2}$ and purple $CeBr_6^{-2}$ have intense electron-transfer bands at 26 600 and 19 200 cm^{-1}, respectively (76). The same is true for the uranyl ion UO_2^{+2} which has a very complicated photochemistry (77). The first excited state was shown by *Becquerel* to live for 10^{-4} sec, and it has even parity, an odd linear combination of oxygen 2 pπ orbitals having transferred an electron to the empty 5f shell. Mixed-oxidation state colours (78) due to electron transfer from one atom to another of the same element may also occur in lanthanides, as suggested by the black colour of Eu_3O_4 and Pr_6O_{11} as compared with red-brown PrO_2. The famous case of dark blue $Ce_xU_{1-x}O_2$ (79) seems to be electronically ordered Ce(IV) and U(IV) in the groundstate of the fluorite, whereas the excited state corresponds to $4f^1$ Ce(III) and $5f^1$ U(V), an electron being transferred from uranium to cerium as in the silver(I) and thallium(I) salts of 5d group hexahalides (80).

Nevertheless, the close connections between atomic spectra of M^{+3} and excited levels of M(III) made many rare-earth spectroscopists unreceptive to the possibility of electron transfer bands. Actually, it was reported (81) that $YbCl_3$, $6H_2O$ and $Yb(H_2O)_9(C_2H_5SO_4)_3$ have broad bands in the ultra-violet. If this is not due to traces of iron(III) or other impurities, the only reasonable assignment is electron transfer.

Banks et al. (82) were the first to suggest that the absorption edges in the near ultra-violet of Eu(III) and Yb(III) in molten chlorides are due to electron transfer, since they are the two lanthanides most readily reduced to M(II). The general difficulty of observing electron transfer bands in this case is that most strongly reducing ligands have intense absorption bands in the ultra-violet. Thus, $M(S_2CN(C_2H_5)_2)_3$ is orange for M = Eu and yellow for M = Yb, whereas the electron transfer bands at higher wave-numbers are masked by internal transitions in the ligands in the other 4f group diethyldithiocarbamates. The first comparative study (83) showing electron transfer bands for M = Sm, Eu, Tm and Yb (and $4f \rightarrow 5d$ transitions for M = Ce, Pr and Tb) was of the solution of bromides in almost anhydrous ethanol. In hydroxyl-free solvents such as aceto-nitrile, it is possible to study octahedral MCl_6^{-3} and MBr_6^{-3} (76) and MI_6^{-3} (84). The first electron-transfer band of the hexa-iodide complexes occurs at M =

$$\text{Sm } 24900 \qquad \text{Eu } 14800 \qquad \text{Tm } 28000 \qquad \text{Yb } 17850 \qquad (2)$$

in cm^{-1}. It is worth emphasizing that these octahedral complexes are not at all typical for the 4f group which normally has N = 8, 9, 10 or 12, and actually the pure electronic transitions are extremely weak (except when allowed as magnetic dipole radiation) and surrounded by vibrational structure, as in certain d group hexafluorides. The electron-transfer spectra of oxides (8, 12, 85, 86) and complexes of oxygen-containing ligands such as sulphate (87, 88) have been extensively discussed. A rather extreme case is green $Yb(C_5H_5)_3$ L, where fifteen carbon atoms from the regular pentagonal cyclopentadienide ligands and a sixteenth ligating atom from the molecule L already have a weak electron transfer band in the red (89). The low wave number is connected with strong ligand-ligand repulsion in the most loosely bound M. O. (15). Quite generally, the molar extinction coefficients ε (between 50 and 500) of electron-transfer bands of 4f group M(III) complexes are some ten to a hundred times lower than typical d group complexes. This must be connected with a very small overlap integral (or, more exactly, coexisting electronic density in our space) between the 4f shell with small average radius and the ligand orbitals. This argument does not apply to cerium(IV) complexes with stronger electron-transfer bands, nor to UI_6^{-2}, $NpCl_6^{-2}$, $NpBr_6^{-2}$ $PuCl_6^{-2}$ and $PuBr_6^{-2}$ (90). The electron-transfer bands of californium(III) and einsteinium(III) bromide complexes in ethanol have been determined (91) for comparison of the oxidizing character with 4f group M(III).

III. Refined Spin-Pairing Energy Theory

A. Applications to Spectra

The writer remarked in 1956 that, for a given configuration l^q, the energy difference between the baricenter of all the states belonging to $(S-1)$ and the baricenter of all the states having S can be written as $2\,DS$ where D is a definite linear combination of the integrals of interelectronic repulsion in the *Slater-Condon-Shortley* theory (42). The corresponding spin-pairing energy theory involving a quantity $-DS(S+1)$ for a given value of S was subsequently developed (15, 18, 56) and a general proof given by *Slater* (92). It may be worthwhile to make a parenthetical comment about a recent paper in which *Katriel* (93) argued that term differences, as between ^1D and ^3P of $1s^2\,2s^2\,2p^2$ of the isoelectronic series starting with the carbon atom, are due to larger nuclear attraction in the lower term. In the writer's opinion, this is one more example of the virial theorem paradox (15) that, when a system with the eigenvalue $E_0 = -T_0$ with T_0 the kinetic energy, and having the potential energy $V_0 = -2\,T_0$ suffers a potential perturbation with the diagonal element E_1, it is true to the first order that the new eigenvalues is $E_0 + E_1$ but, in order to satisfy the virial theorem, the wave function is slightly expanded (if E_1 is positive) and the final value of the kinetic energy is $T_0 - E_1$ and of the potential energy $V_0 + 2\,E_1$ to the first approximation. Thus, the change of the potential energy of an ionic crystal (such as LiF) is *twice* the conventional *Madelung* energy (94). In quantum mechanics, it is sometimes difficult to tell whether A is the cause of B or B is the cause of A, and it is frequently more satisfactory to consider $(A+B)$ as a situation without analysing its causes. But it is not the normal use of the wording to answer: "Why is ^3P below ^1D?" thus: "Because of stronger nuclear attraction in ^3P", though it is true that the average value $<r^{-1}>$ is slightly larger for ^3P than for ^1D. The point is that the origin of the differentiation which allows a slightly contracted wave function for ^3P is the smaller $<r_{12}^{-1}>$, the electrons in the 2p shell avoiding each other slightly more successfully in ^3P. In a way, it is interesting to the experimentalist to know that the origin of term distances in the case of a partly filled shell is interelectronic repulsion, whereas the separation of the eigenvalues E in kinetic and potential contributions is not observable. The problem is rather similar to the question of whether covalent bonding is intrinsically connected with changes of the kinetic energy in the bond region (15, 95), in apparent disagreement with the virial theorem.

A partly filled f^q shell has a ground term which is not the only one with maximum S in the case of $q = 2, 3, 4, 5$ [and the values $q = 9, 10, 11, 12$ related to $(14 - q)$ by the *Pauli* relation (15)] and the energy below the baricenter for maximum S can be written as a definite multiple of E^3 introduced by *Racah* (43). The stabilization by decreased interelectronic repulsion relative to the baricenter of the *whole* configuration f^q is:

$$f^2, f^{12} \; ^3H: \quad -\frac{8}{13}D - 9\,E^3$$

$$f^3, f^{11} \; ^4I: \quad -\frac{24}{13}D - 21\,E^3$$

$$f^4, f^{10} \; ^5I: \quad -\frac{48}{13}D - 21\,E^3$$

$$f^5, f^9 \; ^6H: \quad -\frac{80}{13}D - 9\,E^3 \tag{3}$$

$$f^6, f^8 \; ^7F: \quad -\frac{120}{13}D$$

$$f^7 \quad\;\; ^8S: \quad -\frac{168}{13}D$$

whereas it is zero for ^1S of the closed shells f^0 and f^{14} and for the unique term ^2F of f^1 and f^{13}.

For $0 < q < 7$, where the ground level has $J = L - S$ according to one of *Hund's* rules, the first-order stabilization due to spin-orbit coupling is $-(L+1)\zeta_{nl}/2$, whereas for $7 < q < 14$ the ground level has $J = L + S$ and the first-order stabilization is $-L\zeta_{nl}/2$. In the 4f group this is a relatively minor effect; as first pointed out by *Gobrecht* (48) ζ_{4f} varies smoothly from 643 cm^{-1} in Ce^{+3} to 2950 cm^{-1} in Yb(III) and the variation for a given ion in chemical compounds is negligible and considerably smaller than the nephelauxetic effect.

The coefficients to D in eq. (3) when plotted as a function of q look like a trumpet with the mouthpiece downward forming a V-shaped minimum for $q = 7$. Since E^3 is known from absorption spectra to vary smoothly from 460 cm^{-1} for Pr(III) to 630 cm^{-1} for Tm(III) and D is close to 6500 cm^{-1} (by the way, it is $\frac{9}{8}E^1$ in the *Racah* notation), the total expression (3) plots like a tulip. The most frequent use of eq. (3) is to describe the consequences of adding an electron $4f^q \rightarrow 4f^{q+1}$ as was first done (83) in studying electron-transfer spectra. Then, one has to add a linear variation of the one-electron energy across the f group $W - q$ $(E - A)$, where W is a zero point of reference, and the difference $(E - A)$ is introduced in regard to the increasing nuclear attraction diminished by the parameter of interelectronic repulsion A (18) having the coefficient $q(q-1)/2$ closely related to E^0 [by the way, *Racah* (43) choose to write $E^0 = A - \frac{8}{13}D$ giving the simplest expression for maximum S for $q < 8$, as discussed by *Johnson* (96), but needing the addition of $9(q-7)E^1$ in the second half of the shell]. However, it should not be construed that $(E - A)$ is exactly the difference between two calculable quantities, since we did not consider the total energy but only the consequences of adding an f electron, and since we neglect the second-order effects of modifying the parameters slightly in going from one element to the next. Actually, the refined spin-pairing energy theory is normally used with a fixed value of D. If $W - q(E - A)$ is added to the difference of eq. (3) for $(q + 1)$ and q, two zig-zag curves showing a strong discontinuity between $q = 6$ and $q = 7$ are obtained. If only the coefficients to D were considered, the differences would constitute two parallel straight line segments jumping $8 D$ when the half-filled shell is crossed. It is noted that if $7(E - A)$ coincides with $48 D/13$, the electron transfer band occurs at the same wave number (disregarding spin-orbit coupling) for f^6 and f^{13}. This is close to the observed behaviour. If $D = 6500$ cm^{-1} is assumed, the electron-transfer bands of bromide complexes in ethanol (83) indicate $(E - A) = 3000$ cm^{-1} and of hexahalides in acetonitrile (76, 84) $(E - A) = 3200$ cm^{-1}. Quite generally, the more covalent complexes tend toward higher $(E - A)$ values, but only the cyclopentadienides (89) seem to have crossed the critical value 3430 cm^{-1}, producing lower wave numbers for the electron-transfer bands of Yb(III) than of Eu(III). However, the sterical conditions of these squeezed ligands may contribute to this phenomenon (15).

When $4f^q \rightarrow 4f^{q-1}$ 5d transitions are considered, it is feasible to add $W/ + q(E - A)$ to the difference taken the other way of eq. (3) and obtain satisfactory agreement with the same values of D and E^3 (derived from the internal transitions in the partly filled 4f shell) but with $(E - A)$ around 5000 cm^{-1}. It is not astonishing that $(E - A)$ is larger; the interelectronic repulsion between a 4f and a 5d electron is still strong, though it is smaller than between two 4f electrons. As also discussed by *McClure* and *Kiss* (73), it is surprising that the complicated complete expressions for interelectronic repulsion in $4f^{q-1}$ 5d do not produce stronger deviations from this theory. Also, one might have expected more spectacular effects of the "ligand field" splitting of the 5d shell known to be about 10 000 cm^{-1} for M(II) in CaF$_2$ (18) and presumably larger for M(III).

Almost simultaneously, *Nugent* and *Vander Sluis* (97), *Brewer* (98, 99) and *Martin* (100) pointed out that a similar treatment is applicable to the energy difference between the lowest *level* of $4f^{q-1}$ 5d 6s^2 and $4f^q$ 6s^2 of the neutral atom, and between $4f^{q-1}$ 5d 6s and $4f^q$ 6s of M$^+$, and between $4f^{q-1}$ 5d and $4f^q$ of gaseous M^{+2}, as was already known for M(II). Table 1 gives experimental values for gaseous species (100) with all figures in cm^{-1}, whereas the values ending in two zero have been calculated by *Martin*. The theoretical values (97) derived from $(E - A) = 3580$ cm^{-1} have also been included for comparison with the neutral atoms. Recently, *Sugar* and *Kaufman* (*29*) reported results for gaseous M^{+3} which are given in Table 1 in the same way as *Martin's* results. These authors also calculate that the 4f→5d excitation energy smoothly increases from 115052 cm^{-1} for Pr^{+4} (*28*) to 156500 cm^{-1} for Tb^{+4} and then from 120600 cm^{-1} for Dy^{+4} to 155700 cm^{-1} for Hf^{+4}. It may finally be noted that *Brewer* (*99*) gives the value 10594 cm^{-1} for Dy^{+}.

Table 1. *Energy difference between the lowest level of the configuration containing* $4f^{q-1}$ *5d and the lowest level of the configuration containing* $4f^q$ *in* cm^{-1}

q =		M^0 calc. (*97*)	M^0 (*100*)	M^+ (*100*)	M^{+2} (*100*)	M(II) in CaF$_2$ (*18*)	M(III) in CaF$_2$ (*72*)		M^{+3} (*29*)
1	La	-14300	-15300	-12253	$- 7195$	negative	Ce	32500	49737
2	Ce	$- 4000$	$- 4763$	$- 1472$	$+ 3277$	$- 7100$	Pr	45600	61171
3	Pr	$+ 4200$	$+ 4300$	$+ 8200$	$+12847$	4000	Nd	55900	70100
4	Nd	$+ 5400$	$+ 6764$	$+11310$	$+16000$	7000	Pm	—	73300
5	Pm	$+ 5900$	$+ 8000$	$+12300$	$+17100$	—	Sm	59500	74400
6	Sm	$+15600$	$+15500$	$+19600$	$+24500$	15000	Eu	68500	81800
7	Eu	$+26800$	$+25100$	$+29000$	$+33900$	25000	Gd	>78000	91200
8	Gd	-11900	-10947	$- 7992$	$- 2000$	negative	Tb	46500	55300
9	Tb	$- 200$	$+ 286$	$+ 3400$	$+ 9600$	negative?	Dy	58900	66900
10	Dy	$+ 8400$	$+ 7565$	$+10800$	$+17200$	10400	Ho	64100	74500
11	Ho	$+ 7200$	$+ 8000$	$+11400$	$+18100$	11100	Er	64200	75400
12	Er	$+ 5900$	$+ 7177$	$+10600$	$+17100$	10900	Tm	64000	74400
13	Tm	$+14100$	$+13120$	$+16567$	$+22897$	17000	Yb	70700	80200
14	Yb	$+27000$	$+23188$	$+26759$	$+33386$	27500	Lu	>80000	90432

It is seen from Table 1 that the two zig-zag curves are displaced in a fairly regular fashion when the charge is increased, about 3500 cm^{-1} from M^0 to M^+, and 5000 to 6000 cm^{-1} from M^+ to M^{+2}. This illustrates the highly external nature of the 6s electrons, when compared to the shift 57000 cm^{-1} between isoelectronic M^{+2} and M^{+3}. A characteristic feature of eq. (3) is the plateau where the removal of an electron from f^q needs almost the same energy for q = 3, 4, 5 and, in particular, for q = 10, 11, 12. These plateaux are less pronounced in the 5f group (*97*, *98*) because of the smaller parameters of interelectronic repulsion, and larger $(E-A)$.

It is possible to define the *conditional oxidation state* written in square brackets (*18*) from the number of electrons in the partly filled shell which have the distinctly smallest average radius in the system. Thus, in lanthanides containing $4f^q$ in the preponderant electron configuration, the conditional oxidation state is $(Z-54-q)$. The two columns to the right of Table 1 correspond to M[III] and the rest of the entries in Table 1 to M[II] when no 5d electron occurs, whereas they correspond to M[IV] and M[III], respectively, when a 4f electron is replaced by a 5d electron. Hence, the ground states of Ce^0, Ce^+, Ce(II) and Ce(III) have the conditional oxidation state Ce[III] containing one 4f electron, whereas the ground state of Ce^{+2} is Ce[II] containing two.

Table 1 gives the impression that the fractional charge of M(III) in CaF_2 is somewhere between 2.7 and 2.8, and of M(II) in CaF_2 close to 1.9 by comparison with M^{+2} and M^{+3}. It is obvious that one may modify the central field $U(r)$ in such a way as to imitate this behaviour, as has been elaborated by *Briffaut* (*101*), but the numerical statements about the fractional charges have to be mitigated by considerations of the smaller average radii of the 4f than of the 5d shell. The situation is analogous to *Orgel* (*102*) comparing the nephelauxetic effect in manganese(II) compounds with the moderate effect of 4s electrons on the parameters of interelectronic repulsion in the half-filled 3d shell, the term distance between ^4G and ^6S being 26846 cm^{-1} in $3d^5$ of Mn^{+2}, 25279 cm^{-1} in $3d^5 4s^2$ of the neutral atom, 25300 cm^{-1} in solid MnF_2, 25000 cm^{-1} in $Mn(H_2O)_6^{+2}$ and smaller values, down to about 21000 cm^{-1} in MnSe, in more covalent compounds (*18*). This extent of the nephelauxetic effect is twice as large as in Pr(III) but three times weaker than in typical d group central atoms such as Cr(III), Fe(III) and Rh(III).

B. Metallicity of Undiluted Compounds

The physical properties of the metallic elements from lanthanum to lutetium vary in a monotonic way (increasing melting and boiling points, decreasing molar volume etc.) with the striking exception of europium and ytterbium, which are much more similar to barium with low melting and boiling points, low density and much higher chemical reactivity. *Klemm* (*103*) pointed out that the magnetic properties of lanthanide alloys ($S = \frac{7}{2}$ for Eu[II] and diamagnetic $S = 0$ for Yb[II] but otherwise normally the values expected for M[III]) indicate the number q of 4f electrons. Metallic cerium is a somewhat exceptional case. At normal pressure, the low-temperature modification has lower molar volume corresponding to Ce[IV] and another allotropic modification stable above $-120\,°C$ has higher molar volume and seems to be Ce[III]. The unexpected feature is that both crystal forms are strictly isotypic, just differing in the unit cell parameters. In other words, the cerium metal decides collectively which of the two conditional oxidation states to adopt. However, *Jayaraman* (*104*) showed that a critical temperature exists at higher pressure where the two phases coalesce.

Seen from the point of view of the chemist, metallicity may seem a somewhat accidental property. Thus, stoichiometric compounds exist which are metallic, though they have almost the same unit cell parameters as semi-conducting analogous compounds formed by neighbour elements.

All the lanthanides MS, MSe and MTe crystallize in the NaCl type, but only M = Sm, Eu and Yb are semi-conductors [some doubts have been expressed about the thulium(II) chalcogenides] whereas the rest are metals. The most ready explanation (23, 105) is that when the ground state belongs to $4f^q$, the conditional as well as the conventional oxidation state is M[II] but when the ground state belongs to $4f^{q-1}$ 5d ,the metallic material turns out to have the conditional oxidation state M[III], as seen from the magnetic properties. The conduction band may be formed from one 5d electron per atom. It is seen from Table 1 that the condition for M[III] runs some 14000 cm^{-1} lower than for the gaseous atom or some 23000 cm^{-1} below the limit for M^{+2}. By the same token, several of the di-iodides are metallic, though *Corbett* and his collaborators have made strongly coloured, but non-metallic, di-chlorides and di-iodides of Nd(II), Sm(II), Dy(II) and Tm(II). Here, the limit seems to run 7000 cm^{-1} lower than M^0, or 15000 cm^{-1} lower than M^{+2}. The corresponding condition for M[III] in the metallic element would be even lower than in the sulphides MS, but the origin would no longer be the same, the question now being whether (the more voluminous) M[II] plus two conduction electrons is more stable than M[III] plus three conduction electrons. High-pressure experiments sometimes allow a reversible change of the conditional oxidation; thus, Yb Te becomes Yb[III] and metallic above 150000 atm. (106).

The textbook explanation of metallicity in terms of partly filled energy bands encounters serious difficulties in lanthanide compounds. Not only all halides MX$_3$ but also dark green NdCl$_2$ or deep purple NdI$_2$ are high-energy-gap semiconductors in spite of the partly filled 4f bands. This cannot be explained away as due to metals with vanishing mobility; a closer analysis (18) shows that the large difference between electron affinity and ionization energy of the partly filled shell inhibits the coherent use of the word "one-electron energy" and the definite conditional oxidation states are connected with the minimized interelectronic repulsion when fixing an integral number of 4f electrons to each atom. It is far more appropriate to have a definite *SLJ*-level of M[III] as the ground state surrounded by other electrons assuring the metallicity. *Brewer* (98) commented on the relations between atomic spectroscopy and the thermodynamic properties of metallic lanthanides. *Gschneider* (107) proposed two kinds of 4f electrons, the integral number corresponding to M[III] having ionization energies 3.2 to 8.5 eV higher than the work function indicating the *Fermi* level (obtained from photo-electron spectra as discussed in section IVB) and a fractional number between 0.1 and 0.6 of delocalized 4f electrons in the conduction band. It is necessary to analyze what orbital occupation numbers mean in M.O. theory (15, 66). In typical non-metallic d- and f-group compounds, the filled M.O. with largest ionization energy I are delocalized on the ligands with a small contribution on the central atom. Hence, in a Mulliken population analysis, they contribute a small occupation number of the central atom nl-shell considered. On the other hand, the partly filled shell is localized to an extent between 60 and 98 percent on the central atom and the rest delocalized as an anti-bonding L.C.A.O. with nodal surfaces between the central atom and the ligands. The partly filled shell containing q electrons contributes q multiplied by the factor 0.6 to 0.98 (indicating the square of the amplitude of the central atom A.O.) to the population analysis. It cannot be emphasized too much that the spectroscopic oxidation state (18) is based on the integer q and that the fractional occupation number of the nl-shell does not have exactly the connotation it would have to a spectroscopist. However, in the special case of 4f group compounds, it is known from photo-electron spectra that I of the most loosely bound ligand orbitals is comparable to or *lower* than $I(4f)$ of M(III). Hence, we may have to accept the apparent monstrosity that the weakly anti-bonding 4f-like orbitals [that they are not bonding can be seen from the agreement with the angular overlap model (60, 64)] have higher I than the corresponding bonding M.O. It is perhaps worthwhile to state that, even when the well-defined M.O. configuration is a good approximation, the *total energy* of the system is minimized but not necessarily the individual one-electron energies. Just as the conduction electrons with lower I do not invade the partly filled 4f shells in metals, the non-metallic 4f group compounds may be better off by containing an integral number of 4f electrons, exploiting to the maximum the fact that the coefficient to A is $q(q-1)/2$ rather than $q^2/2$ as in a classical charge density (108) or, expressed colloquially, an electron does not repel itself.

Returning to metals, the proposal by *Gschneider* (107) would indicate that the 4f shell is bonding with respect to the conduction electrons (this does not exclude that it may be anti-bonding with respect to some other orbitals of the neighbour atoms) and that the weak 4f character (most pronounced in La and the early lanthanides) corresponds to some extent to anti-bonding of the (perhaps empty) conduction orbitals. The physical significance of *Pearson's* concept of hardness and softness has been discussed earlier in this series (109, 110) and it has been realized (15) that deformation orbitals belonging to the continuum (with positive E) may explain characteristic details of the chemistry of silver(I), palladium(II), mercury(II), thallium(III), iodide, sulphur-containing ligands, phosphines and hydride which are not readily incorporated in a simple L.C.A.O. description. Since the metals are intrinsically soft, it would not be surprising if the partly filled 4f shell were to be slightly stabilized by the conduction orbitals.

C. Standard Oxidation Potentials

Connick (111) pointed out that the major difference between the 4f group and the 5f group chemistry in aqueous solution is that the standard oxidation potential E_0 for the aqua ions of one oxidation state to the next are far more negative in the 5f group and cannot even be measured for M(II)/M(III). Even the M(III)/M(IV) standard oxidation potentials are too negative (hydrogen evolution) to be measured for thorium and protactinium. Further on, the difference between consecutive standard oxidation potentials is smaller [as seen from the existence of Np(III), Np(IV), Np(V), Np(VI) and Np(VII)] if the linear dioxo complexes MO_2^+ and MO_2^{+2} are accepted for comparison. For a given central atom, increasing pH varying between -1 and 15 induces tendencies aqua $\rightarrow$ hydroxo $\rightarrow$ oxo complexes, and if no complexing anions are bound, the ionic charge per central atom either stays constant or decreases monotonically, as when vanadium(V) changes from VO_2^+ (or $V(OH)_4^+$) via $V_{10}O_{28}^{-6}$ and $V_3O_9^{-3}$ to $HOVO_3^{-2}$ and VO_4^{-3} as a function of increasing pH. Technically speaking, one can always think about E_0 for forming a deprotonated complex as being decreased relative to the aqua ion in a way calculated from *Nernst's* law. However, the deprotonation corresponds to enormous changes in free energy when forming tetroxo complexes *(18)* of ruthenium(VI), (VII) and (VIII), and this is the major difficulty for predicting high oxidation states accurately even if the consecutive ionization energies I_n of gaseous M^{+n-1} are known *(15)*. Keeping pH constant (especially acidic solutions around 1), the tendency toward deprotonation increases as a function of the oxidation state divided by the ionic radius *(112)*. Thus, the small phosphorus(V) has $P(OH)_4^+$ (known in perchloric acid solution) deprotonating to PO_4^{-3} with the four consecutive pK values close to $-3, 2, 7$ and 12, whereas the larger protactinium(V) is the least acidic. By the same token, thorium(IV) aqua ions have $pK = 4$ and uranium(IV) close to 1, whereas other quadrivalent central atoms such as cerium(IV) and hafnium(IV) have aqua ions so acidic that pK must be negative. This whole argument is not as clear-cut in the case of the uranyl ion or the vanadyl(IV) ion $VO(H_2O)_4^{+2}$ where oxide ligands (having lost their proton affinity even in strong perchloric acid) coexist with coordinated water molecules.

The standard oxidation potentials E_0 given in the literature for M(II)/M(III) aqua ions are:

$$\text{Sm} - 1.53 \text{ V} \qquad \text{Eu} - 0.43 \text{ V} \qquad \text{Yb} - 1.15 \text{ V} \qquad (4)$$

The value for Sm is determined by polarography *(113)*. On the other hand, E_0 for Ce(III)/Ce(IV) in 1 molar $HClO_4$ is $+1.71$ V, so strictly speaking, no lanthanide should occur in any other oxidation state than M(III) in this solvent if equilibria with respect to hydrogen and oxygen evolution are established. On the other hand, Ce(IV) is far more stable and Pr(IV) and Tb(IV) reasonably stable in solid oxides (12). There is no direct connection between the E_0 of F^- to F_2 close to $+2.9$ V and the existence of Cs_3 Nd F_7 (114) and Cs_3 Dy F_7 (115), which have the internal transitions characteristic for $4f^2$ Nd(IV) and $4f^8$ Dy(IV) and are both orange, having their first electron-transfer band at 26000 cm^{-1}. However, it is striking that the corresponding praseodymium(IV) and terbium(IV) fluorides having their first electron-transfer band at higher wave-numbers are colourless.

Until now, we have only been talking about the relative ease of removing an 4f electron (by $4f \rightarrow 5d$ excitation or by forming metallic compounds, or by E_0) or adding an 4f electron (electron transfer spectra) across the lanthanides. However much the spectroscopic details of internal $4f^q$ transitions are subtle, and however much useful information we have extracted from the broader absorption bands, there is one kind of information still lacking in the puzzle, i.e. the absolute ionization energy $I(4f)$ relative to vacuo in compounds. As we see below, photo-electron spectrometry has supplied exactly this result though with only a moderate precision around 1 eV. We remember 1 eV is equivalent to 8067 cm^{-1} and it can be argued (15, 18) that the ionization energy of the standard hydrogen electrode is 4.5 eV. This statement neglects certain uncertainties related to the entropy, and it is not more precise than 0.1 eV. But the *chemical ionization energy* $I_{chem} = (4.5 + E_0)$ eV is *adiabatic*, allowing the internuclear distances to vary. Hence, we expect the corresponding values of I_{chem} in eV of aqua ions

$$\text{Ce(III) 6.2} \quad \text{Sm(II) 3.0} \quad \text{Eu(II) 4.1} \quad \text{Yb(II) 3.4} \tag{5}$$

to be smaller than $I(4f)$ determined by spectroscopic measurements obeying *Franck* and *Condon's* principle of unchanged internuclear distances, as discussed in section IV D.

The E_0 values of Eq. (4) follow the refined spin-pairing energy theory with $(E-A) = 2900$ cm^{-1}, almost the same value as found for electron-transfer spectra of M(III) complexes. E_0 for M(III)/M(IV) in both the 4f and the 5f groups (116) has been calculated. The characteristic difference is that $(E-A)$ is only 0.62 eV $= 5000$ cm^{-1} in the 4f but 1.07 eV $=$ 8600 cm^{-1} in the 5f group. A chemical corollary of the higher $(E-A)$ in the 5f group is that low oxidation states become stable for high Z, as already suggested by the electron-transfer spectra (97) of Cf(III) and Es(III), and E_0 of mendelevium(II) aqua ions ($Z = 101$) has been shown by ingenious tracer experiments to have $E_0 = -0.1$ V, and nobelium(II) aqua ions ($Z = 102$) are almost impossible to oxidize and have E_0 around

$+1.5$ V. This forms a most striking difference with americium(III) which cannot be reduced in aqueous solution, and can be explained (117) by the combination of larger $(E-A)$ and smaller D in the 5f group. There is this analogy between the 3d and 4f groups with rather invariant oxidation state and the 4d and 5f groups with more variable oxidation numbers that, at the beginning, they are higher in the 4d and 5f groups (cf. Mo, Tc, Ru and Cr, Mn, Fe; or Th, Pa, U, Np with Ce, Pr, Nd, Pm) whereas they are lower at the end [compare Cu(II) and Ag(I) or Yb(III) and No(II) as the most common oxidation states] (21).

The *hydration difference* (15, 18, 21) is the difference between I_n of gaseous M^{+n-1} and I_{chem} of the corresponding aqua ion. It turns out to be approximately $(2n-1)\varkappa$ where $\varkappa$ is a parameter dependent on the nl-shell (and undoubtedly related to decreasing ionic radii) and is about 44000 cm^{-1} or 5.5 eV for beryllium and aluminium and 42500 cm^{-1} or 5.3 eV in the 3d group. I_4 of Ce^{+3} is 36.7 eV (71) and, when compared with Eq. (5), the hydration energy parameter $\varkappa$ turns out to be $(36.7 - 6.2)/7 = 4.36$ eV or 34000 cm^{-1}. This value is also approximately valid for I_{chem} of 4f group M(II) since *Johnson* (96) estimates $I_3 = 24.9$ eV for Eu^{+2} with the result that $\varkappa = (24.9-4.1)/5 = 4.16$ eV. The spectroscopic $I_3 = 25.2$ eV for Yb^{+2} combined with Eq. (5) gives the similar value $\varkappa = 4.36$ eV. In the 5f group (18), $\varkappa$ seems to be 31000 cm^{-1} or 3.8 eV.

D. Tetrad Effect

It is well-known that the 3d group shows important "ligand field" effects on the heat of formation of solid compounds, complex formation constants and ionic radii. When 3d^5 systems have $S = 5/2$, each of the five 3d-like orbitals contain exactly one electron. Hence, it is possible to make an interpolation between 3d^0 systems such as calcium(II), the half-filled shell manganese(II) and the 3d^{10} systems such as zinc(II) (118—120) and, in general, the other systems are slightly more stable. According to present understanding (15) the main contribution to this "ligand field" stabilization is that the other d^q systems contain a lower number of σ-anti-bonding d-like electrons than the statistical average, $(q/10)$ times their number in the closed shell.

It is not surprising that a similar effect was sought in f^q systems and, as a matter of fact, *Schwarzenbach* and *Gut* (121) collected convincing evidence that a discontinuity occurs in the formation constants of trivalent lanthanides with multidentate ligands such as synthetic aminopolycarboxylic acids, such that the gadolinium(III) complexes are slightly weaker than interpolated from the neighbour elements. However, when the small one-electron energy differences in the 4f shell were evaluated (60) and their almost statistical distribution in the sub-levels of the ground J-level were recognized, it became clear that this mechanism is at least one order of magnitude too weak to explain the experimental effect. This is further attenuated by the *Boltzmann* population of sub-levels at room temperature ($kT = 210$ cm^{-1}).

Irena Fidelis and *Siekierski* (122, 123) pointed out that irregularities just above the limits of experimental error also occur at the quarter-filled shells, between q = 3 and 4 and between q = 10 and 11. *Peppard* (124) further studied this situation (calling it the *tetrad effect*) and asked why it occurs. *Nugent* (125) and the writer (126) independently suggested a variation of the third decimal of the nephelauxetic effect, the complex considered having slightly lower phenomenological parameters of inter-electronic repulsion than the aqua ion. Normally, one does not consider the huge stabilization $-(168/13) D = 84000$ cm^{-1} of gadolinium(III) according to Eq. (3) as more of an experimentally verifiable quantity relative to a hypothetical 4f^q baricenter system than, for instance, the explosive nature of a lithium atom if *Pauli's* exclusion principle were suddenly to be suspended and it rearranged to 1s^3. However, it is clear that if D is 1 percent smaller in a given complex than in the aqua ion, the corresponding energy difference 840 cm^{-1} (equivalent to a factor of 50 in the complex formation constant K at room temperature related by RT ln K) is more than needed to explain the gadolinium(III) irregularity even when it is spread out between Pr(III) and Tm(III) according to Eq. (3).

However, in order to obtain good agreement with experience, one has to assume (126) that E^3 is influenced four to five times more than E^1 (and hence D). This is not without connection with the general question (15) whether the nephelauxetic effect in lanthanide compounds is more pronounced for E^3 than for E^1. If this is the case, we have exactly the opposite situation of the correlation effects described as a "*Coulomb* hole" which, anyhow, is a bad approximation since the *Watson* effect is particularly pronounced for the differences between terms having the maximum value of S (56). By the way, the distribution of L values for maximum S is also symmetric around quarter-filled shells (f^q and f^{7-q}) by a kind of *Van Vleck* relation (15).

It is not excluded that other effects may contribute to the tetrad effect. If the covalent bonding is stronger when 4f^{q-1} 5d has low excitation energy, an irregularity would be introduced by the stronger effect in Tb(III) than in Eu(III) and Gd(III) whereas the opposite [stronger effects in Sm(III) and Eu(III) than in Gd(III)] stabilization would occur if the position of the electron-transfer bands were important. No such effect has been detected with certainty in the electric polarizabilities derived from measurements of refractive indices (127, 128). Nevertheless, it is possible that a part of the tetrad effect is connected with "ligand field" stabilization involving the 5d shell or other empty orbitals of the central atom.

The weak, but perceptible, gadolinium(III) irregularity in the ionic radii is difficult to explain *via* Eq. (3) since the virial theorem should make the average radius of the 4f shell smaller in Gd(III), and its slightly increased ionic radius cannot be explained that way unless the outer shells (5s and 5p) expand as a second-order effect of the 4f contraction.

IV. Photo-Electron Spectrometry

A. Inner Shells and Satellites

The ionization energy I_0 of an inner shell or a penultimate delocalized M. O. can be determined from peaks in the probability distribution of the kinetic energy $E_{kin} = h\nu - I$ of the electrons ejected from a sample with monochromatic photons having the energy $h\nu$. Since 1963, this technique has been applied to gaseous molecules using the helium resonance line at 584 Å corresponding to 21.2 eV (sometimes, the 2p → 1s emission line of He$^+$ at 304 Å or 40.8 eV is used) and the penultimate M. O. predicted in diatomic and triatomic molecules have been confirmed, and many polyatomic molecules have been studied (*129*). The resolution is very good, about 0.01 eV, and the vibrational structure is conspicuous and can help in many identifications, though it may be a little too complicated (as in ultra-violet spectra) in polyatomic molecules. On the other hand, only volatile compounds can be studied, and only I values below 21 (or 40) eV can be determined.

A group of physicists in Uppsala (*130, 131*) have developed photoelectron spectrometry using soft X-rays with wavelengths slightly below 10 Å. In actual practice, the photons used originate in 2p → 1s transitions in an anti-cathode consisting of metallic magnesium (1253.6 eV) or aluminium (1486.6 eV). Though such X-rays penetrate about 3000 Å of a typical solid sample, the ejected electrons with full E_{kin} (not having suffered inelastic scattering) arrive from the outermost 20 to 50 Å. Hence, the yield of counted electrons is very low, and an intense background occurs below the peaks, especially for small E_{kin}. It is customary to plot the spectra with E_{kin} increasing toward the right with the result that I increases toward the left. The resolution is not much better than 1 eV. This regrettable fact is connected with *Heisenberg's* uncertainty principle that, when the half-life of an excited state is sufficiently short, the double-sided half-width 2δ is:

$$10^{-14} \text{ sec: } 0.046 \text{ eV} \qquad 10^{-15} \text{ sec: } 0.46 \text{ eV} \qquad 10^{-16} \text{ sec: } 4.6 \text{ eV} \tag{6}$$

The original instrument in Uppsala (*130*) and a home-made instrument in Berkeley (*132*) use magnetic deflection (like a β-spectrograph) whereas the instruments commercially available (from at least five companies) use electrostatic deflection, allowing a much more compact design. Normally, the monochromatic photons are obtained by filtering in a thin film as this less transparent for adjacent photon energies, but one company attempts direct passage in a crystal monochromator allowing a better resolution at the expense of much lower intensities.

Photo-electron spectrometry on solid samples tends to fall into two categories having each their separate problems. Metallic samples are readily brought into electric contact with the rest of the apparatus; and it is argued that the kinetic energy $E_{kin} = h\nu - I^*$ refers to I^* measured relative to the *Fermi* level of the metal. Thus, if I^* of $4f_{7/2}$ of metallic gold is 83.8 eV and the work function 4.8 eV, I *relative to vacuo* is 88.6 eV. The advantage of metallic samples is that small variations of I^* can be measured (*133*) but only in the case of Au is it certain that no superficial oxide, hydroxide or carbonate occurs. However, it is fairly easy to see both Al and the superficial oxide simultaneously on good-quality aluminium, as well as metallic Ni, Ga, Rh, Pd, Ag, Ir and Pt. There has been much discussion of *charging effects* on non-conducting samples (*134, 135*). During the bombardment with X-rays, a quasi-stationary equilibrium is rapidly established where the sample becomes positive because it loses photo-electrons, but electrons slowly diffuse in from the surroundings to replace them. It has been argued (*134*) that the charging effects on typical non-conductors such as $BaSO_4$ amount to almost 2 eV lower E_{kin} than expected. We have performed experiments with a mixture of powdered $BaSO_4$, ThF_4 and Au on scotch tape where the intensity of the signals (presumably approximately proportional to the X-ray intensity) is varied by a factor of 6 by choosing various combinations of X-ray potential (9000 and 6000 V) and filament current (70, 50 and 30 mA) on our Varian IEE-15 instrument. The I^* values recorded vary parallel (to the extent of 0.5 to 0.7 eV higher values for the highest photon intensity) for both Au and the F, Ba and Th signals of the non-conductors, whereas I^* of carbon 1s of scotch tape only varies 0.2 eV. However, the charging effects show some kind of hysteresis in the sense that, when returning to standard conditions (9000 V and 70 mA) after 10 hours of measurement, the I^* did not fully return to their original values. It is worthwhile to note that ThF_4 alone or $BaSO_4$ alone on scotch tape have only marginally higher I^* values; the admixed gold powder does not modify the electric conditions. However, powdered gold alone on scotch tape ($I^* = 83.2$ and I, according to the definition below, 88.6 eV of the Au $4f_{7/2}$ signal) has much lower I^* values than Au in our experiment ($I^* = 86.25$ to 86.6 eV, I close to 91.2 eV). We doubt whether it is a good idea to put thin layers of a sample on a support of metallic gold. The main conclusion of this experiment is that the charging effects may be as large as 3 eV, but then, they occur already, for a threshold X-ray intensity lower than used by us, and the writer still believes that the *internal standard* is a more reliable technique for non-conducting samples. In particular (*136*) the carbon 1s signal of scotch tape can be defined to have $I = 290.0$ eV relative to vacuo in analogy to other hydrocarbons (*130, 131*) if one-sided scotch tape (of which the cohesive side is the aliphatic hydrocarbon polymerized isoprene) and not double-sided scotch tape (a poly-ether of the type of polymerized ethyleneglycol) is used. Thus, the correction C_{st} in each experiment is given as the difference between 290 eV and the lowest I^* value (usually between 284.5 and 286 eV) of carbon 1s recorded by the apparatus, and all the other signals then have I evaluated as $I^* + C_{st}$. The chemical shift for an element in a definite oxidation state is from 2 to 8 eV according to our measurements (*137*) of 600 compounds containing 77 elements, i.e. all elements which are neither noble gases nor strongly radioactive. 40 of these compounds have been re-measured with some months' interval, and the average deviation is 0.7 eV. However, this is not a fair comparison in the sense that such measurements have most frequently been repeated where the position of the carbon 1s signal is suspect. The writer is convinced that the reproducibility under normal conditions is better than an average deviation of 0.4 eV. However, another interpretation is that the C1s signal at 1.5 to 4 eV higher I^* is representative of the positive charging in a quasi-stationary equilibrium on non-metallic samples, and that one should introduce I' as I diminished by the difference between the two carbon signals. The chemical shift dI' is about two-thirds of dI and the reproducibility of I' is 0.2 eV (*137*).

What is colloquially called the *stone age* of this subject is the opinion which prevailed around 1967 that the chemical shift dI is mainly determined by the oxidation state (*138, 139*) as suggested by a shift 9.6 eV between europium(II) and (III) and a more moderate shift 5.3 eV between I^- and IO_3^-. However, it was pointed out (*130, 139*) that the *Madelung* potential (*15, 18*) of the surrounding ions produces a considerable contribution to dI, counteracting the influence of the charge z of the corresponding gaseous ion M^{+z}. The *bronze age* was initiated in 1969 by *Kramer* and *Klein* (*140*) comparing $I^*(3p)$ values of iron compounds, where iron(III) shows a distribution over 4.3 eV, the highest value (for K_3FeF_6) being identical with that of the ferrate(VI) K_2FeO_4 whereas highly covalent dithiocarbamates (dtc$^-$ is $(C_2H_5)_2NCS_2^-$) such as Fe dtc$_3$ and Fe dtc$_2$ Cl show ionization energies lower than those of many iron(II) complexes such as $[Fe(H_2O)_6]SO_4$ and even the iron(0) compound $Fe(CO)_5$. The most plausible explanation of the "bronze-age" opinions is that dI is proportional to the fractional atomic charge (15, 41) attenuated by the effects of the *Madelung* potential. However, it is always dangerous to make too many measurements, and the *Byzantine epoch* started with the realization that $I(\text{K2p}_{3/2})$ of 64 potassium salts (137, 141) varies from 301 eV for fluorine-containing compounds such as K_2BeF_4, K_2SiF_6, KPF_6, KBF_4, $KNiF_3$ and K_2HfF_6 over intermediate values, e.g. for oxygen-containing compounds such as KIO_4, KIO_3, $KClO_4$, $KReO_4$ and KNO_3, down to 298 eV for hexa-complexes of heavy halides and pseudo-halides such as $K_3Cr(NCS)_6$, K_2OsCl_6, $K_2Pt(SCN)_6$ and $K_3Fe(CN)_6$ as well as KSeCN itself. This variation almost coincides with the classification of the anions according to their hard and soft character according to *Pearson*, with exception of KI where I is 301.0 eV. We return to the possible explanation of this surprising regularity in section IV E. Anyhow, it cannot be connected with the *Madelung* potential which would decrease I the most for small anions, though the order of magnitude 8 eV (136) is valid for the *Madelung* contribution.

Not only is it not possible to determine the oxidation number of a given element from the I values of inner shells unless comparison is made with other compounds containing analogous ligands, but in the meantime three cases of anti-stone-age behaviour have been noted: Tl_2O_3 has *lower* I than all measured thallium(I) compounds and PbO_2 lower I than all measured lead(II) compounds (136) and generally cobalt(III) has lower I than cobalt(II) (137).

The I values for each element have been tabulated from X-ray spectra by *Bearden* and *Burr* (142) and supplemented with photo-electron values (130). If the variation from one element to the next is plotted as a function of the atomic number Z, several shells (in particular in the lanthanide region) show inexplicable irregularities amounting to some 5 eV. Fig. 1 shows the variation dI/dZ where the empty circles derive from the Uppsala tables (130) and the filled circles from our measurements of oxides. In spite of the inevitable experimental uncertainties producing the "noise" in the curves, there is a definite effect already emphasized by *Coster* (20) that, in the transition groups, and in particular in the lanthanides, the change dI/dZ is smaller than outside the transition groups. From a theoretical point of view, this quantity is, to the first approximation, the difference between $<\text{r}^{-1}>$ for the nl-shell being ionized and for the shell to which the most loosely bound electron is added going from Z to $(Z+1)$. The latter shell has so large an average radius outside the transition groups that $<\text{r}^{-1}>$ is rather negligible, whereas it is seen from Fig. 1 that it seems to be 5 eV in the 5d group and about 12 eV in the 4f group, perhaps increasing slightly from cerium to lutetium. This phenomenon is called "external screening" in X-ray

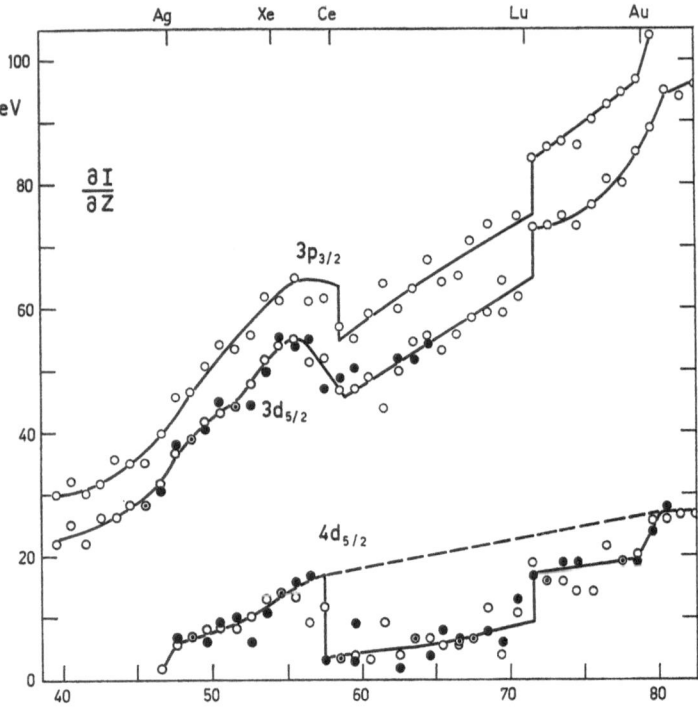

Fig. 1. The change of the ionization energy I from one atomic number Z to the next. The empty circles are differences between entities in the Uppsala tables (130) and the filled circles differences between our measurements on solid compounds, mainly oxides

spectroscopy in sofar as the added electron is in a larger orbital than the electron ionized.

If the ground state of the molecule studied has a positive value of the total spin quantum number S, it is sometimes possible to detect two adjacent photo-electron signals, the lower I corresponding to the ionized system having the increased value $(S+1/2)$ and slightly lower inter-electronic repulsion, whereas the higher I corresponds to $(S-1/2)$. The intensity ratio between the two signals is the ratio $1+(1/S)$ between the degeneracy numbers of the ionized systems. This effect was discovered for the 1s signals of the gaseous molecules $O_2 (S=1)$ and $NO(S=1/2)$ in Uppsala (130, 131) and then for solid fluorides and oxides of manganese(IV) $(S=3/2)$ and manganese(II) and iron(III) $(S=5/2)$ in Berkeley (143). It has been identified in the (otherwise weak) 4s and 5s signals of lanthanide fluorides (144). To the first approximation, the distance between the two signals is $(2S+1) K_{av} (nl, n' l')$ where K_{av} is the average value of the exchange integral of the two-electron operator (18) between the partly filled (nl) shell and the ($n'\, l'$) shell ionized. Hence, K_{av} is closely related to the spin-pairing energy parameter D for a single partly filled shell [which is $(2l+3)\, K_{av}/(2l+2)$] but K_{av} for two partly filled shells can be very small and is, to the first approximation, $<1/r>$ for the "squared overlap density" $\psi_1^2\,\psi_2^2$. However, this analysis is only valid when at least one of the two shells is an s orbital; the number of states for two positive l-values is distributed on a large number of levels, and it might be expected that several signals would have comparable intensities, as is definitely true for the five or six components of the 4d signal of ytterbium(III) compounds (137, 145). The situation is complicated by the strong deviation from *Russell-Saunders* coupling, the two values of $j = l + 1/2$ (having the lower I) and $j = l - 1/2$ of the shell being strongly separated in energy. The main consequence (42) is that S is no longer a good quantum number for most of the levels. The writer believes that the four signals in the 2p region of high-spin ($S = 1$) nickel(II) compounds represent some of the 270 (six times 45) states of the electron configuration $2p^5\,3d^8$ of the ionized system separated by this effect. It is known that the distribution of levels of such configurations containing two partly filled shells and showing strong deviations from *Russell-Saunders* coupling can be quite wide (146). It has been argued (147) that the second and the fourth signals (in order of increasing I) are satellites in the sense discussed below for copper(II). Though the separation around 6 eV between the first and the second signals is considerably larger than the value of 3 K_{av} (2p, 3d) calculated by *Watson*, the writer still believes, in effects of interelectronic repulsion, in part because diamagnetic ($S = 0$) nickel(II) complexes show only two signals belonging to $2p_{3/2}$ at lower and $2p_{1/2}$ at higher I, and in part because their separation 17.2 eV is smaller than the distance 17.7 eV between the first and the third signals of ($S = 1$) complexes which would be hard to reconcile with the hypothesis of satellites because the distance of I between the two j-values corresponding to a positive l-value is constant in different compounds of the same element better than 0.1 eV (148).

The satellites at 8 to 9 eV higher I than the main signals belonging to $2p_{3/2}$ and $2p_{1/2}$ of copper(II) compounds were described by *Novakov* (149). It was thought for a while that certain copper(I) compounds such as Cu_2O, CuCl and CuI also exhibit satellites, but this seems to be due to superficial oxidation, since CuCN and [Ni en$_2$] (CuI$_2$)$_2$ shows only two signals (137, 150) like the isoelectronic closed-shell systems such as zinc(II) and gallium(III). One may imagine two explanations for the specific occurrence of satellites in copper(II), either electron transfer from reducing ligands, as discussed below in the case of lanthanum(III) or simultaneous ejection of two electrons, a 3d-like one as well as one from the inner shell. The latter alternative might be connected with the presence of one strongly σ-anti-bonding, somewhat delocalized electron in the orbital $(x^2 - y^2)$ directed toward the ligands. In such a case, it might also be argued that high-spin octahedral nickel(II) complexes contain two σ-anti-bonding (but less delocalized) electrons in contrast to diamagnetic Ni(II). The position of the satellites is some two to three eV lower than expected for $I(3d)$ (disregarding the additional stabilization of 3d by the hole in the inner shell) so an attractive alternative to "shake-off" is *intra-atomic shake-up* (147) as well known from noble gases (130, 131) where the central atom is left in an excited configuration such as $3d^8\, 4s$. For instance, satellites to the 4s signal of rubidium(I) salts (151) have been ascribed to [Ar] $3d^{10}\, 4s^2\, 4p^4\, 5s$ of Rb^{+2} where the primary ionized state corresponding to [Ar] $3d^{10}\, 4s\, 4p^6$ rearranges. However, the difficulty with the shake-up explanation is that it is not easy to understand why [Ar] $3d^3$ chromium(III) with relatively more extended 3d orbitals do not show conspicuous satellites but only the shoulders corresponding to the $(S \pm {}^1/_2)$ mechanism. Actually, copper(I) would be one of the best candidates for showing low-energy intra-atomic shake-up, but it does not show perceptible satellites.

Table 2. *The four photo-electron signals of lanthanum (III) compounds in the 3d region. The ionization energies I in eV relative to vacuo are determined from the scotch tape correction C_{st} defined in the text. Shoulders are given in parentheses*

	C_{st}	I (La 3d)			
LaF$_3$	4.7	(868)	863.9	(851.4)	847.0
La(IO$_3$)$_3$	5.0	863.8	860.7	847.0	843.9
La$_2$(SO$_4$)$_3$	5.0	863.6	860.4	846.6	843.4
La$_2$(CO$_3$)$_3$	4.0	863.5	860.1	846.4	843.2
La As O$_4$	4.7	863.2	859.9	846.5	843.1
La O Cl	4.4	862.8	859.1	845.5	842.4
La Cr O$_3$	4.5	862.1	858.6	845.2	841.5
La$_2$ O$_3$	4.6	861.9	857.2	844.5	840.9

Lanthanum(III) is an almost unique closed-shell system (152) having four signals in the 3d region. Table 2 shows the four I values of eight La(III) compounds showing a chemical shift $dI = 6$ eV between the fluoride and the oxide. The second and fourth signals are ascribed to inter-atomic shake-up as a situation where an electron has been transferred from the ligands to the empty 4f shell which has become highly stabilized in the ephemeral La(IV) by the absence of one of the ten 3d electrons, but still overlaps to some extent with the ligands, as is known from the high intensity of electron-transfer bands (76) of complexes of the comparable cerium(IV). This suggestion is corroborated by the

Table 3. *The photo-electron signals of lanthanide compounds in the 3d region. Notation as in Table 2*

	C_{st}	I (M 3d)
Ce F$_4$, H$_2$O	4.9	926.1(IV), 913.1(III), 907.6(IV), 894.4, (890.7) (III)
Ce F$_3$	4.8	(915.5), 912.2, (897.1), 893.5
Ce O$_2$	4.8	924.2(IV), 915, 909(III), 905.8(IV), 896.3, 890.1(III)
Th$_{0.6}$ Ce$_{0.2}$ U$_{0.2}$ O$_2$	5.9	909, 906.4(IV), 896, 890.4(III)
Th$_{0.8}$ Ce$_{0.15}$ U$_{0.05}$ O$_2$	4.8	915.4, 908.7(III), 906.0(IV), 896.2, 890.4(III)
Ce$_2$(CO$_3$)$_3$	4.8	911.6, 908.2, 892.8, 890.0
Ce$_2$(SO$_4$)$_3$, 8 H$_2$O	4.7	909.9 asym., 889.9 asym.
Ce$_{0.5}$ U$_{0.5}$ O$_2$	6.1	908(III), 906.1(IV), 895, 889.8(III)
Ce(CrO$_4$)$_2$	4.8	923.8, 914.5, 908.6, 905.4, 895.5, 889.9
Pr F$_3$	4.9	(968.4), 962.1, 943.7 broad
Pr(IO$_3$)$_3$	4.6	962.2, (957), 9421, (937)
Pr$_6$ O$_{11}$	6.5	962.3, 941.6, (937.2)
Th$_{0.9}$ Pr$_{0.1}$ O$_2$	5.5	—, 941.2, (937.5)
Pr$_2$(SO$_4$)$_3$, 8 H$_2$O	4.2	960.8, 940.2
Pr$_2$(C$_2$O$_4$)$_3$	3.7	(963.5), 959.3, 938.8
NdF$_3$	4.7	1013.2, (1005), 990.3 asym.
Nd(IO$_3$)$_3$	4.7	1013.4, (1010.3), 990.8, (988.5)
Nd As O$_4$	4.5	1012.6, (1008), 990.0, (985.5)
Nd$_2$(SO$_4$)$_3$, 8 H$_2$O	4.6	1012.7 asym., (1009), 990.0, (985)
Nd$_2$ O$_3$	5.1	1012.4, 989.8, (985)
Mg$_3$ Nd$_2$(NO$_3$)$_{12}$, 24 H$_2$O	4.6	1011.3, (1005), 988.8, (984)
Sm F$_3$	5.0	(1122), 1118.5, 1091.6
Sm(IO$_3$)$_3$	4.6	1118.6, 1091.4
Sm$_2$ O$_3$	5.3	(1121), 1117.6, 1090.3
Eu F$_3$	4.8	(1177), 1173.6, (1146), 1143.8
Eu(IO$_3$)$_3$	4.9	1172.6, 1143.0
Eu$_2$(SO$_4$)$_3$, 8 H$_2$O	4.5	1171.5, 1141.8
Eu$_2$ O$_3$	4.5	1171.0, 1141.3
Eu SO$_4$	4.4	1171.7(III), 1160.6, 1141.4(III), 1130.8
GdF$_3$	4.4	1129.0, 1196.6
Gd(IO$_3$)$_3$	5.0	1229, 1195.3
Gd Ta O$_4$	5.0	1227.7, 1194.9 asym.
Gd$_2$(SO$_4$)$_3$, 8 H$_2$O	4.8	1227.8, 1194.8
Gd$_2$ O$_3$	4.0	1227.3, 1193.0
Tb O$_2$	5.4	1296, 1284.5(III), 1260.0, 1249,4(III)
Tb(OH)$_3$	5.1	1284.0, 1248.9
Tb Cl$_3$	4.5	1283.2, 1248.1
Dy$_2$O$_3$	4.3	1342.5, 1303.6

slightly higher separation of the satellites and their weaker intensity in LaF_3 containing less reducing fluoride ligands. It must be emphasized that the isoelectronic systems iodide, caesium(I) and barium(II) do not show 3d satellites. In the 5f group, thorium(IV) does not show satellites, whereas uranyl(VI) compounds have weak shoulders to the left of the $4f_{5/2}$ and $4f_{7/2}$ peaks.

However, as seen from Table 3, the subsequent lanthanides show complicated behaviour. The cerium(III) compounds show shoulders only on the two 3d signals which may be ascribed to the $(S\pm^1/_2)$ effect, but CeO_2 most unexpectedly shows six signals, of which four are due to a most unexpected impurity of cerium(III) present in constant proportion. This photo-electron spectrum has been repeated several times, also on freshly calcined samples (to avoid carbonate impurities) and on mixed oxides also containing thorium(IV) and uranium(IV). Madame *Christiane Bonnelle* asked the question whether the electron-transfer satellites of lanthanum(III) might not rather correspond to the first and third signals, having the conventional signals to the left. This is not probable for several reasons (*e.g.* the behaviour of LaF_3) but it is rather striking that the weak signals occur on the right-hand side (as broad shoulders) in all six neodymium(III) compounds studied. It is not entirely excluded that the many states of $3d^9\ 4f^3$ may contain a few low-lying energy levels to which the ionization process has a low probability, but it is conceivable that the electron transfer constitutes a lower total energy than the ephemeral Nd(IV) just lacking one 3d electron. The photo-electron spectra of Sm(III), Eu(III), Eu(II), Gd(III), Tb(IV) and Tb(III) are only weakly structured in the 3d region mainly suggesting $(S\pm^1/_2)$ effects of interelectronic repulsion. The disappearance of the satellites may be caused by the far smaller average radius of the 4f shell no longer overlapping significantly with the electronic density of the ligands.

The presence of a partly filled 4f shell produces complicated X-ray spectra originating in the other inner shells. *Sakellaridis (153, 154)* reports structures of the 2s and 2p absorption edges of the oxides stretching over 100 eV and multiplets of the emission spectra (surprisingly enough stretching over 50 eV) which have been further discussed by *Fischer* and *Baun (155)*. The electron configuration $4d^9\ 4f^{q+1}$ corresponding to absorption lines in the ultra-soft X-ray region typically stretching over 20 eV has been studied experimentally and theoretically *(156, 157)*. As already mentioned, the 4d region of the photo-electron spectra *(137, 145)* shows two broad bands in the case of Eu(III) and Gd(III) and only one, very broad band (with $\delta(-)$ around 3 eV) in Dy(III), Ho(III), Er(III) and Tm(III). On the other hand, YbF_3 has $I = 216, 209, 202$ and 194.7 eV and $Yb_2\ O_3\ I=210, 203.3, 196$ and 189.2 eV.

$Lu_2 O_3$ has only two signals, as expected for a closed-shell system, at $I = 213,9$ and 203.9 eV and $Lu(IO_3)_3$ at $I = 214.8$ and 204.8 eV. It is obvious that the main part of the broadening and structure in the earlier lanthanides is due to interelectronic repulsion in the two partly filled shells $4d^9 4f^q$. However, *Fadley* and *Shirley* (*158*) pointed out that the two 4d signals are perceptibly broadened in LuF_3. This was ascribed to short radiative half-life acting *via* Eq. (6) because the 4f electrons very rapidly jump down into the 4d vacancy. Under our conditions of 100 eV analyzer energy (giving the strongest signals but not the best resolution) $\delta(-)$ is between 0.95 and 1.0 eV for signals having no significant width such as 4f in elements heavier than tungsten (*148*), whereas it is 2.1 to 2.2 eV for $4d_{5/2}$ of lutetium(III) and all heavier elements. Quite generally, ns signals in elements with Z above 20 are broad and have $\delta(-)$ at least 2 eV, and 4s of the lanthanides (*144*) have $\delta(-) = 3$ eV. Whereas the 4p signals is rather narrow until molybdenum, it suddenly increases to about 5 eV between rhodium(III) and cadmium(II) containing many 4d electrons having a high probability of transfer to the 4p vacancy shortening its radiative half-life to 10^{-16} sec, but the width decreases to $\delta(-)$ around 2 eV in barium(II) and metallic gold.

The question of relative intensities of photo-electron signals is rather intricate. If we disregard the question of the probability of an ejected electron escaping without inelastic scattering and consider the simplified problem of the probability of ionizing a definite nlj-sub-shell of a given element with a beam of monochromatic X-ray photons, the relative probability is a product of three factors: the number $(2j+1)$ of electrons in the relativistic sub-shell, the average value $<r^{-2}>$ and a third factor which can be approximated by the square of an effective nuclear charge or by I (*159*). Hence, the high I-values generally give the strongest signals (though not always the best signal/noise ratio) but the unusually small average radius of the 4f shell produces fairly strong signals because of the dependence on $<r^{-2}>$.

B. Partly Filled Shells

One of the major advantages of X-ray-induced photo-electron spectra is that I of the partly filled shell of transition-group compounds and the corresponding filled shell of the immediately following post-transition-group compounds can be measured (*136*). At this point, the intensity distribution containing the factor $<r^{-2}>$ is favourable, and on the whole, the valence electrons give far weaker signals. Thus, each iodate anion in Table 4 (giving results for non-metallic lanthanide, hafnium and tantalum compounds) contains 26 valence electrons and each of the four mandelate anions in hafnium(IV) mandelate contains 58 valence electrons.

Nevertheless, the fourteen 4f electrons give a much stronger signal, as also seen from Fig. 2.

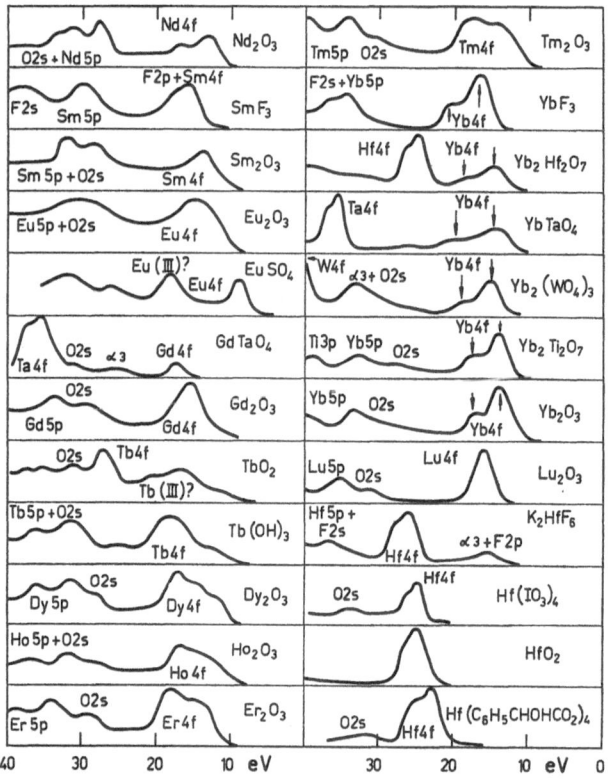

Fig. 2. Photo-electron spectra of compounds of lanthanides, hafnium and tantalum in the region from I between 0 and 40 eV. The origin of the signals from various shells is indicated; the intensities are not on the same scale going from one compound to another, and an arbitrary height of background has been removed. Signals due to $K\alpha_3$ photons with 9.8 eV higher energy than the main exciting line are marked α_3

Independently of our measurements, *Wertheim, Rosencwaig, Cohen* and *Guggenheim (160)* pointed out that the fluorides MF_3 have a signal growing in intensity from Ce(III) to Lu(III) and showing a doublet structure from Tb(III) (q = 8) to Yb(III) (q = 13). This is exactly what one would expect for the second half of the shell, because the ionization leaving between 7 and 12 electrons has the choice between increasing S by half a unit (the lower I) and decreasing S by half a unit (the baricenter of these states has $2 DS + D$ higher energy) whereas the ionization for q below 8 always decreases S by half a unit. The authors at *Bell* (*160*) did not want to pronounce themselves about the absolute ionization energy relative to vacuo, because it is difficult to estimate the charging effects, but it is important to note that the I(M 4f) signal coincides with the I(F 2p) for lanthanides around Sm(III) whereas it occurs at lower I in PrF$_3$. Using one-sided scotch tape as internal standard and the 290 eV definition, it is known (*136*) that the most loosely bound p shells of halides have I close to $(1 + 3.7\ \varkappa_{opt})$ eV where $\varkappa_{opt}$ is the *optical electronegativity* derived from electron transfer spectra (*75*) and where 3.7 eV is the constant 30000 cm^{-1} introduced in order to have $\varkappa_{opt} = 3.9$, 3.0, 2.8 and 2.5 for fluoride, chloride, bromide and iodide. The corresponding I-values 15.4, 12.1, 11.4 and 10.3 eV are close to I of the most loosely bound M. O. of gaseous halides (*129*) though variations around 1 eV occur. *Nugent, Baybarz, Burnett* and *Ryan* (*116*) calculated what is technically the uncorrected optical electronegativity for 4f group M(IV) which is the appropriate quantity to consider since the excited state of the electron transfer spectra of M(IV) corresponds to the reduced M(III) Fig. 3 gives I(4f) of various compounds where the highest value for a given oxidation state usually is represented by the fluoride (square) and the lowest by the oxide (empty circle). The range of chemical shift dI is 3.1 eV for Yb(III), as seen from Table 4, about half the chemical shift for the inner shells. The black curves are the values of $(1 + 3.7\ \varkappa_{uncorr})$ eV calculated for maximum S of 4f^{q-1} and the additional contribution of spin-pairing energy for $(S - 1)$; it is a fortunate accident that the curves join at the experimental value for Lu$_2$O$_3$. On the whole, the agreement is unexpectedly good, since the ionized system 4f^{q-1} have many SL-terms, and this may explain the intensity distribution which is normal in Yb(III) insofar as ^{3}H and ^{3}F are a more probable product of ionization of ^{2}F$_{7/2}$, whereas the secondary peak at 3.8 eV higher I almost corresponds to the position of ^{1}D of the 4f^{12} system Tm(III) (*53*) and a considerably larger energy difference than simply $2 D = 1.6$ eV. By the same token, the broad but less intense structures of I(4f) from the ground level $^4I_{15/2}$ of the 4f^{11} system Er(III) may correspond to the many terms ^{5}I, ^{5}F, ^{5}S, ^{5}G and ^{5}D of 4f^{10} of which the four first are distributed over 3 eV in Ho(III). The shoulders of the photo-electron spectra in the 4f region of Nd(III) and Sm(III) (however much they are suspected of being due to valence electrons) may also correspond to the various terms having maximum S of the ionized system. It should be possible to predict relative intensities using fractional parentage (*44, 45*).

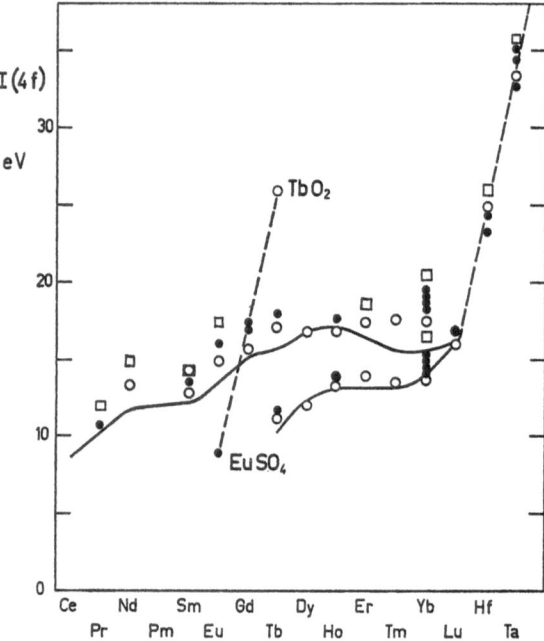

Fig. 3. The ionization energy of the 4f shell in oxides (empty circles), fluorides (squares) and other compounds (filled circles) of lanthanides, hafnium and tantalum. The effects of spin-pairing energy are compared with values of I derived from the calculated uncorrected optical electronegativities (*116*) for M(IV) and given as full curves appropriate for the ionization of M(III). The dashed lines connect the iso-electronic $4f^7$ and $4f^{14}$ systems with differing oxidation state.

Another source of separations in the I(4f) signals is spin-orbit coupling. We already mentioned that the two J-levels of $4f^{13}$ Yb(III) are separated 1.3 eV. What is more important for our purposes is that the same difference in Lu^{+4} is (*29*) 1.46 eV. In our photo-electron spectra, a minor asymmetry can be perceived in the Lu(III) signals, and a well developed shoulder in Hf(IV) representing $4f_{5/2}$ having I 1.6 eV higher than $4f_{7/2}$. This separation is 1.8 eV in Ta(V) showing two distinct signals and increases smoothly (*148*) to 5.3 eV in bismuth ($Z = 83$). On our scale of resolution, it is clear that spin-orbit coupling in partly filled 4f shells only plays a minor role. Thus, the width of ^7F (known to be 0.5 eV in $4f^6$ Eu(III)) contributes to a certain broadening of the strong photo-electron peak of $4f^7$ systems (*158*).

Stig Hagström and collaborators (*161, 162*) went to ultra-high vacuo (10^{-10} mm rather than 10^{-6} to 10^{-7} mm used by us) to detect photo-electron spectra of metallic lanthanides induced by 21 eV or lower energy ultra-violet photons. Interestingly enough, the 4f signals are far less distinct than in the X-ray induced spectra. Great care was taken (161) to use freshly evaporated samples which had not yet developed an oxygen 1s signal, and the $I*$ (4f) were measured relative to the *Fermi* level. The two metals with conditional oxidation state[II] have very low $I*$, 2.2 eV for europium and 1.6 and 2.9 eV (separated by spin-orbit coupling) for ytterbium. Since the work function for these two metals must be close to 2 eV, Eu has I(4f) some 4.7 eV lower than our $EuSO_4$, which is a quite normal chemical shift from a metallic element to its compounds. Gaseous europium atoms (158) have this signal at 12.0 eV. Metallic gadolinium (161) had $I*$(4f) = 7.8 eV, and the work function (163) is 3.2 eV with the result that I(4f) = 11.0 eV, again a reasonable chemical shift when compared with the Gd(III) compounds in Table 4. Metallic dysprosium has two signals with $I* = 3.8$ and 8.0 eV, undoubtedly separated by the $(S \pm 1/2)$ effect of Fig. 3 like $I = 16.8$, (13.8) and (12) eV in Table 4 for $Dy_2 O_3$. By the same token, metallic erbium (161) has $I* = 5.0$ and 8.5 eV.

These results are important for solid-state physics insofar as they demonstrate the presence of an integral number of 4f electrons having a larger I than the most loosely bound conduction electrons in the metals, and for M. O. theory insofar as they show the paradoxical result that I(4f) in almost ionic compounds is larger or comparable with I of the most loosely bound delocalized M. O. consisting, for instance, of oxygen or fluorine 2p orbitals. Whereas gaseous VCl_4 has I(V 3d) = 9.41 eV below the various I(Cl 3p) between 11.75 and 15 eV (164) the vapours of tris(hexafluoroacetylacetonates) have I(M 3d) for M = Ti, V and Cr as a sharp peak below a characteristic set of M. O. ionization energies also observed for M = Al and Sc, whereas M = Fe and Co have the 3d signal inside this set (165). By the way, this observation confirms the opinion (166) that the diagonal elements of 3d energy vary in a regular way with Z in such a series, as suggested by other d group electron-transfer spectra (75). Anyhow, the theoretical challenge from the photo-electron results is the renewal of the question (15, 56) whether models of the *Hückel-Wolfsberg-Helmholz* type have diagonal elements representing the ionization energies I(M) and I(X) of atoms in a highly heteronuclear bond MX, or whether one should consider the *Mulliken* electronegativity (the average of the ionization energy and the electron affinity) as the appropriate diagonal elements, or whether, finally, one should compare the electron affinity of M and the ionization energy of X as a measure of the extent of weak covalent bonding. In calculations, these quantities should, of course, include the *Madelung* potential (167).

C. K. Jørgensen

C. Difference between Electron Affinity and Ionization Energy

Metals which are freely conducting in all three dimensions (or anisotropically conducting in two dimensions, such as graphite) have the same electron affinity and ionization energy, which is called the work function. The situation is entirely different in a solidified noble gas or in a stoichiometric alkali halide crystal with very low electron affinity and high ionization energy. This closed-shell behaviour is easy to understand. What is more fascinating is the existence of partly filled shells with low electron affinity and high ionization energy because of effects of interelectronic repulsion. Depending on the exact shape of the radial function, the theoretical value of the integral (18) $J_{av}(nl, nl)$ is between 0.6 and 0.7 times $<r^{-1}>$ of the nl-shell (in atomic units of 1 bohr $= 0.528$ Å and 1 hartree $= 27.2$ eV). Because of the *Watson* effect $(15, 56)$ the spectroscopic value is expected to be close to $0.55 <r^{-1}>$. For gaseous Gd^{+3}, the 4f shell of Dr. Watson's Hartree-Fock function has $<r^{-1}> = 1.61$ bohr^{-1}. Hence, a phenomenological $J_{av}(4f, 4f)$ close to 24 eV is expected. This is slightly more than the difference (of which ~ 6 eV is due to spin-pairing energy) between the ionization energy $I_4 \sim 43$ eV of Gd^{+3} and $I_3 \sim 21$ eV of Gd^{+2}, but this discrepancy is caused by the readaptation of the 54 electrons of the xenon core. Such adaptation is much more important in chemical compounds, such as hydration energy or Madelung energy, and it cannot be argued that the standard oxidation potentials give more than a highly attenuated value of J-integrals. As a matter of fact, disproportionating oxidation states such as antimony(IV) have an apparently negative $J(5s, 5s)$. Whereas $Sb^{III}(Sb^V O_4)$ is colourless, halide complexes such as $K_4^+(Sb^{III} Cl_6) (Sb^V Cl_6)$ are dark blue with Sb(IV, IV) as the *excited* state of the electron-transfer band (78). Nevertheless, the difference of 5 to 6 V between the extrapolated standard oxidation potential E_0 of hypothetical cerium(II) aqua ions and E_0 of cerium(III) aqua ions is far larger than usual in the d groups (for vanadium, the two E_0 values are -0.25 V and $+0.34$ V, the latter referring to vanadyl ions as the oxidized form).

It is seen from Table 4 that $I(4f) = 8.9$ eV for $EuSO_4$ (which could not be obtained free of a minor superficial impurity of Eu(III) even in freshly prepared samples) is 7 eV lower than the europium(III) sulphate (the corresponding dI for the 3d signal in Table 3 is 11.0 eV) and it is 7 to 8 eV lower than the isoelectronic gadolinium(III) compounds. This large shift made us look for $I(4f)$ of terbium(IV). The dark brown $Tb_4 O_7$ contains equal amounts of Tb(III) and Tb(IV) and shows a signal at 27.0 eV not seen in terbium(III) compounds. However, it was slightly suspected of being due to oxygen 2s which is known to give I values between 28 and 32 eV in many oxygen-containing compounds,

Table 4. $I(4f)$ of lanthanide, hafnium and tantalum compounds. Notation as in Table 2

	C_{st}	$I(4f)$		C_{st}	$I(4f)$
PrF_3	4.9	13.7 (F 2p?), (12.0)	$Er(IO_3)_3$	4.3	17.5, (16), (13.8)
$Pr_2(SO_4)_3$, 8 H_2O	4.2	10.8 weak	Er_2O_3	4.7	17.5, (14)
NdF_3	4.7	(17), 14.9	$Tm(IO_3)_3$	4.3	17.2, (13.8)
$Nd\ AsO_4$	4.5	(16.5), 13.5	Tm_2O_3	5.1	17.7, (13.6)
$Nd_2(SO_4)_3$, 8 H_2O	4.6	(16), 13.2	YbF_3	4.7	20.5, 16.7
Nd_2O_3	5.1	(16.6), 13.4	$Yb(IO_3)_3$	4.7	(20), 16.0
SmF_3	5.0	(15.8), 14.1	$YbTaO_4$	4.7	(19.4), 15.7
$Sm(IO_3)_3$	4.6	17.3 (iodate?), (14)	$Yb_2 (WO_4)_3$	4.3	19.1, 15.1
Sm_2O_3	5.3	(16), 13.7	$Yb_2 Hf_2O_7$	4.7	18.7, 15.0
EuF_3	4.8	17.5	$Yb_2(SO_4)_3$, 8 H_2O	4.2	18.4, 14.6
$Eu_2(SO_4)_3$, 8 H_2O	4.5	16.0	$Yb\ As\ O_4$	4.2	(18.8), 14.6
Eu_2O_3	4.5	14.9	$Yb\ VO_4$	4.9	18.2, 14.3
$EuSO_4$	4.4	8.9	$Yb_2 Ti_2 O_7$	4.6	17.5, 13.7
GdF_3	4.4	18.9	$Yb_2 O_3$	3.8	(17.6), 13.6
$Gd_2(SO_4)_3$, 8 H_2O	4.8	17.5	LuF_3	5.5	20.5
$Gd\ TaO_4$	5.0	17.4	$Lu(IO_3)_3$	4.5	16.8
$Gd_2 O_3$	4.0	15.6	Lu_2O_3	4.8	16.0
TbO_2	5.4	27.8, 18(III), (12) (III).	$K_2 Hf F_6$	5.0	(27.5), 26.0
Tb_4O_7	5.1	27.0(IV), 17.2, 11.1	$Hf(IO_3)_4$	5.0	(27.3), 25.6
$Tb(OH)_3$	5.1	18.1, (11.5)	$Hf\ O_2$	4.5	(26.5), 25.0
$Tb\ Cl_3$	4.5	17.7, 11.0	$Yb_2 Hf_2 O_7$	4.7	(26), 24.4
Dy_2O_3	3.3	16.8, (13.8), (12)	$Hf(C_6 H_5\ CHOHCO_2)_4$	5.0	(25), 23.2
$Ho(IO_3)_3$	4.2	17.8, (15.4), (13)	$K_2 TaF_7$	4.5	(37.5), 35.7
$Ho_2 O_3$	4.9	16.9, (13)	$Yb\ TaO_4$	4.7	(36.3), 34.6
$Er\ F_3$	4.5	18.7, (15.7)	$Ta_2 O_5$	4.9	35.1, 33.3

oxides at the lower end (*e.g.* MgO 29.6 eV) and sulphates and other oxygen-containing anions (*168*) at the higher limit of this interval. Considerations of relative intensities and comparison with other oxides have removed this doubt. Professor *Georg Brauer*, University of Freiburg im Breisgau, was so kind as to supply us with a sample of brickred TbO_2 obtained by a special exhaustive extraction of Tb_4O_7 with acetic acid (*169*) and having the correct cubic unit-cell parameter and chemical analysis. Whether this material reacts with the scotch tape or contains superficial Tb(III) is not known, but we could not obtain spectra differing from two-thirds Tb(IV) and one-third Tb(III), even from a sample cautiously washed with cold 1 molar hydrochloric acid. Anyhow, Table 3 shows $dI = 10.6$ eV for the 3d signals, and the $I(4f) = 27.8$ eV was confirmed for the untreated sample. This is the *highest I of any chemical system containing a partly filled shell* (closed-shell systems, such as beryllium(II), may have much larger values, here 119 to 123 eV, for I of the most loosely bound shell). The regularities seen on Fig. 3 clearly show that $I(4f)$ of praseodymium(IV) would be close to 22 eV. Unfortunately, the photo-electron signal of this single 4f electron not having a particularly large $<r^{-2}>$ is so weak that it has not yet been observed. However, since its position is related to the chemical stability of Pr(V) (an oxidation state proposed in certain mixed oxides by *Prandtl*) it can be said, as in the song by *Juliette Greco* "Une fourmi de dix-huit mètres, cela n'existe pas", that the electron affinity of Pr(V) is certain to be far superior to be compatible with any neighbour atoms.

Apparently, all our cerium(IV) compounds contain some Ce(III) on the surface. The most plausible interpretation involves a chemical shift of the 3d signals about 16 eV between these two oxidation states, as suggested in Table 3. This large value is a most striking difference from elements not being lanthanides, Professor *Cornelius Keller*, Kernforschungszentrum Karlsruhe, was so kind as to send us a sample of LuF_3. The $I(4f) = 20.5$ eV, but the distance between the two carbon 1s signals is unusually large, so I' is only 15.3 eV like for GdF_3.

It is seen from Table 4 that hafnium(IV) has $I(4f)$ some 9 eV higher than analogous lutetium(III) compounds, and tantalum(V) approximately 9 eV higher than Hf(IV). This almost constant value of dI/dZ for the $4f_{7/2}$ signal remains 9 eV up to gold. Though we have not been able to measure any ytterbium(II) compound, there is no doubt that it would be similar to $EuSO_4$. Hence, it is seen from Fig. 3 that if any thulium(I) compound existed, and it had the [68] configuration $[Xe]4f^{14}$, it would have $I(4f)$ close to 1 eV. This is another way of expressing why Tm(I) does not exist, and why the almost constant trivalency of the lanthanides corresponds to the range of I_{chem} where the electron affinity is not so large as to oxidize most ligands, and where the ionization energy is not too low. The latter condition is more severe in solvents (such as water evolving hydrogen) than in crystals (such as CaF_2 having only a very weak tendency to form Ca^+, F^{-2} or "F centers" consisting of an electron on an anion vacancy).

D. Difference between Vertical and Adiabatic Ionization Energy

It is well known from spectra in the visible and the ultra-violet that the width of absorption bands indicates the difference in the bonding character of the excited state and the ground state. Thus, the potential surfaces of the excited $4f^q$ states of lanthanide complexes, or of some of excited $3d^5$ states of manganese(II) complexes (15, 102) have the same internuclear distances as the ground states, whereas in other cases, when the excited states prefer either much shorter or much longer internuclear distances, the absorption bands (and emission bands of luminescence) are broad.

Whereas the typical one-sided half-width of broad absorption bands is 0.2 eV, the photo-electron spectra of gaseous molecules (129) frequently have $\delta(-)$ from 0.5 to 1 eV when a strongly bonding M. O. is depleted. Probably, these co-excitations of vibrations are not very important for X-ray induced spectra of solid samples. The observed $\delta(-)$ can be considered as the square root of the sum of squared contributions from various sources. The radiative half-life of the 2 p→1s transition in the anti-cathode corresponds to $\delta = 0.6$ eV and the apparatus contributes $\delta = 0.6$ eV under our standard conditions producing $\delta(-) = 1.0$ eV. The 1s signal of lithium(I) has $\delta(-) = 1.3$ eV corresponding to an intrinsic δ of 0.8 eV, whereas the 1s signal of beryllium(II) is marginally broadened to $\delta(-) = 1.1$ eV. This broadening has probably a vibrational origin, since the theoretical expression (42) for the half-life of one-electron atoms performing 2 p→1s transitions is $11.10^{-10} \sec/Z^4$ giving perceptible δ only for Z above 8.

A strong discrepancy between the "vertical" ionization energies obtained from photo-electron spectra and adiabatic values obtained from chemical equilibria are the I_{chem} of Eq. (5) such as 4.1 eV for europium(II) aqua ions, whereas we find $I = 8.9$ eV for $EuSO_4$. Though it can be argued that this crystal is particularly stable, a difference of at least 4 eV has to be explained. First, it may be argued that the adiabatic process corresponds to the right-hand foot of the photo-electron signal and not to its maximum. A Gaussian error curve rapidly fades away and has 1/256 of its maximum height 3 δ before the maximum and 16 ppm at 4 δ before the maximum. The observed $\delta(-)$ for $I(4f)$ of $EuSO_4$ correcting for the background is 1.4 eV [it is 1.6 eV in Gd(III)] and three times δ around 1 eV might very well explain a part of the discrepancy. It might also be argued that charging effects have modified the I value (I' defined above would be 6.9 eV), but the writer does not believe that this is the major contribution. Though E_0 for VCl_4 to VCl_4^- is not known, it cannot be very far from $+0.5$ V corresponding to I_{chem} around 5 eV, but the gaseous molecule VCl_4 has $I = 9.4$ eV (164) and the solvation energy of the anion VCl_4^- in a perfect dielectric is 1.7 eV, so again, the adiabatic I_{chem} is much smaller than I for a spectroscopic process obeying the principle of *Franck* and *Condon*. The $I(3d)$ values for many iron(III), nickel(II) and copper(II) complexes (137) are between 10 and 12 eV, far beyond the conceivable I_{chem} values below 8 eV. Finally, CeF_3 has $I(4f)$ on our scale close to 10 eV by comparison with the other fluorides (160), again 4 eV higher than I_{chem}. This may be compared with $I(5p)$ of crystalline iodides (135) close to 11 eV and the (not strictly relevant) E_0 of I^- in aqueous solution to form I_2 corresponding to $I_{chem} = 5.0$ eV which would be changed to 5.8 eV if free iodine atoms are formed. However, I' (5p) are between 8 and 9 eV (137).

E. The *Manne* Effect and the Inner Structure of the Continuum

One may ask whether a principle of *Franck* and *Condon* operates on the motion of the other electrons, when a given electron is ionized. It was a dogmatic truth in the Copenhagen interpretation of quantum mechanics that the quantum jumps do not need any time. This is a non-relativistic statement; since light takes 10^{-18} sec to cross an atom having a diameter of 3 Å, the primary process of the photo-electron ionization cannot conceivably be more rapid. At one time, it was becoming a paradox that the main photo-electron signal corresponds to adaptation of the other electrons to the new, more negative central field U(r) in the ionized system, whereas the idea of complete adaptation can be reduced *ad absurdam*. *Manne* and *Åberg* (170) resolved this apparent contradiction by a careful study of the ten-electron systems methane and neon, where *Hartree-Fock* calculations give $I(1s)$ 15 and 20 eV too high, respectively, than the experimental $I = 290.8$ and 870.2 eV. However, if the eight outer electrons are allowed to contract their radial functions in the modified central field having almost the same form as if an additional proton were situated on the nucleus, very good agreement is obtained. Thus, the "frozen" *Hartree-Fock* orbitals of neon (171) give $I = 891.7$ eV whereas the energy difference between the solution constrained to have the electron configuration $1s \, 2s^2 \, 2p^6$ and of the ground state solutions gives $I = 868.6$ eV. To this can be added the three obvious sources of deviations, $+0.8$ eV for the relativistic effect on a 1s electron, $+1.1$ eV for the correlation between the two 1s electrons in the ground state, and $+0.3$ eV for other correlation effects, giving a total of $I = 870.8$ eV.

The interesting point is that only 80 percent of the intensity of the 1s region of the neon photo-electron spectrum occurs in the main peak at 870.2 eV whereas 20 percent is distributed on some 50 weak peaks at higher I (130, 131). These peaks are due to "shake-up" to states such as 1s 2s^2 2p^5 3p or 1s 2s 2p^6 3s and to "shake-off" where two or more electrons are ionized away to states such as 1s 2s^2 2p^5 or 1s 2s 2p^5. The baricenter of the whole structure (stretching up to above $I = 1000$ eV) and of the main peak occurs at 886 ± 1 eV, close to the "frozen" orbitals according to *Koopmans* yielding 892 eV (170). In other words, we are in the presence of a *Franck-Condon* structure of the unusual character that the main peak at lowest I contains most of the intensity followed by many minor peaks, but corresponding to the "sudden" approximation of rapidly removing a 1s electron and then letting the total wave function of the nine other electrons, which is no longer an eigenfunction of the new Hamiltonian operator, collapse into its linear combination of eigenstates, of which the "adapted" solution 1s 2s^2 2p^6 has a much larger coefficient than the others. The small peaks at higher I can only be studied in practice for gaseous samples, whereas they are masked to a large extent in solids by the much stronger background due to inelastically scattered electrons. However, we see a weak, broad structure close to 311 eV following the main peak at 290 eV of scotch tape, probably due to shake-off of the most loosely bound valence electron, and many other intense lines have such weak hills at the left at a distance of some 20 eV.

The writer believes that the ordering of K 2p$_{3/2}$ signals of 64 potassium salts (137, 141) following the "soft" and "hard" character of the anion (with exception of the iodide) is due to a similar *Manne effect*, and also that the electronic density of the neighbour atoms adapts in the ionized system producing the main peak in apparent disagreement with *Franck* and *Condon's* principle. It is quite conceivable that this interatomic *Manne* effect decreases I to the extent of some 10 eV (methane and neon may be somewhat exceptional by having so relatively many external electrons) explaining why the *variation* of the *Madelung* potential in potassium salts is overcompensated by the variation of the *Manne* effect, the hard anions BeF_4^{-2} and BF_4^- adapting their electronic density to a smaller extent than soft anions such as $Pt(SCN)_6^{-2}$ and $SeCN^-$. The exceptional behaviour of iodide [quite generally, most iodides and caesium salts tend to have I-values 1 to 2 eV larger than expected (136, 137)] can be connected with an effect pointed out quite early (139) that in crystals with high electric polarizability, the ionization of an electron is made intrinsically more difficult because the polarization of the ions surrounding the electron stabilizes the groundstate.

Seen from a chemical point of view, the inter-atomic *Manne* effect can be regarded as an ephemeral highly noble metal being produced by the removal of an electron from an inner shell of a hard cation. Already the ground state of K(II) belonging to the configuration $1s^2\, 2s^2\, 2p^6\, 3s^2\, 3p^5$ is exceedingly oxidizing, about to the same extent as the hypothetical praseodymium(V) discussed above, since $I(K\ 3p)$ varies between 25 and 22 eV (137, 141). However, this is nothing compared with the excited state $1s^2\, 2s^2\, 2p^5\, 3s^2\, 3p^6$ having the electron affinity between 301 and 298 eV. Obviously, we are speaking about a very short-lived species (longer than $3\cdot 10^{-15}$ sec as seen from the marginal broadening of the signal, but probably not a much longer half-life) and *Franck* and *Condon's* principle prevents the internuclear distances from contracting from the values in the ground state corresponding to the ionic radius of K^+. However, otherwise, the ephemeral potassium(II) is a kind of extreme caricature of a noble metal M(II).

The picture of molecules obtained by visible spectroscopy refers to a *time-scale* of about 10^{-13} sec. It is well known (15) that questions of symmetry of a given molecule or polyatomic ion can have highly different answers in cases such as *Jahn-Teller* unstable copper(II) or *Gillespie* unstable tellurium(IV) and lead(II) for the time-average picture obtained from X-ray or neutron diffraction of a crystal, whereas experimental techniques with an intermediate timescale, such as electron or nuclear magnetic resonance, may give differing results, which may also depend on the temperature at which the experiment is performed. There is little doubt that the primary process of the photo-electron ejection is an even closer approximation to giving an instantaneous picture, a reasonable guess of the order of magnitude being 10^{-17} sec. This raises the interesting question of *electronic ordering* of mixed oxidation states of a given element. Thus, the two kinds of antimony atoms in dark blue $Cs_4^I(Sb^{III}Cl_6)\,(Sb^V\,Cl_6)$ kindly provided by *Peter Day* (78) produce two sets of distinct signals, though dI is smaller than between SbF_6^-, $SbCl_6^-$ and $Sb(OH)_6^-$, having differing ligands. We have been trying to find distinct signals from non-equivalent atoms in compounds such as $Fe_3\,O_4$ (which is metallic at room temperature), $Pr_6\,O_{11}$ and $Pb_3\,O_4$, but without much success as yet. We continue studies of interesting cases of niobium(II, III) ruthenium(III, IV) and iridium(III, IV) complexes. Nevertheless, the writer believes in the fundamental electronic ordering of all such compounds (78) on the time scale of 10^{-17} sec, though the experimental results may happen to be only compatible with this idea rather than confirming it positively.

A completely extreme case of electronic ordering on an ultra-short time scale may occur in metallic transition-group elements. *Baer et al.* (172) studied X-ray-induced photo-electron spectra in the region of I^* (relative to the *Fermi* level) below 10 eV containing the signals from the most loosely bound d shell, as well as regions corresponding to inner shells. Copper has a sharp signal at $I^* = 3.1$ eV corresponding to ionization of the closed shell $3d^{10}$, whereas silver and gold show two signals, probably separated by combined "ligand field" and spin-orbit coupling effects. However, the preceding metallic elements Ni, Co, Fe and Pd, Rh, Ru and Pt, Ir and Os show asymmetric broad signals decreasing slowly in the direction of high I^* values. Though most solid-state physicists would think about this phenomenon in terms of the density of states in the filled energy bands (multiplied by a presumably roughly constant factor expressing the probability of ionization), it might also be argued that on the very short time scale of photo-electron spectrometry, metals such as iron are distributed on different conditional oxidation states, say a high concentration of Fe[0] and lower concentrations of Fe[I] and Fe[II] having preponderant electron configurations containing $3d^8$, $3d^7$ and $3d^6$, respectively.

Seen from the point of view of pure quantum mechanics, both the older X-ray spectrometry and the photo-electron spectrometry study excited states constituting singularities (as regards transition probabilities, half-life of producing *Auger* electrons by auto-ionization, *etc.*) *inside the continuum*. To chemists accustomed to vibrational (infra-red and *Raman*) and conventional electronic (visible and ultra-violet) spectroscopy, this is an entirely new feeling. It must be remembered, however, that any system not confined in a small volume has a continuous spectrum of translational energy starting already at the ground state, so the concept of discrete states always need some amplification; it is always an approximation in one way or another.

V. Conclusions — What is so Peculiar About the 4f Group?

The value $l = 3$ certainly contributes to specific properties of the 4f group. Thus, the ionization energy I can have the same order of magnitude as ligand valence electrons (as seen from Fig. 3) and at the same time, the shell can have a small average radius (contributing to the relatively strong photo-electron signals and to the almost monatomic character of the $4f^q$ J-levels in compounds) because of the angular contribution $l(l+1) <r^{-2}>/2$ to the one-electron kinetic energy (15). This pseudo-potential of centrifugal action was recognized by *Maria Goeppert-Mayer* (173) as the main explanation of the f group characteristics. However, these arguments already apply to a smaller degree to the 5f group having much more variable oxidation states. It has even been argued (117) that only the $(n = l + 1)$ cases such as 3d, 4f, 5g, 6h show this tendency, so all the elements from $Z = 123$ to 140 have the almost invariant oxidation state M(IV) containing from one to eighteen 5g electrons, and the question as to why $Z = 164$ prefers the oxidation state M(II) with $7d^8$ is closely related to differences between consecutive ionization energies I_n calculated for the gaseous ions, and the parameter $\varkappa$ of hydration difference (174), and the chemistry of $Z = 184$ is determined by the question as to whether the 6g group unexpectedly belongs to the invariant class, or whether a variation between $6g^8$ M(IV) and $6g^0$ M(XII) occurs (15, 34).

Coster (20) believed that the lanthanides intrinsically end with $Z = 71$. It is seen from Fig. 3 that this is more a question of preferred oxidation state; one might have suspected that the almost invariant oxidation state of the 4f group would have been M(II), in which case the lanthanides would end with Yb(II) exactly as the 5f group ends with No(II). The black curves on Fig. 3 can be moved 9 eV higher up along the high-slope dashed line to represent the variation from Pr(IV) to Hf(IV) or they can be moved 9 eV down in which case they would have represented the development from La(II) to Yb(II) if the former system had not had a negative I. The 4f group is characterized (probably together with the 5g group) by unusually small values of $(E - A)$ combined with large values of D. One important consequence is that the $4f^q$ and $4f^{7+q}$ have comparable I. Thus, it is easier to oxidize Yb(II) than Eu(II), and it is seen from Fig. 3 that Gd(III) and Lu(III), or TbO_2 and HfO_2, have closely similar I (4f). If such an event had happened in the 3d group, it would have been easier to oxidize zinc(II) to Zn(III) than it would be to oxidize manganese(II) to Mn(III). Another difference from the 4f group is that I (3d) is between 27.6 and 29.5 eV for Ga(III) compounds (137) much higher than for Fe(III).

It is interesting to note that the refined spin-pairing energy theory is applicable not only to spectroscopic excitations where a 4f electron is removed (4f→5d) or added (electron transfer spectra), and to the corresponding standard oxidation potentials, but also to the actual ionization energies I (4f) obtained from photo-electron spectra.

A book (175) by Sinha about rare earth complexes in general, and another book (176) about europium, illustrate well the two main epochs through which the interest in lanthanides has passed. For the chemist, it is useful to have a smoothly varying series from La(III) to Lu(III) in which Y(III) is inserted between Dy(III) and Ho(III), where the chemical properties essentially vary as a function of the ionic radius, and it is also interesting to study the exceptional deviations in direction of M(IV) or M(II). The chemists early developed a devotion to half-filled shells which is now explained by interelectronic repulsion in the partly filled shell, and which got its quantitative expression as spin-pairing energy. For the spectroscopist, each M(III) is interesting as an individual, and the solutions and solids containing $4f^q$ offer an unusual opportunity of studying atomic line spectra in the condensed states of matter, which is otherwise only presented by isolated atoms in cool matrices, such as solid argon condensed at 4°K. The third aspect which is now making the lanthanides fascinating is the recent studies of photo-electron spectra. Though the resolution obtained is not brilliant at present, they have added information which it was not possible to obtain previously.

Acknowledgements: The writer is grateful to Dr. *Hervé Berthou* and Dr. *Lucette Balsenc* for their careful work with the photo-electron spectrometer which was purchased with a grant from the Swiss National Science Foundation.

VI. References

1. *Weeks, M. E.:* Discovery of the Elements (6. Ed.). Easton: Journal of Chemical Education 1960.
2. *Heller, A., French, K. W., Haugsjaa, P. O.:* J. Chem. Phys. *56*, 2368 (1972).
3. *Urbain, G.:* Ann. Chim. Phys. (Paris) *18*, 289 (1909).
4. *Jørgensen, C. K., Judd, B. R.:* Mol. Phys. *8*, 281 (1964).
5. *Ropp, R. C., Gritz, E. E., Haberland, P. H., Ridout, D. E. U.:* Electrochem. Technology *4*, 24 (1966).
6. *Brecher, C., Samelson, H., Lempicki, A., Riley, R., Peters, T.:* Phys. Rev. *155*, 178 (1967).
7. *Sovers, O. J., Yoshioka, T.:* J. Chem. Phys. *51*, 5330 (1969).
8. *Jørgensen, C. K., Pappalardo, R.: Rittershaus, E.:* Z. Naturforsch. *20a*, 54 (1965).
9. *Van Uitert, L. G., Lida, S.:* J. Chem. Phys. *37*, 986 (1962).
10. *Blasse, G.:* Philips Res. Reports *24*, 131 (1969).
11. *Haissinsky, M.:* Pseudo-decouvertes dans l'histoire de la radioactivité. XIII Congrès International d'Histoire des Sciences. Moscou: Naouka 1971.
12. *Jørgensen, C. K., Rittershaus, E.:* Mat. fys. Medd. Danske Vidensk. Selskab *35*, no. 15 (1967).
13. *Rydberg, J.:* Lunds Universitets Årsskrift *9*, no. 18 (1913).
14. — J. Chim. Physique *72*, 585 (1914).
15. *Jørgensen, C. K.:* Modern Aspects of Ligand Field Theory. Amsterdam: North-Holland Publishing Co. 1971.
16. — Mol. Phys. *2*, 96 (1959).
17. *Kauffman, G. B.:* J. Chem. Educ. *46*, 128 (1969).
18. *Jørgensen, C. K.:* Oxidation Numbers and Oxidation States. Berlin-Heidelberg-New York: Springer 1969.
19. *Bohr, N.:* Naturwiss. *11*, 606 (1923).
20. *Coster, D.:* Naturwissenschaften *11*, 567 (1923).
21. *Jørgensen, C. K.:* Chimia (Aarau) *23*, 292 (1969).
22. *Hund, F.:* Linienspektren und Periodisches System der Elemente. Berlin: Springer 1927.
23. *Jørgensen, C. K.:* Mol. Phys. *7*, 417 (1964).
24. *Gruen, D. M.:* Progr. Inorg. Chem. *14*, 119 (1971).
25. *Moore, C. E.:* Nat. Bur. Stand. Special Publ. no. 306—4. Washington D. C. (1969).

26. *Klinkenberg, P. F. A.:* cited in Ref. 97 and 98.
27. *Sugar, J.:* J. Opt. Soc. Am. *55*, 1058 (1965); *61*, 727 (1971).
28. *Kaufman, V., Sugar, J.:* J. Res. Natl. Bur. Std. *71A*, 583 (1967).
29. *Sugar, J., Kaufman, V.:* Phys. Rev. *B*, inpress (1972).
30. *Becquerel, H.:* Ann. Chim. Phys. (Paris) *30*, 5 (1883); *14*, 257 (1888).
31. *Urbain, G.:* Compt. Rend. *140*, 1233 (1905).
32. *Gielessen, J.:* Ann. Physik *22*, 537 (1935).
33. *Ephraim, F., Mezener, M.:* Helv. Chim. Acta *16*, 1260 (1933).
34. *Jørgensen, C. K.:* Angew. Chem., in press.
35. *Hofmann, K. A., Kirmreuther, H.:* Z. Phys. Chem. *71*, 312 (1910).
36. *Ephraim, F., Bloch, R.:* Ber. Deut. Chem. Ges. *59*, 2692 (1926); *61*, 65 and 72 (1928).
37. *Boulanger, F.:* Ann. Chim. (Paris) *7*, 732 (1952).
38. *Jørgensen, C. K.:* Mat. fys. Medd. Danske Vidensk. Selskab *30*, no. 22 (1956).
39. *Misumi, S., Kida, S., Isobe, T., Inazumi, A.:* Bull. Chem. Soc. Japan *41*, 25 (1968).
40. *Caro, P., Derouet, J.:* Bull. Soc. Chim. France *1972*, 46.
41. *Jørgensen, C. K.:* Helv. Chim. Acta, Fasc. extraord. Alfred Werner, 131 (1967).
42. *Condon, E. U., Shortley, G. H.:* Theory of Atomic Spectra (2. Ed.). Cambridge: University Press 1953.
43. *Racah, G.:* Phys. Rev. *76*, 1352 (1949).
44. *Wybourne, B. G.:* Spectroscopic Properties of Rare Earths. New York: Interscience (John Wiley) 1965.
45. *Judd, B. R.:* Operator Techniques in Atomic Spectroscopy. New York: McGraw-Hill 1963.
46. *Jørgensen, C. K.:* Solid State Phys. *13*, 375 (1962).
47. *Ellis, C. B.:* Phys. Rev. *49*, 875 (1936); *55*, 1114 (1939).
48. *Gobrecht, H.:* Ann. Phys. *28*, 673 (1937); *31*, 181 and 755 (1938).
49. *Mukherji, P. C.:* Indian J. Phys. *11*, 123 (1937).
50. *Jørgensen, C. K.:* Mat. fys. Medd. Danske Vidensk. Selskab *29*, no. 11 (1955).
51. *Satten, R. A.:* J. Chem. Phys. *21*, 637 (1953).
52. *Bethe, H., Spedding, F. H.:* Phys. Rev. *52*, 454 (1937).
53. *Jørgensen, C. K.:* Acta Chem. Scand. *9*, 540 (1955).
54. *Judd, B. R.:* Proc. Roy. Soc. (London) *A 228*, 120 (1955).
55. *Wybourne, B. G.:* J. Chem. Phys. *32*, 639 (1960); *34*, 279 (1961).
56. *Jørgensen, C. K.:* Orbitals in Atoms and Molecules. London: Academic Press 1962.
57. *Dieke, G. H.:* Spectra and Energy levels of Rare Earth Ions in Crystals. New York: Interscience 1968.
58. *Carnall, W. T., Fields, P. R., Rajnak, K.:* J. Chem. Phys. *49*, 4412, 4424, 4443, 4447 and 4450 (1968).
59. *Trees, R. E.:* J. Opt. Soc. Am. *54*, 651 (1964).
60. *Jørgensen, C. K., Pappalardo, R., Schmidtke, H. H.:* J. Chem. Phys. *39*, 1422 (1963).
61. *Schäffer, C. E., Jørgensen, C. K.:* Mol. Phys. *9*, 401 (1965).
62. — Pure Appl. Chem. *24*, 361 (1970).
63. *Burns, G.:* Phys. Letters *25A*, 15 (1967).
64. *Kuse, D., Jørgensen, C. K.:* Chem. Phys. Letters *1*, 314 (1967).
65. *Harnung, S. E., Schäffer, C. E.:* Struct. Bonding *12*, 257 (1972).
66. *Jørgensen, C. K.:* Chimia (Aarau) *25*, 109 (1971).

67. *Herzberg, G.:* Electronic Spectra and Electronic Structure of Polyatomic Molecules. Princeton: Van Nostrand 1966.
68. *Freed, S.:* Phys. Rev. *38*, 2122 (1931).
69. *Jørgensen, C. K., Brinen, J. S.:* Mol. Phys. *6*, 629 (1963).
70. *Geier, G., Jørgensen, C. K.:* Chem. Phys. Letters *9*, 263 (1971).
71. *Lang, R. J.:* Can. J. Res. *14A*, 127 (1936).
72. *Loh, E.:* Phys. Rev. *147*, 332 (1966).
73. *McClure, D. S., Kiss, Z.:* J. Chem. Phys. *39*, 3251 (1963).
74. *Alig, R. C., Kiss, Z. J., Brown, J. P., McClure, D. S.:* Phys. Rev. *186*, 276 (1969).
75. *Jørgensen, C. K.:* Progr. Inorg. Chem. *12*, 101 (1970).
76. *Ryan, J. L., Jørgensen, C. K.:* J. Phys. Chem. *70*, 2845 (1966).
77. *Rabinowitch, E., Belford, R. L.:* Spectroscopy and Photochemistry of Uranyl Compounds. Oxford: Pergamon Press 1964.
78. *Robin, M. B., Day, P.:* Advan. Inorg. Radiochem. *10*, 248 (1967).
79. *Hofmann, K. A., Höschele, K.:* Ber. Deutsch. Chem. Ges. *48*, 20 (1915).
80. *Jørgensen, C. K.:* Acta Chem. Scand. *17*, 1034 (1963).
81. *Freed, S., Mesirow, R. J.:* J. Chem. Phys. *5*, 22 (1937).
82. *Banks, C. V., Heusinkveld, M. R., O'Laughlin, J. W.:* Analyt. Chem. *33*, 1235 (1961).
83. *Jørgensen, C. K.:* Mol. Phys. *5*, 271 (1962).
84. *Ryan, J. L.:* Inorg. Chem. *8*, 2053 (1969).
85. *Blasse, G., Bril, A.:* J. Chem. Phys. *45*, 3327 (1966).
86. *Ropp, R. C.:* J. Electrochem. Soc. *112*, 181 (1965).
87. *Barnes, J. C., Day, P.:* J. Chem. Soc. *1964*, 3886.
88. *— Pincott, H.:* J. Chem. Soc. (A) *1966*, 842.
89. *Pappalardo, R., Jørgensen, C. K.:* J. Chem. Phys. *46*, 632 (1967).
90. *Ryan, J. L., Jørgensen, C. K.:* Mol. Phys. *7*, 17 (1963).
91. *Nugent, L. J., Baybarz, R. D., Burnett, J. L.:* J. Phys. Chem. *73*, 1177 (1969).
92. *Slater, J. C.:* Phys. Rev. *165*, 655 (1968).
93. *Katriel, J.:* Theoret. Chim. Acta *23*, 309 (1972).
94. *Calais, J. L., Mäkilä, K., Mansikka, K., Petterson, G., Vallin, J.:* Physica Scripta (Stockholm) *3*, 39 (1971).
95. *Ruedenberg, K.:* Rev. Mod. Phys. *34*, 326 (1962).
96. *Johnson, D. A.:* J. Chem. Soc. (A) *1969*, 1525 and 1528.
97. *Nugent, L. J., Vander Sluis, K. L.:* J. Opt. Soc. Am. *61*, 1112 (1971).
98. *Brewer, L.:* J. Opt. Soc. Am. *61*, 1101 (1971).
99. — J. Opt. Soc. Am. *61*, 1666 (1971).
100. *Martin, W. C.:* J. Opt. Soc. Am. *61*, 1682 (1971).
101. *Briffaut, J. P.:* Phys. Status Solid. *44*, 769 (1971).
102. *Orgel, L. E.:* J. Chem. Phys. *23*, 1819, 1824 and 1958 (1955).
103. *Klemm, W., Bommer, H.:* Z. Anorg. Chem. *231*, 138 (1937).
104. *Jayaraman, A.:* Phys. Rev. *137A*, 179 (1965).
105. *Hulliger, F.:* Helv. Phys. Acta *41*, 945 (1968).
106. *Chatterjee, A., Singh, A. K., Jayaraman, A., Bucher, E.:* Phys. Rev. Letters *27*, 1571 (1971).
107. *Gschneider, K. A.:* J. Less-Common-Metals *25*, 405 (1971).
108. *Jørgensen, C. K.:* Chimica Teorica, VIII, Corso Estivo di Chimica Milano 1963, p. 63. Fondazione Donegani, Accademia Nazionale dei Lincei, Rome, 1965.
109. — Struct. Bonding *1*, 234 (1966); *3*, 106 (1967).
110. *Ahrland, S.:* Struct. Bonding *5*, 118 (1968).

111. *Connick, R. E.:* J. Chem. Soc. (Suppl.) *1949*, 235.
112. *Jørgensen, C. K.:* Inorganic Complexes. London: Academic Press 1963.
113. *Timnick, A., Glockler, G.:* J. Am. Chem. Soc. *70*, 1347 (1948).
114. *Varga, L. P., Asprey, L. B.:* J. Chem. Phys. *49*, 4674 (1968).
115. — — J. Chem. Phys. *48*, 139 (1968).
116. *Nugent, L. J., Baybarz, R. D., Burnett, J. L., Ryan, J. L.:* J. Inorg. Nucl. Chem. *33*, 2503 (1971).
117. *Jørgensen, C. K.:* Chem. Phys. Letters *2*, 549 (1968).
118. *Bjerrum, J., Jørgensen, C. K.:* Rec. Trav. Chim. *75*, 658 (1956).
119. *Dunitz, J. D., Orgel, L. E.:* Advan. Inorg. Radiochem. *2*, 1 (1960).
120. *George, P., McClure, D. S.:* Progr. Inorg. Chem. *1*, 382 (1959).
121. *Schwarzenbach, G., Gut, R.:* Helv. Chim. Acta *39*, 1589 (1956).
122. *Fidelis, I., Siekierski, S.:* J. Chromatog. *17*, 542 (1965).
123. — — J. Inorg. Nucl. Chem. *28*, 185 (1966); *29*, 2629 (1967).
124. *Peppard, D. F., Lewey, S., Bloomquist, C. A. A., Horwitz, E. P., Mason, G. W.:* J. Inorg. Nucl. Chem. *32*, 339 (1970).
125. *Nugent, L. J.:* J. Inorg. Nucl. Chem. *32*, 4385 (1970).
126. *Jørgensen, C. K.:* J. Inorg. Nucl. Chem. *32*, 3127 (1970).
127. *Salzmann, J. J., Jørgensen, C. K.:* Helv. Chim. Acta *51*, 1276 (1968).
128. *Jørgensen, C. K.:* Rev. Chim. Minérale (Paris) *6*, 183 (1969).
129. *Turner, D. W., Baker, C., Baker, A. D., Brundle, C. R.:* Molecular Photoelectron Spectroscopy. London: Interscience (John Wiley) 1970.
130. *Siegbahn, K., Nordling, C., Fahlman, A., Nordberg, R., Hamrin, K., Hedman, J., Johannsson, G., Bergmark, T., Karlsson, S. E., Lindgren, I., Lindgren, B.:* ESCA-Atomic, Molecular and Solid-State Structures Studied by Means of Electron Spectroscopy. Uppsala: Almqvist and Wiksell 1967.
131. *Siegbahn, K., Nordling, C., Johansson, G., Hedman, J., Hedén, P. F., Hamrin, K., Gelius, U., Bergmark, T., Werme, L. O., Manne, R., Baer, Y.:* ESCA Applied to Free Molecules. Amsterdam: North-Holland Publishing Co. 1969.
132. *Hollander, J. M., Jolly, W. L.:* Accounts Chem. Res. *3*, 193 (1970).
133. *Watson, R. E., Hudis, J., Perlman, M. L.:* Phys. Rev. B *4*, 4139 (1971).
134. *Hnatowich, D. J., Hudis, J., Perlman, M. L., Ragaini, R. C.:* J. Appl. Phys. *42*, 4883 (1971).
135. *Bremser, W., Linnemann, F.:* Chemiker-Ztg. *95*, 1011 (1971).
136. *Jørgensen, C. K.:* Chimia (Aarau) *25*, 213 (1971); *26*, 252 (1972).
137. — *Berthou, H.:* Mat. fys. Medd. Danske Vidensk. Selskab, *38*, no. 15 (1972).
138. *Fadley, C. S., Hagström, S. B. M., Hollander, J. M., Klein, M. P., Shirley, D. A.:* Science *157*, 1571 (1967).
139. — — *Klein, M. P., Shirley, D. A.:* J. Chem. Phys. *48*, 3779 (1968).
140. *Kramer, L. N., Klein, M. P.:* J. Chem. Phys. *51*, 3618 (1969).
141. *Jørgensen, C. K., Berthou, H., Balsenc, L.:* J. Fluorine Chem. *1*, 327 (1972).
142. *Bearden, J. A., Burr, A. F.:* Rev. Mod. Phys. *39*, 125 (1967).
143. *Fadley, C. S., Shirley, D. A., Freeman, A. J., Bagus, P. S., Mallow, J. V.:* Phys. Rev. Letters *23*, 1397 (1969).
144. *Cohen, R. L., Wertheim, G. K., Rosencwaig, A., Guggenheim, H. J.:* Phys. Rev. B *5*, 1037 (1972).
145. *Bonnelle, C., Kartanak, R. C., Jørgensen, C. K.:* Chem. Phys. Letters *14*, 145 (1972).
146. — *Jørgensen, C. K.:* J. Chim. Physique *61*, 826 (1964).
147. *Rosencwaig, A., Wertheim, G. K., Guggenheim, H. J.:* Phys. Rev. Letters *27*, 479 (1971).
148. *Jørgensen, C. K.:* Theoret. Chim. Acta *24*, 241 (1972).

149. *Novakov, T.:* Phys. Rev. *B 3*, 2693 (1971).
150. *Jørgensen, C. K., Berthou, H.:* J. Electron Spectr., under preparation.
151. *Wertheim, G. K., Rosencwaig, A.:* Phys. Rev. Letters *26*, 1179 (1971).
152. *Jørgensen, C. K., Berthou, H.:* Chem. Phys. Letters, *13*, 186 (1972).
153. *Sakellaridis, P.:* Compt. Rend. *236*, 1014, 1244, 1457 and 1767 (1953).
154. — J. Phys. Radium *16*, 271 and 422 (1955).
155. *Fischer, D. W., Baun, W. L.:* J. Appl. Phys. *38*, 4830 (1967).
156. *Dehner, J. L., Starace, A. F., Fano, U., Sugar, J., Cooper, J. W.:* Phys. Rev. Letters, *26*, 1521 (1971).
157. *Sugar, J.:* Phys. Rev. *B 5*, 1785 (1972).
158. *Fadley, C. S., Shirley, D. A.:* Phys. Rev. *A 2*, 1109 (1970).
159. *Jørgensen, C. K., Berthou, H.:* Discussions Faraday Soc., *54*, in press.
160. *Wertheim, G. K., Rosencwaig, A., Cohen, R. L., Guggenheim, H. J.:* Phys. Rev. Letters 27, 505 (1971).
161. *Hedén, P. O., Löfgren, H., Hagström, S. B. M.:* Phys. Rev. Letters *26*, 432 (1971).
162. *Hagström, S. B. M., Brodén, G., Hedén, P. O., Löfgren, H.:* J. Physique (Colloque CNRS no. 196) C 4–269 (1971).
163. *Eckstein, W., Georg, K. F., Heiland, W., Kirschner, J., Müller, N.:* Z. Naturforsch. *25 a*, 1981 (1970).
164. *Cox, P. A., Evans, S., Hamnett, A., Orchard, A. F.:* Chem. Phys. Letters, *7*, 414 (1970).
165. *Orchard, A. F.:* private communication.
166. *Jørgensen, C. K.:* Acta Chem. Scand. *16*, 2406 (1962).
167. — *Horner, S. M., Hatfield, W. E., Tyree, S. Y.:* Intern. J. Quantum Chem. *1*, 191 (1967).
168. *Prins, R., Novakov, T.:* Chem. Phys. Letters *9*, 593 (1971).
169. *Brauer, G., Pfeiffer, B.:* J. Prakt. Chem. *34*, 32 (1966).
170. *Manne, R., Åberg, T.:* Chem. Phys. Letters 7, 282 (1970).
171. *Verhaegen, G., Berger, J. J., Desclaux, J. P., Moser, C. M.:* Chem. Phys. Letters *9*, 479 (1971).
172. *Baer, Y., Hedén, P. F., Hedman, J., Klasson, M., Nordling, C., Siegbahn, K.:* Physica Scripta (Stockholm) *1*, 55 (1970).
173. *Goeppert-Mayer, M.:* Phys. Rev. *60*, 184 (1941).
174. *Penneman, R. A., Mann, J. B., Jørgensen, C. K.:* Chem. Phys. Letters *8*, 321 (1971).
175. *Sinha, S. P.:* Complexes of the Rare Earths. Oxford: Pergamon Press 1966.
176. — Europium. Berlin-Heidelberg-New York: Springer 1967.

Received April 17, 1972.

Structure and Bonding: Index Volume 1-13